POLAR AND MAGNETOSPHERIC SUBSTORMS

ASTROPHYSICS AND
SPACE SCIENCE LIBRARY

A SERIES OF BOOKS ON THE RECENT DEVELOPMENTS

OF SPACE SCIENCE AND OF GENERAL GEOPHYSICS AND ASTROPHYSICS

PUBLISHED IN CONNECTION WITH THE JOURNAL

SPACE SCIENCE REVIEWS

VOLUME 11

SYUN-ICHI AKASOFU

Geophysical Institute, University of Alaska, College, Alaska

POLAR

AND MAGNETOSPHERIC

SUBSTORMS

D. REIDEL PUBLISHING COMPANY

DORDRECHT-HOLLAND

ISBN-13: 978-94-010-3463-0   e-ISBN-13: 978-94-010-3461-6
DOI: 10.1007/978-94-010-3461-6

TO SYDNEY CHAPMAN

who, *unbeknown* to most scientists, has encouraged and
inspired the world's magnetic and auroral observatories
to maintain the essential records upon which our understanding of geomagnetism and the aurora rests.

PREFACE

It has become increasingly clear that the magnetosphere becomes intermittently unstable and explosively releases a large amount of energy into the polar upper atmosphere. This particular magnetospheric phenomenon is called the *magnetospheric substorm*. It is manifested as an activity or disturbance of various polar upper atmospheric phenomena, such as intense auroral displays and X-ray bursts. Highly active conditions in the polar upper atmosphere result from a successive occurrence of such an elementary activity, the *polar substorm*, which lasts typically of order one to three hours.

The concept of the magnetospheric substorm and its manifestation in the polar upper atmosphere, the polar substorm, has rapidly crystallized during the last few years. We can find a hint of such a concept in the term 'polar elementary storm' introduced by Kristian Birkeland as early as 1908. However, we are greatly indebted to Sydney Chapman, who established the basic foundation of magnetospheric physics and has led researches in this field during the last half century. Indeed, the terms '*polar magnetic substorm*' and '*auroral substorm*' were first suggested by Sydney Chapman.

The concept of the substorm was then soon extended by Neil M. Brice of Cornell University, and Kinsey A. Anderson and his colleagues at the University of California, Berkeley, who introduced the term '*magnetospheric substorm*'.

We owe many of these recent developments in magnetospheric physics to the great international enterprise, the International Geophysical Year (IGY) and subsequent international cooperative effort (IGC, IQSY).

However, as ten years have passed since the IGY, it is opportune to examine critically what we have learned, what we should have learned and what should be investigated in the near future. There have already been many excellent review articles in the literature on various polar upper atmospheric phenomena and also many excellent edited books contributed by experts in various fields. What is lacking, in my opinion, is a synthetic and unifying study, by a single author, of various polar upper atmospheric phenomena and associated magnetospheric phenomena on the basis of the concept of the polar substorm and the magnetospheric substorm.

It is quite obvious that for a single author to review studies in such diverse fields is a formidable task. Nevertheless, I have volunteered to do so, since I feel, on the basis of my studies of the polar substorm, that the concept of the polar substorm and of the magnetospheric substorm can help greatly in our understanding of complex polar upper atmospheric phenomena and magnetospheric phenomena.

In this monograph, I have therefore concentrated my effort in organizing available materials in terms of the polar substorm and the magnetospheric substorm. This has enabled me to construct the pattern of the development of substorms for various polar upper atmospheric phenomena. I hope that no significant work has been overlooked in this respect. I am reasonably certain that most of the inaccuracies and obscurities in the constructed patterns presented in this monograph are due to our present lack of knowledge and understanding. Thus, it is my hope that this monograph will prove useful in providing a basis for fruitful and constructive debate toward a better understanding of polar upper atmospheric and magnetospheric phenomena.

It is for this very reason that I believe the monograph should also prove useful in planning future international cooperative studies. By its nature, it is not possible to study the polar substorm without a well-organized network of observatories. A reasonable understanding of both the polar substorm and the magnetospheric substorm also requires a well-planned effort, between 'satellite workers' and 'ground workers'.

I have written this monograph on the assumption that the reader is familiar with at least one of the diverse polar upper atmospheric phenomena. Since it is written for a very specific purpose, the quoted references are rather specialized; however, in order for this book to be useful as an introductory one, I have also included a few references for each chapter under the title 'general references'.

SYUN-ICHI AKASOFU

ACKNOWLEDGEMENTS

I would like to thank Dr Sydney Chapman whose encouragement has made this book possible.

Many colleagues at the Geophysical Institute, University of Alaska (past and present) have contributed directly and indirectly to this book. It would have been impossible to produce a book of this scope without their help. It is a pleasure to be able to record here my gratitude to the former director Dr C. T. Elvey, the present director Dr K. B. Mather, and Dr T. N. Davis for their encouragement. I should like particularly to thank Dr C.-I. Meng for his collaboration in studying the polar sub-storm. I had many stimulating discussions with Messrs. A. E. Belon, F. T. Berkey, R. R. Heacock, J. L. Hook, R. Parthasarathy, G. J. Romick, D. W. Swift, and C. R. Wilson during the preparation of the manuscript.

It is also a great pleasure to thank Drs W. I. Axford, Y. I. Feldstein, E. W. Hones, Jr., G. K. Parks, J. H. Piddington, J. A. Van Allen and J. R. Winckler for their illuminating discussions. My frequent visits to the High Altitude Observatory, NCAR, Boulder, Colorado, have been most useful. I would like to thank the colleagues who have provided me with many of the illustrations used, and also the Editors and Publishers for their kind permission to reproduce these illustrations.

I acknowledge with pleasure the support given to my study by the Atmospheric Sciences Section of the National Science Foundation and the National Aeronautics and Space Administration.

Mr. K. Kawasaki and Mrs. C. Abney read the manuscript with care and found many obscurities. The staff members of the steno, drafting and photo departments of the Geophysical Institute worked hard during the hot (88°!) summer days during the preparation of the manuscript. Special thanks are due to Mrs. J. Lipscomb for her help in the editorial work, and to D. Reidel Publishing Company for its special care in the publication of this monograph. Finally, I would also like to thank my wife who shared with me the 'Arctic' life.

TABLE OF CONTENTS

LIST OF FREQUENTLY USED SYMBOLS

Units	Cgs-emu units are adopted in the book
Coordinates	Cartesian coordinates (x, y, z); spherical coordinates (r, θ, ϕ)
dp lat	dipole latitude ($=$ geomagnetic latitude) [1]
dp long	dipole longitude ($=$ geomagnetic longitude) [1]
a	radius of the earth
a_p	planetary (linear) magnetic index
AE	auroral electrojet index [3]
$\boldsymbol{B}$	magnetic induction
D component	declination component of the earth's magnetic field
$D(a, \theta, \phi)$	magnitude of disturbance field $(H, D, Z) = c_0(\theta) + \sum_n c_n(\theta) \sin(n\phi + \varepsilon_n)$
$Dst(H)$	$c_0(H)$ at $\theta \simeq 70°$
DS	$\sum_n c_n(\theta) \sin(n\phi + \varepsilon_n)$
$\boldsymbol{E}$	electric field
E	energy of a particle
E_0	e-folding energy
e	magnitude of electronic charge
H component	horizontal component of the earth's magnetic field
$\boldsymbol{J}$	electric current density; $J_P =$ Pedersen current; $J_H =$ Hall current
J	flux of particles
J	second integral invariant [5]; $J' = J/\sqrt{2m\mu}$
K	total energy of a particle
K_p	three-hourly planetary magnetic K index
$\sum K_p$	daily sum of K_p
L	McIlwain's L parameter [6]
LT	local time
m	mass of a particle
n	number density of particles
n_e	number density of electrons
r	radial distance
r_e	equatorial radial distance
T	substorm time
T	temperature
t	time
UT	universal time

v	velocity
V_A	Alfvén wave speed $= B/\sqrt{4\pi\rho}$
W	kinetic energy of a particle
$W_\perp$	kinetic energy associated with the perpendicular component of velocity with respect to B
X component	geographic North-South component of the earth's magnetic field
Y component	geographic East-West component of the earth's magnetic field
Z component	vertical component of the earth's magnetic field
γ	10^{-5} gauss
ν	collision frequency
μ	magnetic moment of a particle ($= W_\perp/B$)
σ_1	Pedersen conductivity
σ_2	Hall conductivity
σ_3	Cowling conductivity
ρ	mass density
ω	angular frequency
ω_B	gyro frequency

References

[1] S. CHAPMAN: 1963, 'Geomagnetic nomenclature', *J. Geophys. Res.* **68**, 1174.
[2] G. V. SIMONOW: 1963, 'Geomagnetic time', *Geophys. J.* **8**, 258.
[3] T. N. DAVIS and M. SUGIURA: 1966, 'Auroral electrojet activity index *AE* and its universal time variations', *J. Geophys. Res.* **71**, 785.
[4] S.-I. AKASOFU: 1963, 'The main phase of magnetic storms and the ring current', *Space Sci. Rev.* **68**, 3155.
[5] T. G. NORTHROP: 1963, *The adiabatic motion of charged particles*, Interscience Publishers, New York.
[6] C. E. McILWAIN: 1961, 'Coordinates for mapping the distribution of magnetically trapped particles', *J. Geophys. Res.* **66**, 3681.

CHAPTER 1

INTRODUCTION

1.1. Polar Upper Atmosphere and the Outer Magnetosphere

The polar upper atmosphere is unique in that it is connected to the outer magnetosphere by geomagnetic field lines. The magnetosphere may be divided into two parts:
the inner magnetosphere, where energetic charged particles are temporarily trapped,
and the outer magnetosphere.

There appears to be an almost continuous acceleration of charged particles near
the boundary of the inner magnetosphere and the outer magnetosphere. Some of the
accelerated particles are able to penetrate deep into the polar upper atmosphere,
being guided by the geomagnetic field lines which lie near the boundary, and interact
with atoms and molecules there. The interaction manifests itself in various phenomena

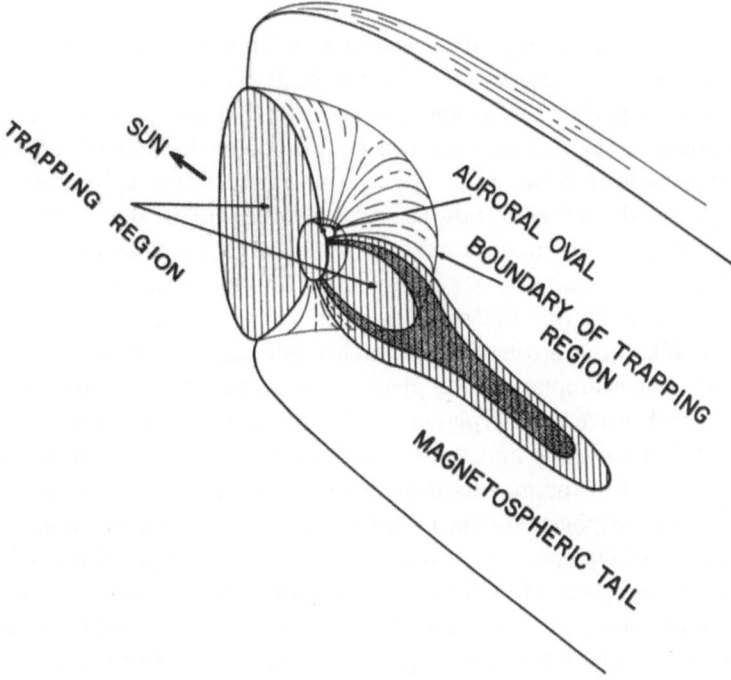

Fig. 1.   Noon-midnight cross-section of the magnetosphere showing the structure of the magnetosphere and its relation to the auroral oval. The auroral oval delineates the projection of the boundary
of the trapping region and the outer magnetosphere onto the polar atmosphere.

and is most readily recognized as a visible glow from the excited or ionized particles in the polar upper atmosphere. From the structure of the magnetosphere shown in Figure 1, it can be seen that the glow appears as an oval band surrounding the dipole pole, delineating approximately the projection of the boundary of the two regions onto the polar upper atmosphere. This luminous oval belt is called the *auroral oval*. The luminosity is provided by visible auroras that lie along the belt. The outer magnetosphere is thus traversed by the geomagnetic field lines which originate in the polar region encircled by the auroral oval.

Because of the great distortion of the outer geomagnetic field lines by the solar wind, the magnetosphere is asymmetric with respect to the dipole axis; the cross-section of the trapping region is thus also greatly asymmetric, as can be seen in Figure 1. Therefore, the auroral oval is not a circle of constant dipole latitude and is eccentric with respect to the dipole pole. Here, the auroral *oval* should not be confused with the auroral *zone* which lies approximately along a circle of dipole latitude 67° (see also Section 1.3).

There is another luminous oval belt which is located just equatorward of the auroral oval. In general, this belt is much more diffuse than the oval belt and contains the hydrogen emissions. For this reason, it is called the *proton aurora* or the *hydrogen aurora*.

1.2. Polar Substorms and Magnetospheric Substorms

The magnetospheric structure and its relation to the polar upper atmosphere described in Section 1.1 may be considered to be the basic frame of reference in describing various magnetospheric phenomena and associated polar geophysical phenomena. This monograph is devoted to disturbed conditions in the magnetosphere and their various manifestations in the polar upper atmosphere. More specifically, by using a full knowledge of these manifestations and available satellite data, an attempt will be made to construct a three-dimensional distribution of charged particles in the magnetosphere and to infer basic processes associated with the magnetospheric substorm.

When the sun is active, the solar wind is variable and causes magnetospheric disturbances. When the disturbance is intense enough, it may be called a *magnetospheric storm*. For example, a solar plasma cloud ejected during intense solar flares generates a shock wave in interplanetary plasma. The magnetospheric storm begins when both the shock wave and the plasma cloud interact with the magnetosphere. Typically, the onset of the magnetospheric storm is marked by the sudden compression of the magnetosphere, owing to the passage of the interplanetary shock wave and of the region of high pressure behind the shock. The compression is often followed by successive occurrences of explosive processes within the magnetosphere. The lifetime of the individual explosive processes is typically one to three hours, much shorter than the lifetime of a typical magnetospheric storm. This process is called the *magnetospheric substorm*. During a single magnetospheric storm, there may occur about ten substorms. We may thus write,

$$\text{Magnetospheric Storm} = \text{Compression} + \sum \text{Magnetospheric Substorm}$$

The process that provides the particles for the auroral oval during quiet conditions is considerably activated during a magnetospheric substorm. As a result, quiet auroral arcs which lie along the auroral oval are activated during the substorm. The activation originates in the midnight sector of the auroral oval, and its effects spread violently in all directions, causing various characteristic displays in different local time sectors. In general, the magnetospheric substorm reaches its peak in a rather short period of about 15 min to 30 min, and then gradually subsides. Auroral activity follows this pattern of the growth and decay of the substorm. This is but one of the manifestations of the magnetospheric substorm in the polar upper atmosphere and is called the *auroral substorm*.

The geomagnetic field is also greatly disturbed by motions of the energetic particles which appear during substorms; in particular, protons of energies of order $1 \sim 50$ keV share a significant portion of the energy associated with the magnetospheric substorm. Electric currents are generated by the gyration, oscillation and drift motions of these protons, resulting in magnetic disturbances in middle and low latitudes, as well as in the magnetosphere. Their asymmetric distribution (with respect to the dipole axis) and the resulting electric field appears to play the key role in the magnetospheric substorm; this electric field in the magnetosphere is communicated to the ionosphere (which is thought to be the base of the magnetosphere) and generates electric currents there. In particular, a concentrated electric current, which is called the *auroral electrojet*, is induced along the auroral oval and causes intense geomagnetic disturbances. The magnetic disturbance generated by the auroral electrojet and the motions of the protons in the magnetosphere is called the *polar magnetic substorm*; this is another manifestation of the magnetospheric substorm.

The polar ionosphere is also greatly disturbed during magnetospheric substorms, resulting in the *ionospheric substorm*. The main feature of the ionospheric substorm is the anomalous ionization in the lower ionosphere by energetic particles coming into the polar ionosphere. The riometer, which measures the intensity of cosmic radio noise, is the most common device used in monitoring changes of this ionization. The ionization in the lower ionosphere is particularly effective in absorbing radio waves there. The deformation of the ionosphere (namely, the redistribution of the ionization in the ionosphere) is another important aspect of the ionospheric substorms. It may be caused by the same electric field which induces electric currents and/or by a heating of the polar upper atmosphere and subsequent motion of the atmospheric gas.

Two types of waves are known to be generated in the ionosphere during the magnetospheric substorm. The first type is called the traveling wave disturbance (TWD) and propagates a great distance. The other type is very closely associated with supersonic motions of active auroras during substorms and can be detected as an infrasonic wave on the ground. Here, both phenomena are called the *atmospheric wave substorm*. Another important manifestation of the magnetospheric substorm in the ionosphere is heating. There is much to be discussed concerning the mechanisms which are responsible for this heating. An important consequence of the heating will be the generation of a large-scale circulating pattern in the polar atmosphere.

In addition to the excitation and ionization of the polar upper atmospheric con-stituents, energetic particles emit or are associated with electromagnetic waves of different frequencies. The energetic electrons emit bremsstrahlung X-rays when they are decelerated by collisions with atmospheric particles. During magnetospheric sub-storms, intense fluxes of X-rays appear not only along the auroral oval, but also along the auroral zone in the morning sector. This phenomenon may be called the *X-ray substorm*.

As mentioned earlier, during magnetospheric substorms the population of low energy protons (1 ~ 50 keV) is increased in the magnetosphere. Some of such protons manifest themselves by disturbing the proton aurora, after penetrating into the polar upper atmosphere. This phenomenon is called the *proton aurora substorm*.

An intense precipitation of energetic particles along the auroral oval and in its vicinity during magnetospheric substorms is also associated with very low frequency emissions, such as hiss and chorus. This phenomenon may be called the *VLF emission substorm*. During magnetospheric substorms, various types of ultra-low frequency electromagnetic waves or the so-called 'micropulsations' are observed. This phenomenon is here called the *micropulsation substorm*. These waves are thought to be generated in the magnetosphere and are propagated in the magnetospheric medium as hydromagnetic waves.

The following chart summarizes various manifestations of the magnetospheric substorm in the polar upper atmosphere. Each phenomenon grows and decays as the magnetospheric substorm develops and subsides. They may be called, as a whole, the *polar substorm*. The *magnetospheric substorm* occurs intermittently and impulsively with a lifetime of order 1 ~ 3 hours. As we shall see in later chapters its intermittent and impulsive characteristics are well displayed in the various features of the polar substorm.

Each manifestation of the magnetospheric substorm in the polar upper atmosphere shows us different aspects of the magnetospheric substorm. Based on the pattern of the development and decay of the auroral substorm, it is possible to determine how the precipitation of electrons of energies 1 ~ 10 keV varies during the substorm

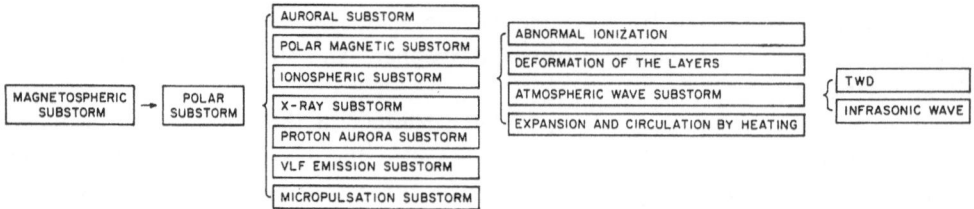

over the entire polar region. The patterns of the ionospheric substorm absorption and the X-ray substorm can tell us how the precipitation of more energetic electrons (~ 50 keV) varies during the substorm. If a possible electric current system could be determined from a study of polar magnetic substorms, it might tell us how the electric field is distributed and how it varies in the magnetosphere during magnetospheric substorms. On the basis of the analysis of these different manifestations, what we hope to establish in this monograph are the self-consistent time and spatial changes of the precipitation of electrons and protons of different energies and the changing distribution of the electric field during the substorm. Then, having an appropriate model of the magnetosphere, such as that schematically shown in Figure 1, we can infer the three-dimensional distribution of the energetic particles and the electric field in the magnetosphere during magnetospheric substorms. In order to construct a reasonable model for the magnetospheric substorm, this study should, however, be combined with *in situ* observations by satellite-borne instruments in the magnetosphere.

1.3. Magnetospheric Substorm and Magnetospheric Storm

As mentioned earlier, the magnetospheric substorm is associated with an increase of the population of protons of energies $1 \sim 50$ keV in the trapping region. Their adiabatic motions induce electric currents and reduce the intensity of the geomagnetic field in low latitudes. If magnetospheric substorms occur frequently enough, these protons tend to accumulate in the trapping region and form an intense ring current or the storm-time radiation belt in the magnetosphere. More specifically, if substorms occur with a time interval of, say, 12 hours or so, many of the protons produced during one substorm will disappear from the trapping region before the next substorm occurs by the charge-exchange process or some other process. However, if the time interval between substorms is less than a few hours, successive substorms tend to accumulate protons in the trapping region, forming an intense radiation belt (or the ring current belt) and causing a significant decrease in the intensity of the geomagnetic field in low latitude regions over the entire longitude range; this is the main phase of a geomagnetic storm.

The magnetospheric storm is thus manifested as the geomagnetic storm which consists of the initial phase and the main phase. The first phase is characterized by a step function-like sudden increase and the second phase by a large decrease of the horizontal component; the first phase results from the compression of the magnetosphere by the interplanetary shock wave. The second phase results from the accumulated effect of the protons; thus:

Magnetospheric storm = Compression + $\sum$ Magnetospheric substorm
$\downarrow$ $\qquad\qquad\qquad$ $\downarrow$ $\qquad\qquad\qquad\qquad$ $\downarrow$
Geomagnetic storm $\quad$ = Initial phase $\ + \sum$ Polar magnetic substorm and
$\qquad\qquad\qquad\qquad\qquad\qquad\qquad\qquad\qquad$ Magnetic field of the ring
$\qquad\qquad\qquad\qquad\qquad\qquad\qquad\qquad\qquad$ current.

It is not difficult to infer from the above discussion that the major phase of geomagnetic storms, namely the main phase, may be characterized entirely by the intensity and occurrence frequency of magnetospheric substorms, since the total number of protons accumulated in the trapping region should be related to the intensity and the occurrence frequency of substorms. Figure 2 shows this situation schematically. The onset time and the intensity of successive substorms are indicated by vertical columns, and the total number of protons which appear during the substorms is also given. If those protons form a uniform belt around the earth (with respect to the dipole axis), the total energy is proportional to the magnitude of the decrease of the geomagnetic field at the equator (DESSLER and PARKER, 1959). Therefore, in this simple situation, the variation of the total energy of the protons in the trapping region is manifested

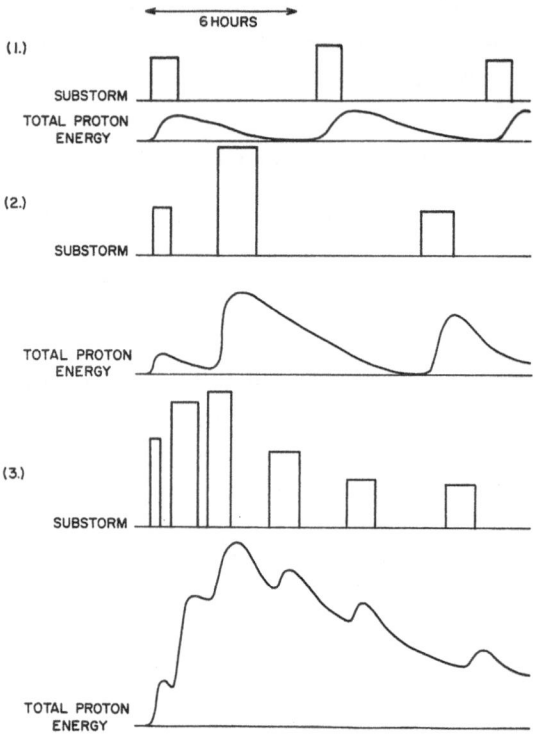

Fig. 2.   Schematic diagram showing how the occurrence frequency and intensity of magnetospheric substorms are related to the total population (or the total kinetic energy) of the protons of energies 1 ∼ 10 keV in the trapping region.

by a uniform depression of the horizontal component, namely the main phase decrease. Thus, the growth and decay of the main phase depends greatly on the occurrence frequency and intensity of substorms. *A study of the major phase of geomagnetic storms, the main phase, is reduced to a study of an elementary and fundamental process: the magnetospheric substorm.*

We shall see in later chapters that the magnetospheric substorm is an internal process within the magnetosphere. It is the magnetosphere which transforms the energy carried away by the solar plasma from the sun into the substorm energy. The mechanism involved in these processes is not known and is one of the most challenging problems.

1.4. Auroral Oval as a Natural Coordinate

Recent studies of the aurora have revealed an important fact, that the auroral arcs tend to lie in a narrow oval belt encircling the dipole pole (Figure 3). The oval is eccentric with respect to the dipole pole and its center is appreciably ($\sim3°$) displaced toward the dark hemisphere (FELDSTEIN, 1963; FELDSTEIN and STARKOV, 1967; KHOROSHEVA, 1962). The geometry of the oval is fixed, as a first approximation, with respect to the sun, and the earth rotates once a day under the oval. The auroral zone is simply the locus of the midnight part of the oval where intense auroral displays are

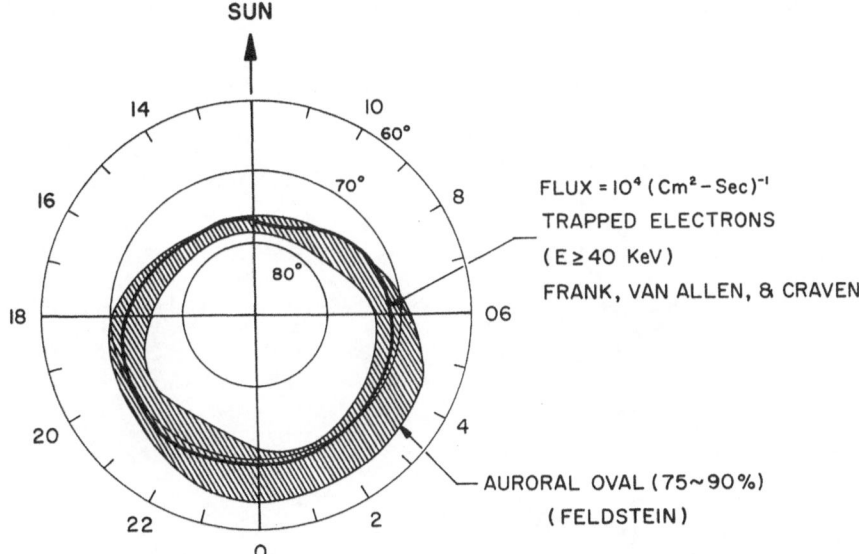

Fig. 3. Auroral oval determined by Feldstein and the outer boundary of the trapping region.

most frequently seen. Figure 3 also shows the line of intersection between the polar ionosphere and the outer boundary of the trapping region, which was determined by FRANK et al. (1964), indicating that the geometry of the auroral oval is closely related to the structure of the magnetosphere (Figure 1).

One of the major difficulties, in fact one of the major sources of confusion in analyzing polar geophysical data, arises because the data are too often discussed relative to geomagnetic (or dipole) coordinates. Since the outer geomagnetic field is greatly distorted by the solar wind and since polar substorms are manifestations of the magnetospheric substorm which originates in the outer magnetosphere, it is clear that the dipole coordinates are not necessarily adequate nor the best coordinates in organizing polar upper atmospheric data. What is important is the relative location of a station with respect to the auroral oval, rather than with respect to the dipole pole. This is particularly the case for studies of auroral phenomena and polar geomagnetic disturbances. Unfortunately, however, the auroral zone, which is a circle of constant dipole latitude of 67°, had long been interpreted as an instantaneous region of the aurora. The instantaneous region of the aurora is the auroral oval and is not the auroral zone. As mentioned earlier, the latter is the locus of the midnight part of the former. Figure 4 shows the approximate location of the auroral oval in the Northern Hemisphere at different UT. It can be seen that typical auroral zone stations such as College and Kiruna, are located well outside the auroral oval in their local daylight hours.

The oval has, however, a much more important meaning than an instantaneous region of the auroras. As we have seen in Section 1.1, its geometry is closely related to the structure of the magnetosphere. The eccentricity of the oval is a manifestation

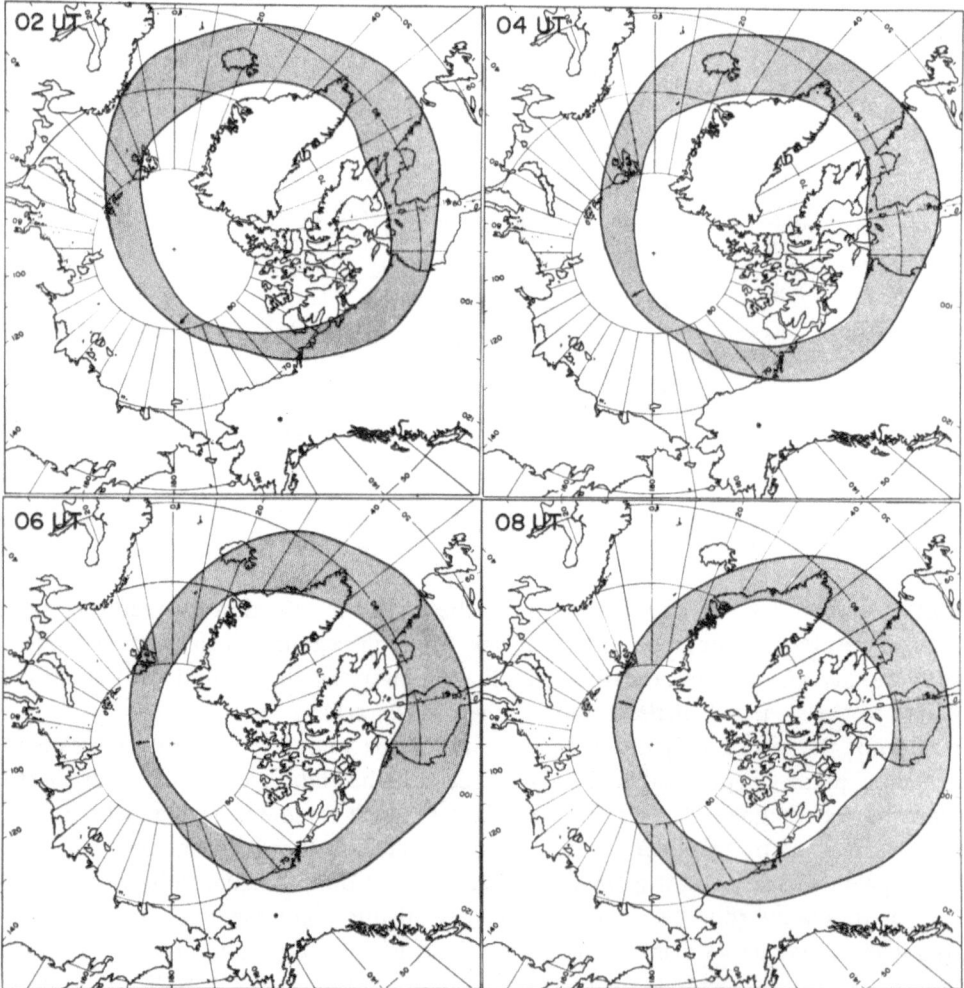

Fig. 4a. Approximate location of the auroral oval in the Northern Hemisphere at different UT hours.

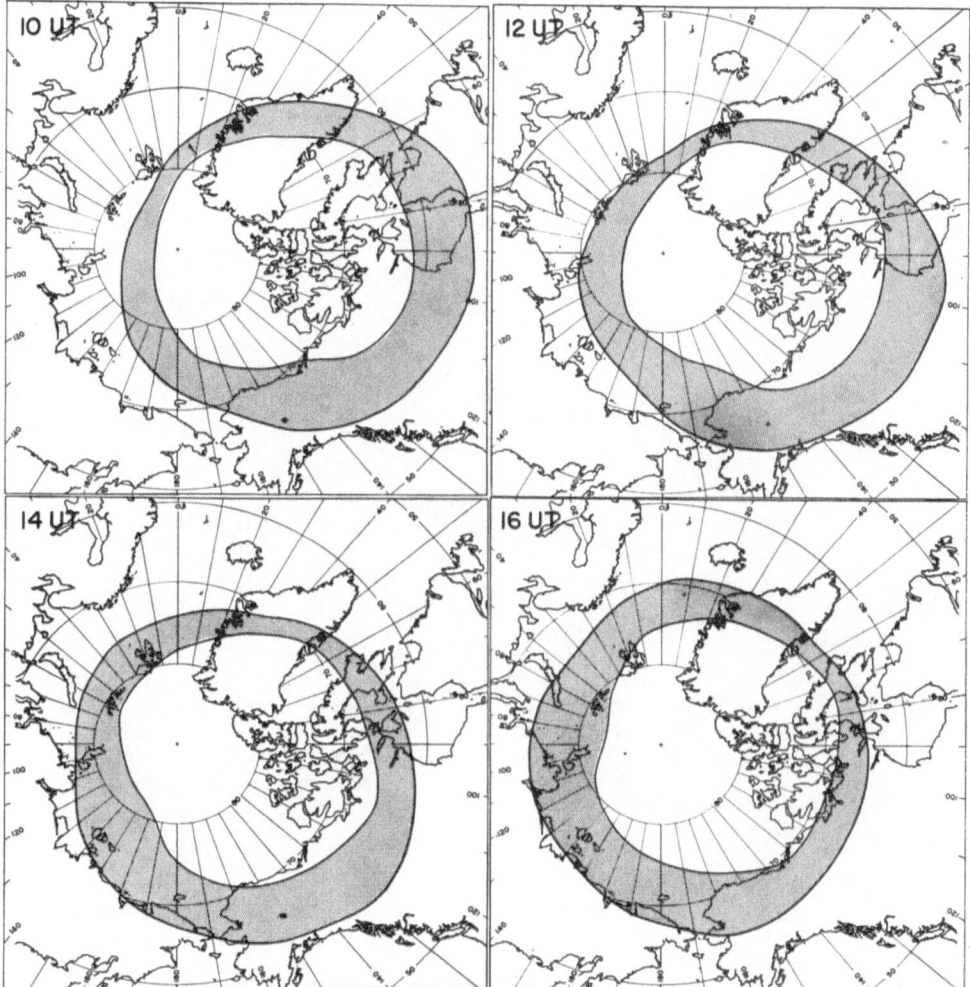

Fig. 4b.

of the day-night asymmetry of the internal structure of the magnetosphere; the outer radiation belt has a marked asymmetry, and the line of intersection between its outer boundary and the ionosphere coincides approximately with the auroral oval (Figure 1). Thus, the concept of the auroral oval has a basic foundation in terms of the structure of the magnetosphere.

This geometrical relationship between the auroral oval and the structure of the magnetosphere suggests that the auroral oval delineates approximately the boundary of the area from which proceed the magnetic field lines that traverse the outer magnetosphere. In the past, the polar cap had been defined as the area within the auroral zone, but it is more natural to regard it as the area inside the auroral oval. A station at dp lat 70° had been regarded as a permanent polar cap station in the past (since it is

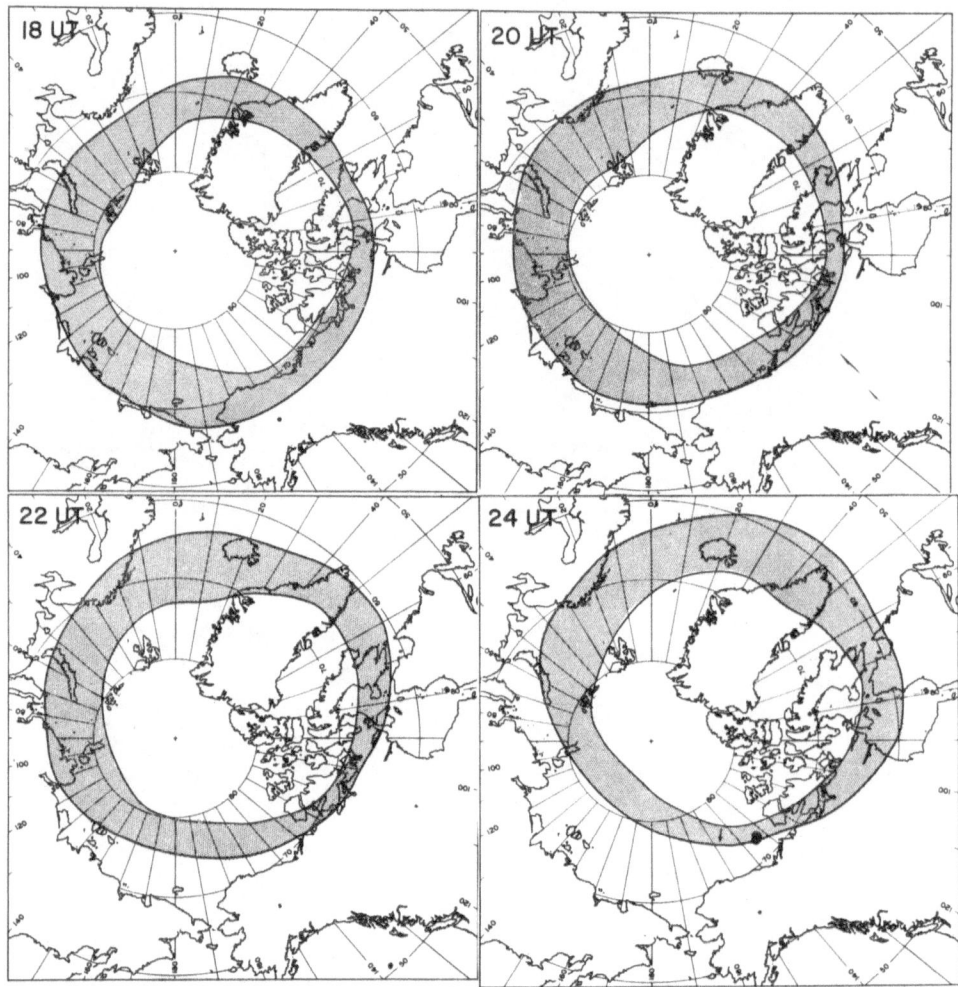

Fig. 4c.

inside the auroral zone), but it is in the newly defined polar cap only in the midnight hours when the station is located inside the oval and when the field line originating at the station lies in the tail region of the magnetosphere; during the rest of the day, the station is not a polar cap station, since it is outside the oval and thus the field line originating there traverses the trapping region.

This is well demonstrated by the fact that the auroral oval delineates approximately the boundary of the area which is constantly bombarded by low energy cosmic rays; Figure 5 shows the location of the boundary of this area (STONE, 1964). These particles have free access to the area bounded by the oval, regardless of their energies. The area is also approximately the region of the polar cap absorption (PCA). A station at dp lat 70° is outside the region of continuous bombardment of low energy cosmic

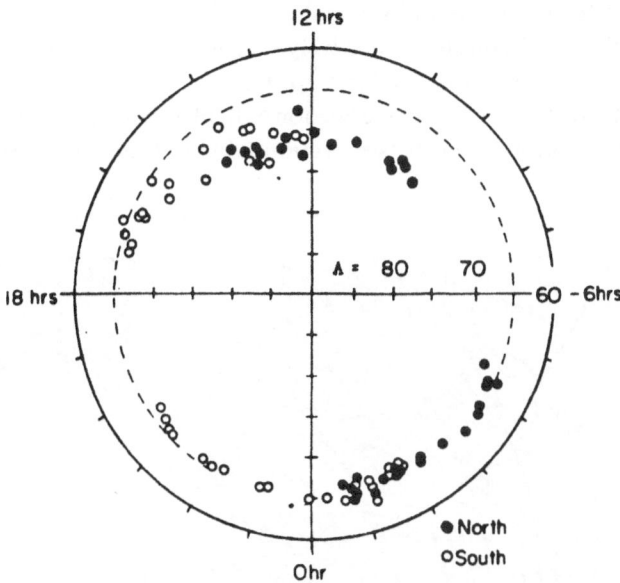

Fig. 5. Location of the boundary for vertically incident 1.5 MeV protons in the dipole latitude and dipole local time coordinates. The protons are observed everywhere inside the boundary
(Stone, E. C.: *J. Geophys. Res.* **69**, 3557, 1964).

rays in the daytime like other stations in middle and low latitudes. On the other hand, it is within the region of bombardment in the night hours, like other stations near the dipole pole.

Apparent complexity of the daily occurrence of some of the polar geophysical phenomena arises from the notable eccentricity of the auroral oval with respect to the dipole pole. For example the daily variation of the occurrence frequency of auroras observed at single stations is characterized by the change of the relative location of the station with respect to the auroral oval, during the course of a day. A typical auroral zone station comes under the oval only in the midnight hours, so that the major polar geophysical phenomena have a single peak of occurrence around midnight, at about 10 UT at College (dp lat 64.7°) and 22 UT at Kiruna (dp lat 65.3°) (see Figure 4). The uniqueness of the auroral zone is due to the fact that the auroral and polar magnetic substorms are most intense in the midnight sector of the oval where the oval and the zone intersect. A station between dp lat 70° and 75° 'crosses' the auroral oval twice a day, so that the occurrence frequency of the aurora has a double peak, one in the evening hours and one in the morning hours. From the geometry of the oval (Figure 3) it can then be inferred that as one advances toward higher latitudes both peaks shift systematically from midnight hours to twilight hours.

When peak times are plotted in the dipole latitude and local time coordinates, they tend to line up along two spiral curves, one in the afternoon-evening sector and the other in the morning-forenoon sector. NIKOLSKY (1947) was the first to obtain the spiral curves by using the daily variation of magnetic activity; the two spirals are named the N and M spirals, respectively. Figure 6a shows the location of the M spiral at different UT hours on a polar map. Later, a number of geophysical phenomena, such as Hα (or Hβ) emission (see Section 6.3), blackout, auroral radar echoes (see Section 4.10) and sporadic E layer (see Section 6.3) are found to have a tendency to

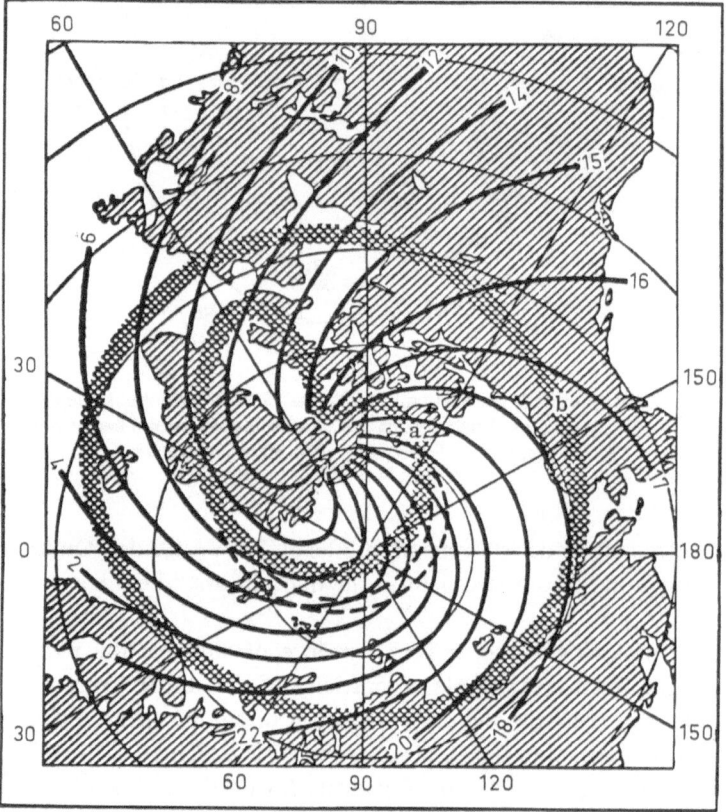

Fig. 6a.   Location of the 'M-spiral' at different UT hours on a polar map. At a particular UT hour (say, 16 UT), geomagnetic disturbances in the morning hours are most intense along the 'spiral' indicated by the number 16 (Nikolsky).

appear along one or both of the spirals. Figure 6b shows a similar map for the blackout (Piggott, 1964; see also Section 4.8). In spite of the very different nature of these phenomena, the two types of spirals are remarkably similar. Figure 6c shows another type of spiral manifested by the storm-type sporadic E (Thomas, 1960). Figure 7 shows spiral curves for various polar geophysical phenomena on the dipole coordinates, constructed by Nagata (1963). By ignoring discrepancies in detail of the curves, however, the geometry of the auroral oval (illustrated in Figure 3) can be easily visualized by combining the spirals.

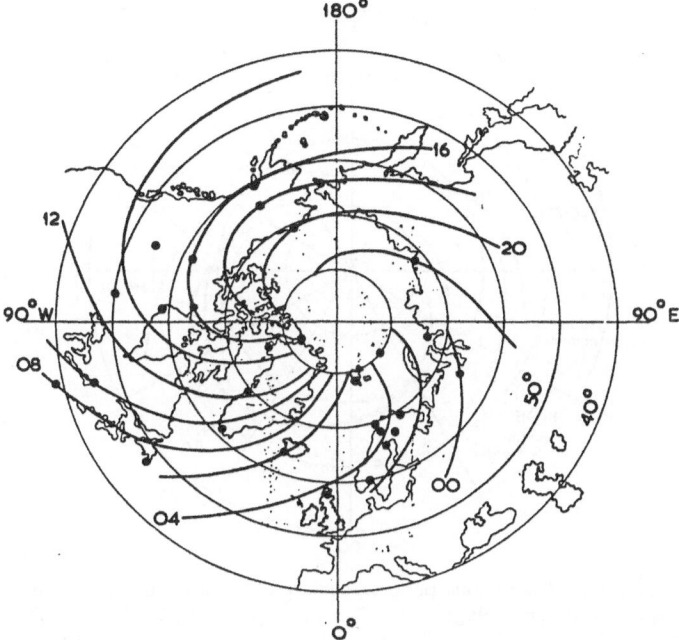

Fig. 6b. Location of the curves along which the polar blackout occurs most frequently at different UT hours (Piggott, W. R.: *Research in Geophysics*, Vol. 1).

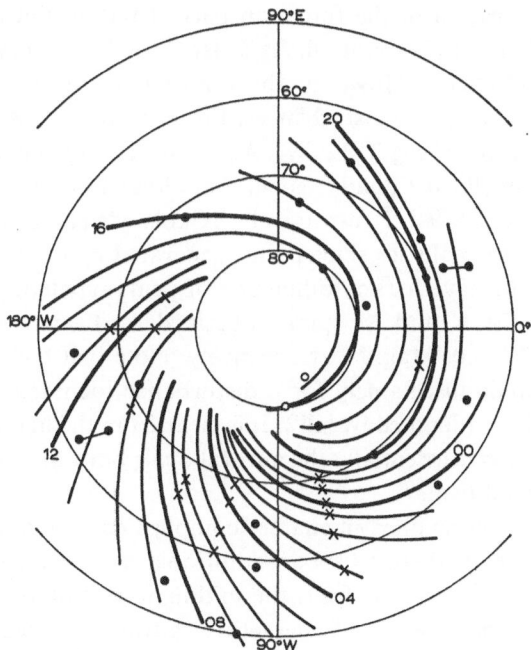

Fig. 6c. Location of the curves along which the storm-type sporadic E occurs most frequently at different UT hours (Thomas, L.: *Some Ionospheric Results obtained during the IGY*).

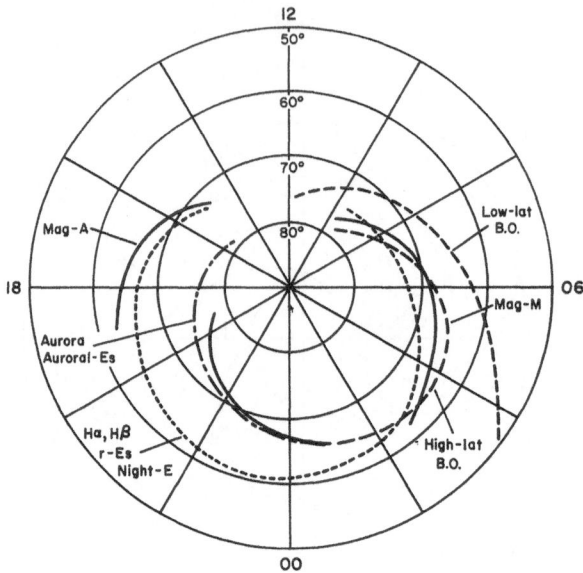

Fig. 7. The 'spiral' curves for various polar upper atmosphere phenomena in dipole latitude and time coordinates (Nagata, T.: *Planetary Space Sci.* **11**, 1395, 1963).

It is not difficult to identify Nikolsky's 'M spiral' as the forenoon part of the oval. Figure 8 shows the location of the forenoon part of the oval at 0, 6, 12 and 18 UT (solid curves) and Nikolsky's spiral curves (dash); the latter curves tend to curl less tightly than the solid curves. However, the solid curves represent a more accurate location of intense geomagnetic disturbances. For example, the K indices at Eskdale-muir (England; geographic long 3° 12′ W), Agincourt (East Canada; geographic long 79° 16′ W), Meanook (West Canada; geographic long 113° 20′ W) tend to peak at 21–24 UT, 0–3 UT, and 6–9 UT respectively. Nikolsky's curves in very high latitudes do not seem to be realistic, since points separated by only a few hundred kilometers are unlikely to have a time difference of peak magnetic activity of 6 hours.

Figure 9 shows the N and M spirals obtained by FELDSTEIN (1963); they are indicated by 'Mag. N' and 'Mag. M', respectively. Note that the N spiral is obtained for two different conditions, namely for a disturbance (denoted by 'disturbed') and for an average condition (denoted by 'all'). It is clear from the figure that the combination of the two spirals give essentially the auroral oval; both M and N spirals are thus segments of the auroral oval.

Polar upper atmospheric phenomena are complicated by the fact that in addition to the narrow oval band, there exists an additional band which extends along the morning half of the auroral zone from the midnight part of the oval. This band is characterized by the precipitation of energetic electrons and their associated phenomena (Chapters 4, 5, 6, 7, 8). The precipitating electrons in this band are more energetic than those in the oval. The poor correlation between such a precipitation and

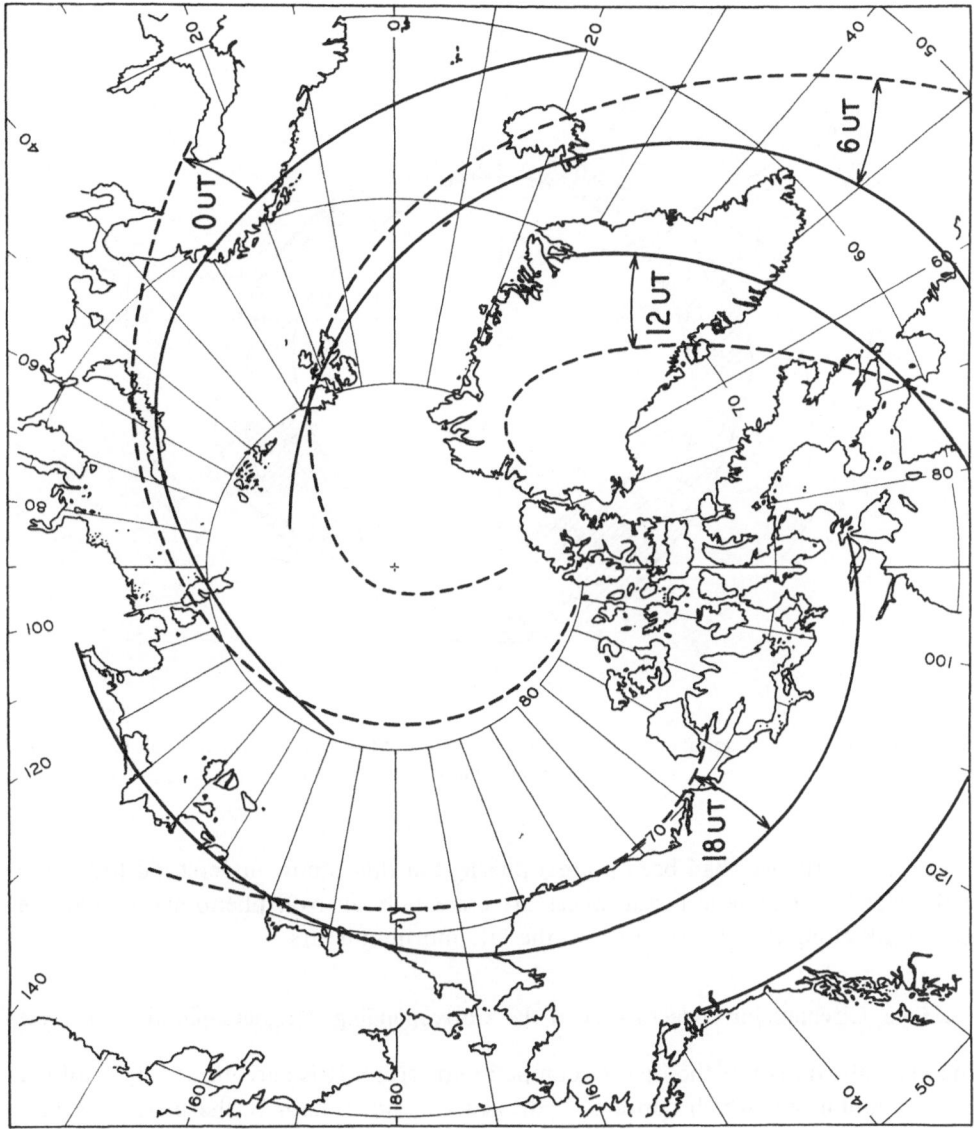

Fig. 8.   Comparison of the locations of Nikolsky's M spiral and the morning half of the oval at different UT hours.

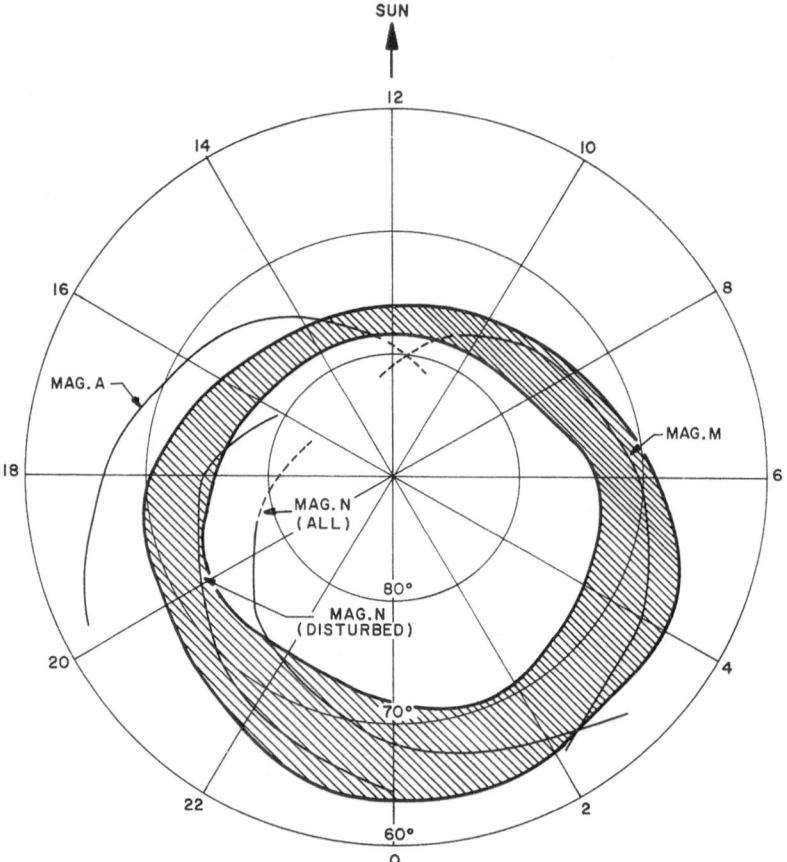

Fig. 9.   The M, N and A spirals determined by Feldstein, and the auroral oval in dipole latitude and time coordinates.

magnetic disturbances had been a great puzzle, but this is now understood to be due to the fact that magnetic disturbances are essentially an oval phenomenon and are quite weak along the auroral zone in the late morning hours.

1.5. Four Circumpolar Structures and the Corresponding Magnetospheric Structures

The auroral oval is not the only circumpolar structure. It is surrounded by a diffuse band of luminosity which contains hydrogen emissions, namely the proton aurora. Protons leaking out from the ring current (or the storm-time belt) may contribute to this luminosity in the midnight sector, particularly during magnetospheric substorms.

The top-side ionospheric sounders carried by satellites have revealed a deep 'trough' of ionization along a circumpolar belt in middle latitudes; this structure is called the mid-latitude trough (MULDREW, 1965; SHARP, 1966). MULDREW (1965) showed that the trough, like the auroral oval, has no constant geomagnetic position, but has its lowest latitude in the midnight sector. Further, the electron density at 1000 km level in the area encircled by the trough is appreciably lower ($<30/cm^3$) than in the surrounding mid-latitude region. Figure 10 shows the locations at which low density was observed (HAGG, 1967). It appears that the mid-latitude trough is the projection of the outer boundary of the ionosphere or the so-called 'plasmapause'

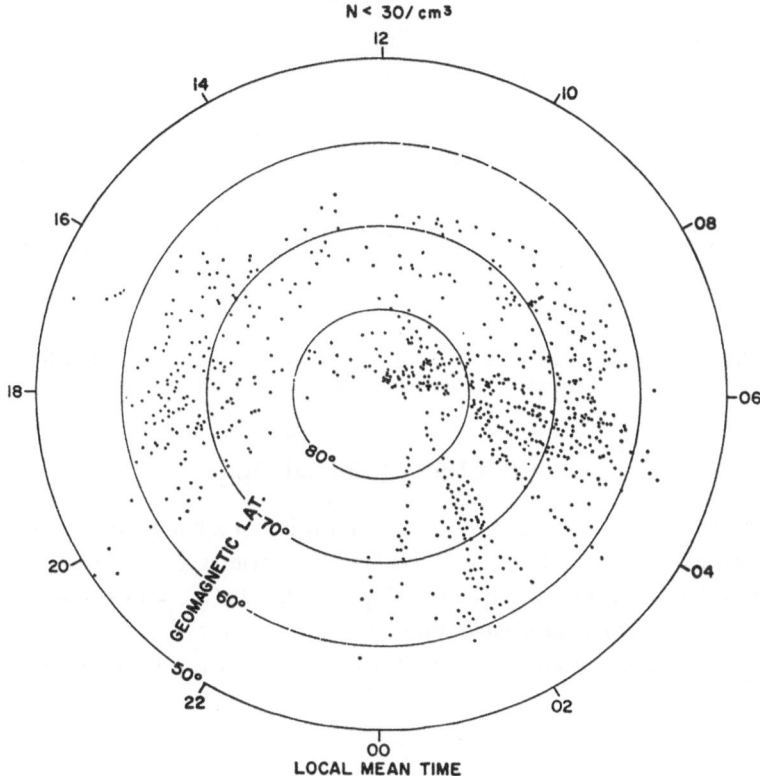

Fig. 10.   Locations (in dipole latitude and time coordinates) where a very low electron density ($<30/cm^3$) was observed at 1500–3000 km level by the Alouette II satellite (Hagg, E. L.: *Canadian J. Phys.* **45**, 27, 1967).

discovered by CARPENTER (1966) in his whistler study. Figure 11 illustrates schematically these three circumpolar structures and their relation to the corresponding structures in the magnetosphere in the midnight meridian.

There appears to be another interesting circumpolar structure during intense geomagnetic storms; it is a rather broad band of luminosity which is characterized by an intense red oxygen line (λ 6300). This luminosity is called the subvisual mid-latitude red arc (cf. ROACH and ROACH, 1963). Its relation to the proton belt and the plasmapause is not well known, but it appears to be the most equatorward structure of the four circumpolar structures.

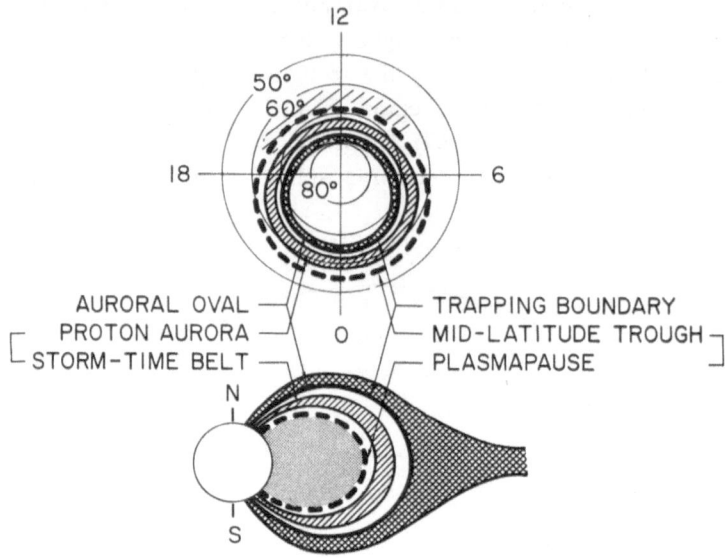

Fig. 11.   Schematic diagram showing the three circumpolar oval structures and their relation to the corresponding magnetospheric structures.

1.6. Changing Auroral Oval

An important feature of the auroral oval is that it is not a fixed curve. Figure 3 shows only the average location of the oval. Its 'radius' varies greatly during geomagnetic disturbances. During prolonged extremely quiet periods ($K_p = 0$ for about 24 hours), the auroral oval contracts poleward, and its location in the midnight sector is beyond dp lat 70°. Auroras also become very faint or invisible in the midnight sector during

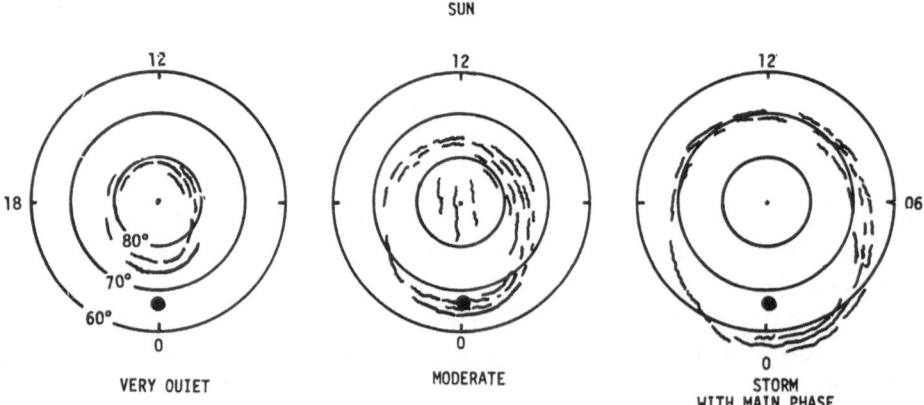

Fig. 12.   Auroral ovals at different magnetic conditions: during a very quiet period, a moderately disturbed period and an intense storm. The location of a typical auroral zone station in the midnight sector is shown by a dot.

such an extremely quiet period (STRINGER *et al.*, 1965); see Figure 12. When the K_p index becomes 1 or 2, the midnight location descends to about dp lat 70° or a little higher; see Figure 12. In this case, a typical auroral zone station, indicated by a dot, is located well outside the auroral oval, so that it becomes temporarily a subauroral zone station. The oval is located at its average location for the K_p value of 3 (STRINGER and BELON, 1967). A further increase of geomagnetic activity is associated with the shift of the oval toward the equator. During intense geomagnetic storms, the oval can descend to as low as dp lat 50° in the midnight sector (AKASOFU and CHAPMAN, 1963). In such a case, a typical auroral zone station, and even a typical subauroral zone station, temporarily becomes a polar cap station, since it is located well inside the oval. A dot in Figure 12 shows the location of an auroral zone station in the midnight sector, illustrating the changes of the relative location of the station with respect to the oval for different conditions. Figure 13 shows the latitude of the equatorward boundary of the auroral oval for different intensities of geomagnetic

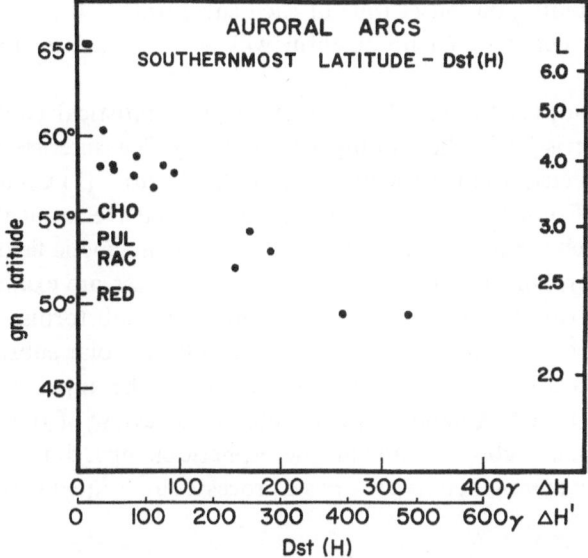

Fig. 13.   Latitude of the equatorward boundary of the auroral oval for different intensities of geomagnetic activity, expressed in term of the *Dst(H)*.

storms; the intensity of geomagnetic storms is given in terms of the magnitude of the magnetic field produced by the ring current (AKASOFU and CHAPMAN, 1963; see also FELDSTEIN and STARKOV, 1968).

Therefore, without monitoring the relative location of stations with respect to the oval, it is impossible to organize in a consistent way the vast amount of complicated polar geophysical data for different degrees of geomagnetic activity and for different periods of the sunspot cycle. This emphasizes the importance of the auroral oval as a natural frame of reference in examining polar upper atmospheric phenomena.

1.7. Scope of the Monograph

In the following eight chapters, we will discuss briefly each manifestation of the magnetospheric substorm in the polar upper atmosphere. The purpose is to construct the pattern of the growth and decay of each polar geophysical phenomenon as the substorm progresses.

For the time frame of reference, we shall choose the onset time of polar magnetic substorms and/or auroral substorms in the midnight sector to be $T=0$. The onset of a polar magnetic substorm is manifested as a sharp onset of the so-called 'negative bay', for details, see Chapter 2. In this respect, we recall that in the classical analyses of geomagnetic storms by Moos (1910) and Chapman (1918), the onset time of storms is chosen to be the time of storm sudden commencements; the growth and decay of geomagnetic storms were then examined as a function of the storm time, reckoned from sudden commencements. In our study, there will be many polar substorms during a single magnetic storm, so that the development of polar phenomena should be examined by finding the onset time of each substorm.

In these eight chapters, each substorm will be discussed in approximately the following order:

(1) Introduction; (2) Typical Daily Variation; (3) Statistical Daily Variation Pattern; (4) Characteristics in the Midnight Sector; (5) Characteristics in the Evening Sector; (6) Characteristics in the Morning and Day Sector; (7) Characteristics in the Geomagnetically Conjugate Areas; and (8) Growth and Decay of the Substorm.

In Chapter 9, changes of the particle fluxes and the magnetic field in the magnetosphere and in interplanetary space during polar substorms are examined. Again, the time is reckoned from the onset time of polar magnetic substorms.

In the first part of Chapter 10, we shall review all the polar substorm phenomena in terms of the precipitation of energetic particles and the appearance of an electric field. The second part of Chapter 10 is devoted to a review of the entire subject by referring to both polar substorms and magnetospheric changes during polar substorms. This will help us in comprehending the basic processes associated with magnetospheric substorms.

References

GENERAL

CHAMBERLAIN, J. W.: 1961, *Physics of the aurora and airglow*, Academic Press, New York.
CHAPMAN, S. and BARTELS, J.: 1940, *Geomagnetism*, The Clarendon Press, London.
HESS, W. N.: 1968, *The radiation belt and magnetosphere*, Blaisdell Pub. Co., Waltham, Mass.

REFERRED TO IN TEXT

AKASOFU, S.-I. and CHAPMAN, S.: 1963, 'The lower limit of latitude (US sector) of northern quiet auroral arcs, and its relation to Dst (H)', *J. Atmospheric Terrest. Phys.* **25**, 9–12.
CARPENTER, D. L.: 1966, 'Whistler studies of the plasmapause in the magnetosphere. 1. Temporal variations in the position of the knee and some evidence on plasma motions near the knee', *J. Geophys. Res.* **71**, 693–709.
CHAPMAN, S.: 1918, 'An outline of a theory of magnetic storms', *Proc. Roy. Soc.* **A95**, 61–83.
DESSLER, A. J. and PARKER, E. N.: 1959, 'Hydromagnetic theory of geomagnetic storms', *J. Geophys. Res.* **64**, 2239–2252.

FELDSTEIN, Y. I.: 1963, 'Some problems concerning the morphology of auroras and magnetic disturbances at high latitudes', *Geomagnetizm i Aeronomiya* **3**, 183–192.

FELDSTEIN, Y. I. and STARKOV, G. V.: 1967, 'Dynamics of auroral belt and polar geomagnetic disturbances', *Planetary Space Sci.* **15**, 209–229.

FELDSTEIN, Y. I. and STARKOV, G. V.: 1968, 'Auroral oval in the IGY and IQSY period and a ring current in the magnetosphere', *Planetary Space Sci.* **16**, 129–133.

FRANK, L. A., VAN ALLEN, J. A., and CRAVEN, J. D.: 1964, 'Large diurnal variations of geomagnetically trapped and of precipitated electrons observed at low altitudes', *J. Geophys. Res.* **69**, 3155–3167.

HAGG, E. L.: 1967, 'Electron densities of 8–100 electrons cm^{-3} deduced from Alouette II high-latitude ionograms', *Canadian J. Phys.* **45**, 27–36.

KHOROSHEVA, O. V.: 1962, 'The diurnal drift of the closed auroral ring', *Geomagnetizm i Aeronomiya* **2**, 696–705.

MOOS, N. A. F.: 1910, *Magnetic observations made at the Government Observatory, Bombay, for the period 1846–1905 and their discussion. Part I. Magnetic data and instruments. Part II. The phenomenon and its discussion*, Government Central Press, Bombay.

MULDREW, D. B.: 1965, 'F-layer ionization troughs deduced from Alouette data', *J. Geophys. Res.* **70**, 2635–2650.

NAGATA, T.: 1963, 'Polar geomagnetic disturbances', *Planetary Space Sci.* **11**, 1395–1429.

NIKOLSKY, A. P.: 1947, 'Dual laws of the course of magnetic disturbances and the nature of mean regular variations', *Terr. Magn. Atmos. Elect.* **52**, 147–173.

PIGGOTT, W. R.: 1964, 'Studies of ionospheric absorption', in *Research in Geophysics*, Vol. 1 (ed. by Hugh Odishaw), The MIT Press, Cambridge, Mass., 277–297.

ROACH, F. E. and ROACH, J. R.: 1963, 'Stable 6300 Å auroral arcs in midlatitudes', *Planetary Space Sci.* **11**, 523–545.

SHARP, G. W.: 1966, 'Mid-latitude trough in the night ionosphere', *J. Geophys. Res.* **71**, 1345–1356.

STONE, E. C.: 1964, 'Local time dependence of non-Störmer cutoff for 1.5-MeV protons in quiet geomagnetic field', *J. Geophys. Res.* **69**, 3577–3582.

STRINGER, W. J. and BELON, A. E.: 1967, 'The statistical auroral zone during IQSY, and its relationship to magnetic activity', *J. Geophys. Res.* **72**, 245–250.

STRINGER, W. J., BELON, A. E., and AKASOFU, S.-I.: 1965, 'The latitude of auroral activity during periods of zero and very weak magnetic disturbance', *J. Atmospheric Terrest. Phys.* **27**, 1039–1044.

THOMAS, L.: 1960, 'The temporal distribution of storm-type sporadic E in the Northern Hemisphere', in *Some ionospheric results obtained during the International Geophysical Year* (ed. by W. J. G. Beynon), Elsevier Publishing Co., New York, 172–179.

CHAPTER 2

AURORAL SUBSTORM AND ASSOCIATED MAGNETIC
DISTURBANCES

2.1. Introduction: Typical Daily Variation

A close relationship between auroral displays and polar magnetic disturbances has
been recognized for many years (HARANG, 1951; HEPPNER, 1954; KHOROSHEVA, 1961).
It is, however, only recently that large-scale auroral displays over the entire polar
region and their detailed relationships with the associated geomagnetic activity have
become clear. In this chapter, we shall examine individual features of auroral displays
and only those geomagnetic disturbances under or near auroras. In Chapter 3, we
shall study geomagnetic disturbances associated with the auroral substorm by examining magnetic records collected from stations widely distributed over the entire
Northern Hemisphere.

A. AURORAL SUBSTORM

The characteristics of auroral displays depend greatly on both universal time and
local time, so that if observed at a particular point on the earth it is very difficult to
distinguish between those that depend on universal time from those that depend on
local time.

When auroral activity during the course of a night is observed at a particular
location, say at an auroral zone station, auroras appear first near the poleward horizon
as a glow; this is because we see only the upper part of an arc (or arcs). As night
progresses, auroras shift equatorward, namely toward the station, so that the lower
border becomes clearly visible. In general, they remain fairly quiet and homogeneous
until about 21 LT, except for the intermittent occurrence of rays and folds. Between
21 and 23 LT, we may see, once or twice, folds which travel westward rapidly along an
arc; in most cases this activity does not last for more than 30 min and auroras are
quickly restored to their quiet forms. The quiet forms in the evening hours are often
referred to as 'the pre-break-up form'.

By 23 LT, arcs may lie close to the zenith. This gradual equatorward shift is now
well understood; it is due to the fact that we are rotating under the auroral oval which
is eccentric with respect to the dipole pole (Section 1.4). Because of the eccentricity
of the oval, the relative distance between the oval and the station decreases until the
midnight hour.

Between 23 LT and 02 LT, auroras may suddenly become active and spread over
a substantial portion of the sky; this phenomenon is the so-called 'breakup'. After
this activity, we get an impression of luminous patchy clouds spreading over the

entire sky. Most of the patches appear to pulsate. This stage is often referred to as the 'post-break-up phase'.

After the 'post-break-up phase', a careful observer may see the formation of a quiet form again. A similar cycle of the change of forms may be repeated – a quiet

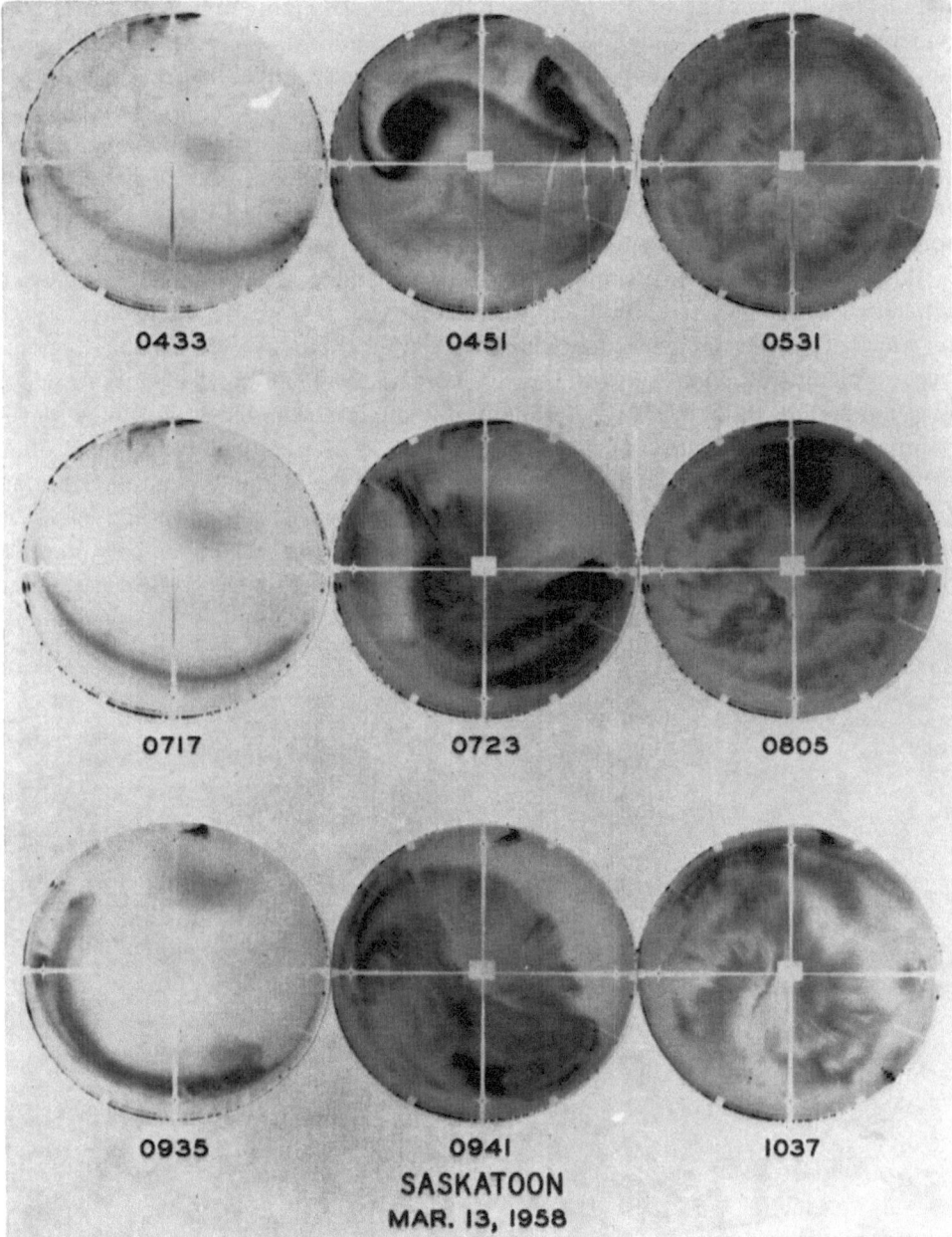

Fig. 14. Example in which three 'auroral cycles' were observed during the course of a night.

form to an active form, then to a patchy form, and finally back to a quiet form. He may see that on many nights such a cycle ('the auroral cycle') may be repeated twice, three times, or even four times during the course of a single night. Figure 14 (in negative) shows an example in which the cycle was repeated three times.

The morphological studies of auroral displays achieved their earliest successes in finding the features of auroral activity that are statistically more prominent during the course of a night at individual observatories. At an auroral zone station, the typical auroral forms are quiet arcs in the evening hours, active rayed bands around midnight, and 'patches' in the morning hours. On this basis, it was long thought that there is a fixed pattern of auroral activity under which the earth rotates once a day; that is to say, auroras in the evening hours are generally of a quiet form, generally of an active form around midnight, and generally of a patchy form in the morning hours. Figure 15 shows schematically this concept of the fixed pattern.

An extensive analysis of simultaneous all-sky camera photographs taken from a number of observatories showed, however, that the concept of a fixed pattern of auroral activity does not accurately describe the auroral activity over the polar regions. Auroras all along the auroral *oval* (not the auroral *zone*) can be of a quiet form for a certain period. But, such a quiet condition is often intermittently disrupted, particularly during geomagnetic storms. The breakdown of the quiet condition is often abrupt, and occurs first near the equatorward edge of the oval in the midnight sector. Soon the active features spread in all directions, poleward, westward and eastward along the oval, and equatorward. After this explosive spread, auroral activity gradually subsides all along the auroral oval. This auroral activity over the entire polar region is described in terms of the auroral substorm (AKASOFU, 1964).

The westward spread is characterized by a specific form of auroral activity which

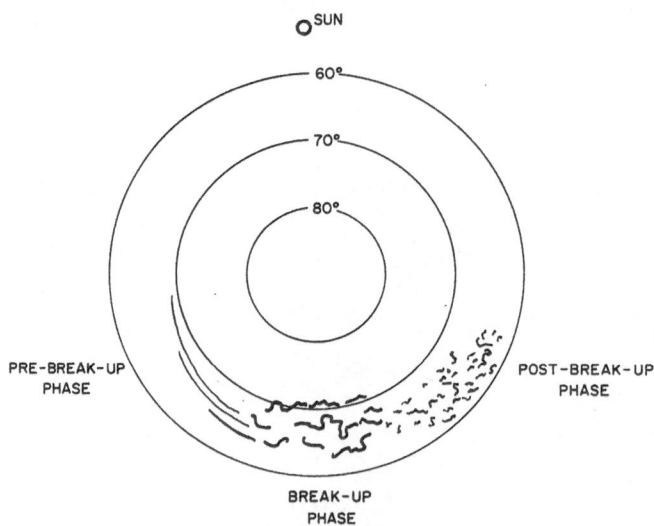

Fig. 15.   Schematic diagram showing the concept of a fixed pattern of auroral activity under which the earth rotates once a day.

is called the westward traveling surge and which travels rapidly along a quiet arc lying in the evening sector of the oval. This active feature does not last long, since the surge passes very rapidly across the sky at any particular station from the eastern horizon to the western horizon. The speed of surges is of order 1 km/sec; after the passage of a surge, quiet conditions are restored. For this reason, this active feature is not *statistically* prominent, and the quiet form is the most persistent feature in the evening sector. Furthermore, since the oval generally lies at least a few hundred kilometers poleward of typical auroral zone stations, such as College (dp lat 64.7°), Kiruna (dp lat 65.3°), and Macquarie Island (dp lat −61.1°), in the evening sector, visual observations (or a single all-sky camera) provide only poor information on active features in the evening hours.

In the morning hours, on the other hand, auroral arcs disintegrate into 'patches' even during a weak substorm. Further, this feature tends to persist well after the maximum epoch of the auroral substorm. For this reason, if the substorms occur very frequently, less than every few hours, auroras may not completely resume their quiet form. For this reason this active feature is *statistically* most prominent.

It is thus not difficult to reconcile apparent disagreements between the concept of the fixed pattern and the concept of the auroral substorm. The concept of the auroral substorm may be roughly expressed by saying that the pattern of active auroral displays shown in Figure 15 appears only intermittently; quiet forms are seen over the entire local time range during the period between two auroral substorms. This situation can easily be recognized by the fact that the fixed pattern (Figure 15) cannot describe auroral activity shown in Figure 14.

The concept of the auroral substorm suggests that the onset of the active period of auroral displays is a universal time dependent phenomenon, while the *characteristics* of auroral activity along the oval during substorms depend strongly on local time. Figure 16 illustrates this concept schematically; it is constructed to show four substorms occurring between 05 and 17 UT, at about 06, 09, 13 and 16 UT respectively.

At 05 UT quiet arcs lie all along the auroral oval. Central Canada is located in the midnight sector at that time. Suppose that the auroral substorm happens to occur a little before 06 UT. A rapid poleward expansive motion of the auroral system is observed in central Canada during the expansive phase. At 06 UT, about 10 min after the onset of the substorm, the northernmost band has attained a dipole latitude of almost 80°. At this time, only a slight increase of brightness of the quiet arcs may be seen in Alaska, but most of the arcs in the morning side disintegrated into 'patches'. The resulting patches drift eastward, namely toward the morning twilight sector. Auroras which occupy the midday part of the oval (namely, off the Siberian Coast), become brighter during this period. At 07 UT the substorm begins to subside and a number of bands return toward their initial location in the midnight sector. In the evening sector, a westward traveling surge generated by the poleward expansive motion in the midnight sector degenerates into an irregular band after traveling a considerable distance; activated bands do not, however, disintegrate into patches. By 08 UT quiet conditions are restored.

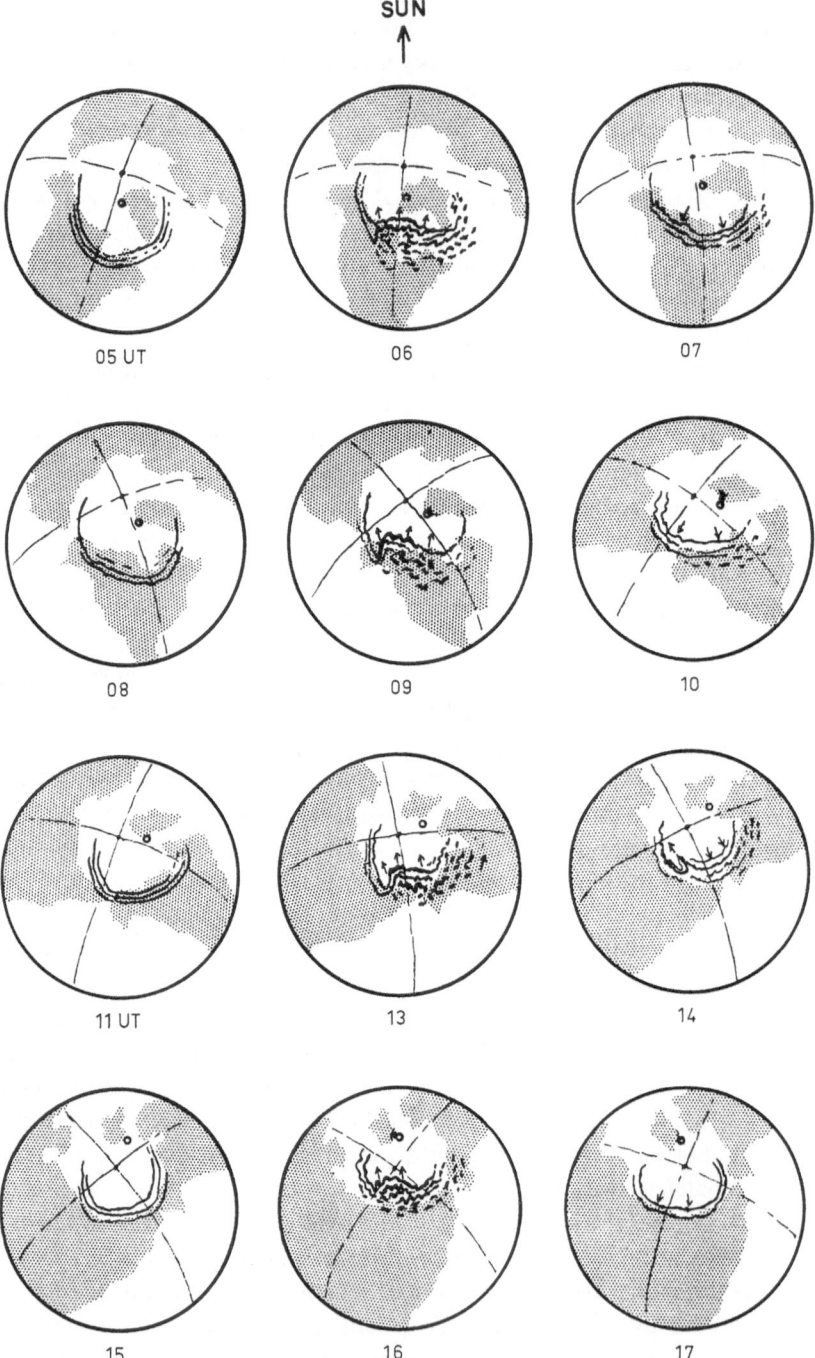

Fig. 16. Schematic diagram to show the concept of the auroral substorm by assuming the situation in which four substorms occur between 05 and 17 UT, at about 06, 09, 13, and 16 UT.

A little before 09 UT another substorm happens to occur. The Alaska-Canada border, where a rapid poleward expansive motion of the auroras is seen, is located in the midnight sector at that time. In central Alaska (which is located in the late evening sector), the poleward expansion is seen first as a mass of light near the eastern horizon and then as a great fold which travels rapidly westward. The fold may be seen later (typically in about 10 to 15 min) as a westward traveling surge over the eastern Siberian sky. The auroras over the American continent are disrupted and are seen as eastward drifting patches. At 10 UT active bands start to return to their initial location from their northernmost location which was attained over the Arctic Sea. Midday auroras over Spitzbergen also become active. By 11 UT the second substorm is essentially over, and three quiet arcs are seen in the dark hemisphere.

A little before 13 UT, a new substorm occurs and the most active display is seen over Eastern Siberia and the Bering Strait which are located in the midnight sector at that time. Irregularly folded bands drift rapidly eastward over Alaska. The arcs over the central Canadian continent disintegrate into 'patches' soon after the beginning of the substorm. At 14 UT the substorm enters into the recovery phase; a group of loops is drifting along the Arctic Sea coast in Siberia. Midday auroras over Greenland also become active. The substorm is essentially over by 15 UT and auroras all along the oval become fairly quiet. Another substorm occurs a little before 16 UT. The most brilliant display is seen in central Siberia and to the west of it during this substorm. Arcs over Alaska disintegrate, and the resulting drifting patches are seen in the twilight sky.

The concept of the auroral substorm has been developed to express systematically such auroral activity. Figure 17 shows the growth and decay of a model auroral substorm $(A \rightarrow B \rightarrow C \rightarrow D \rightarrow E \rightarrow F \rightarrow A)$. The auroral substorm has two characteristic phases; the expansive phase and the recovery phase. The first indication of the substorm is a sudden brightening of one of the quiet arcs lying in the midnight sector of the oval or a sudden formation of an arc (B, $T = 0$–5 min). In most cases, the brightening of an arc or the formation of an arc is followed by its rapid poleward motion, resulting in an 'auroral bulge' around the midnight sector (C, $T = 5$–10 min). The so-called 'break-up' phenomenon occurs in the bulge, but it is not the whole display. As the auroral substorm progresses, the bulge expands in all directions (D, $T = 10$–30 min). In the evening side of the expanding bulge, a large-scale fold appears which travels rapidly westward along an arc, namely, the westward traveling surge. In the morning side of the bulge, arcs appear to disintegrate into 'patches' which drift eastward with a speed of order 300 m/sec.

When the expanding bulge attains its highest latitude, the recovery phase of the auroral substorm begins (E, $T = 30$ min–1 hour). The expanded bulge begins to contract. The westward traveling surge may still travel a considerable distance after the end of the expansive phase, but it degenerates eventually into irregular bands. In the morning sky, eastward drifting patches remain until the very end of the recovery phase (F, $T = 1$–2 hours). At the end of the substorm, the general situation will be similar to that just before the onset of the substorm.

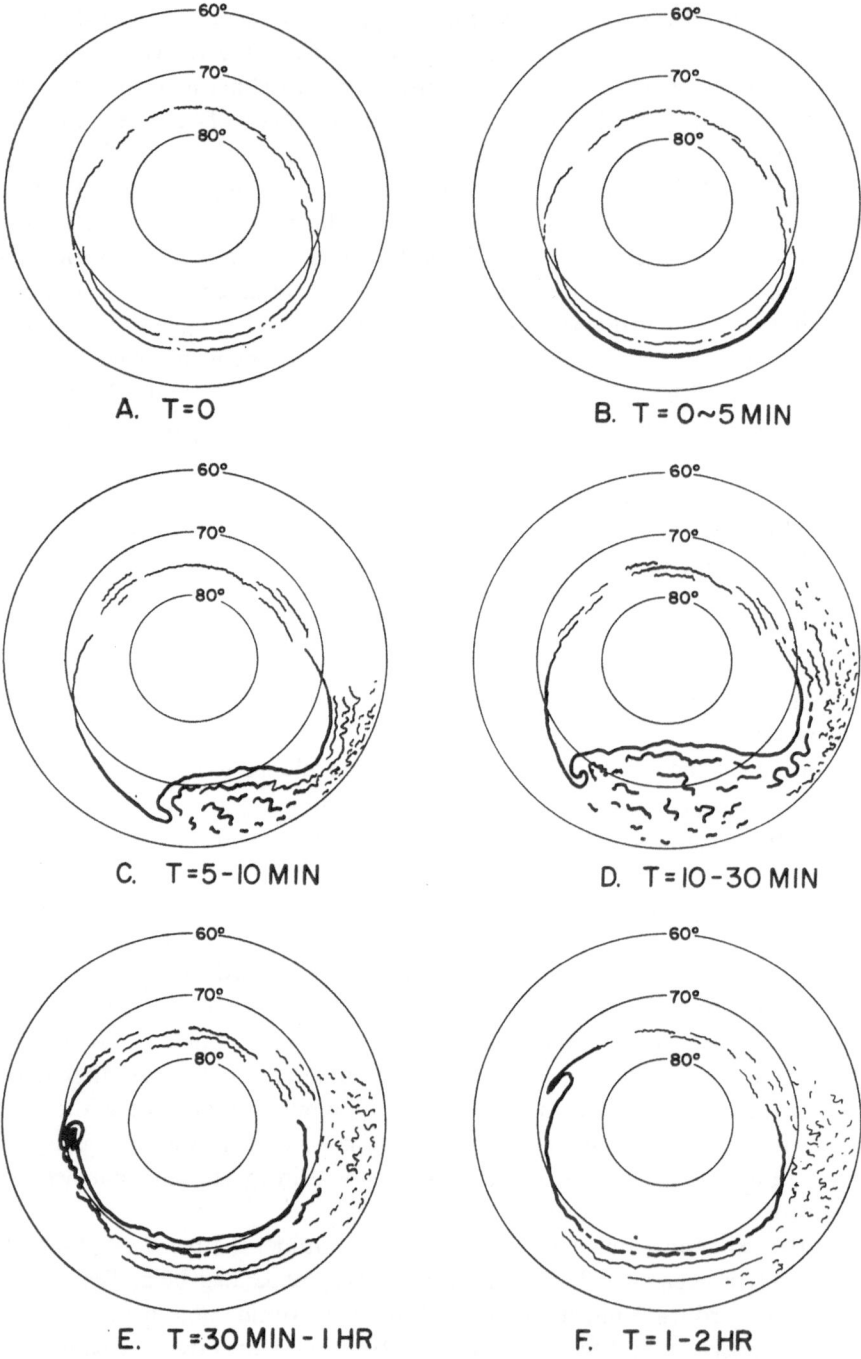

Fig. 17.   Schematic diagram to show the development of auroral substorm.

The concept of the auroral substorm can be tested either by observing the entire polar region by a satellite-borne camera far above the auroral level or by observing auroras in a particular local time sector for several hours. If this concept of the auroral substorm is correct, an observer who remains in the midnight or in the late evening sector observes intermittent activations from the quiet auroral form; that is, quiet arcs intermittently become active rayed bands. In Figure 16, quiet arcs seen in the midnight sector at 05 UT became active at 06 UT. But a quiet form resumed after the substorm was over (08 UT); another substorm disrupted this quiet condition anew at about 09 UT. On the other hand, if the older concept of the fixed pattern is correct, auroras in the midnight or late evening sector would always remain active (like the conditions at 06 or 09 UT).

The availability of a high speed jet aircraft has made it possible to remain for several hours in approximately the same local time sector along the geographic latitude circle 60° or higher. In the following, we shall describe one of the successful constant local time flights which were made during the NASA Airborne Auroral Expedition in the spring of 1968. The flights originated at Churchill (dp lat 68.7°) and ended at Fairbanks (dp lat 64.7°).

The airplane took off from Churchill at 0521 UT on March 3, 1968 ($K_p = 2+, 1+, 3-, 3-, 2+, 3+, 4_0, 4-$) and arrived over Fort Yukon at 0906 UT; the local midnight at the two locations is 0600 UT (CST) and 1000 UT (AST), respectively; the magnetic local midnight at the two places is approximately 0725 UT and 1140 UT, respectively. The flight was made in such a way as to follow approximately along the auroral zone, namely the dipole latitude circle of 67°, so that the airplane was approximately in the late evening sector of the auroral oval for three hours and 45 minutes during this part of the flight. After reaching Fort Yukon, the airplane headed South and arrived over Anchorage at 0955 UT, after passing over Fairbanks at 0924 UT. Then the airplane headed North and landed at Eielson (near Fairbanks) at 1036 UT. The total flight time was 5 hours and 15 min. During the flight between Churchill and Fort Yukon, auroras were seen both to the North and South so that the aircraft was within the oval during this part of the flight.

Between the time of take-off (0521 UT) and 0700 UT, auroras on both sides intermittently showed some active features (see the photographs taken at 0630 and 0634 UT in Figure 18), but remained fairly quiet. At about 0710 UT, auroras to the east suddenly became bright, and an auroral substorm began.

This auroral activity generated an extremely intense westward traveling surge which traveled rapidly westward. The surge caught up to the airplane at about 0713 UT; the airplane was near the western shore of the Great Slave Lake (62° N, 117° W) at that time. (See the photographs taken at 0710, 0712, 0713, and 0715 UT.) Considerable auroral activity then followed until about 0821 UT. By 0900 UT, however, a little before arriving at Fort Yukon, the substorm was practically over.

After 0900 UT, the only noticeable main feature before the onset of a new substorm at 1002 UT was quiet auroras to the South. This was confirmed by flying southward; as can be seen in the photograph taken at 0930 UT, auroras consisted

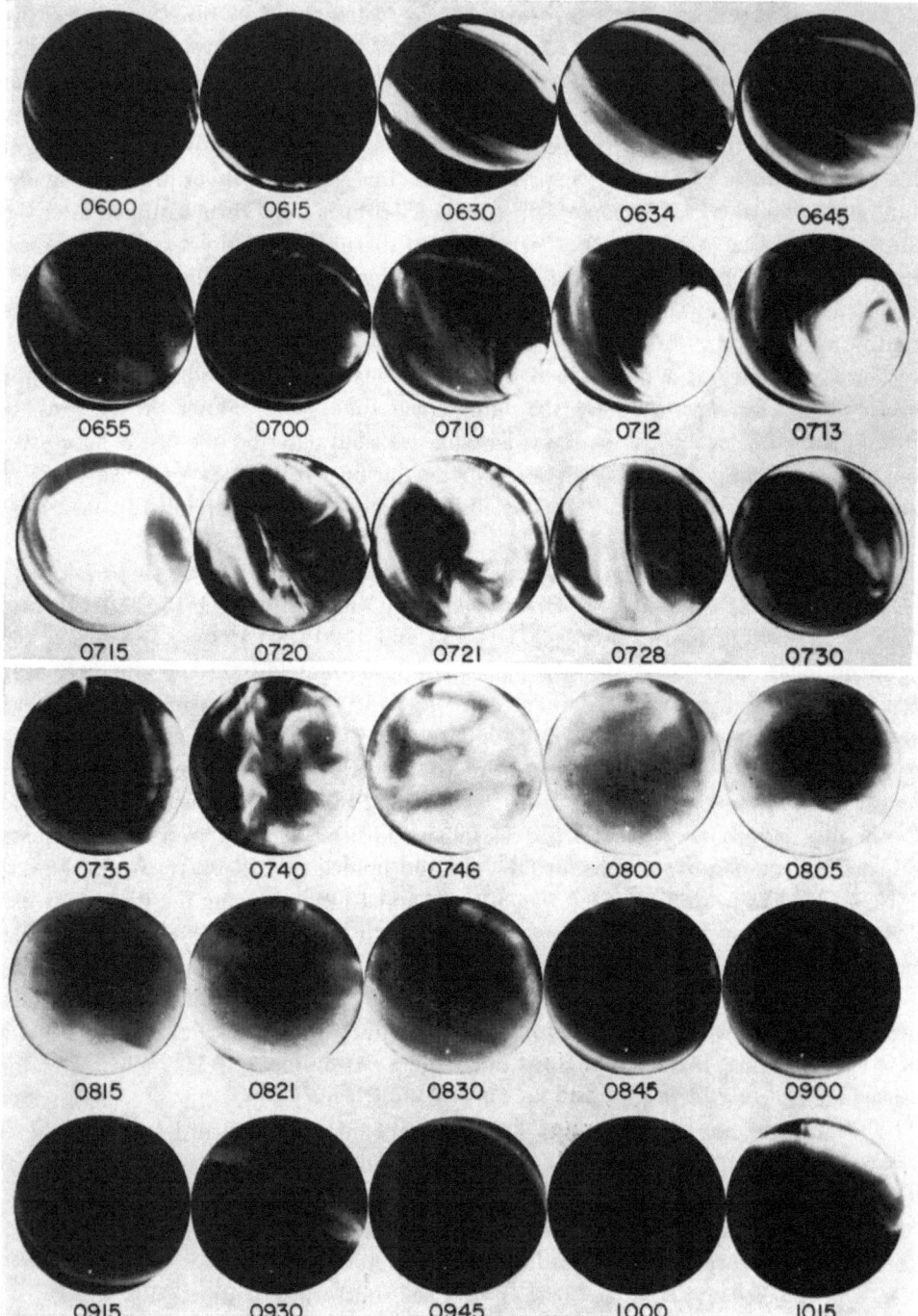

Fig. 18.   All-sky photographs taken during the constant local time flight on March 3, 1968; the plane remained in the late evening sector during the flight from Churchill to Fairbanks.

of several quiet arcs located mostly a little North of Fairbanks. This quiet condition corresponds to that between 07 and 09 UT in Figure 16. The only difference is that the substorm began at about 0710 UT during the flight. At 1000 UT, 5 minutes after heading North, quiet auroras were seen far to the North.

A new activity began along the quiet aurora at about 1002 UT. The photograph taken at 1015 UT shows an intense surge which traveled along the aurora which was very quiet at 1000 UT. The difference in appearance of the two surges is due to the relative location of the plane with respect to the surges. In fact, a typical feature of the surge was recorded at College near the zenith at about 1013 UT. This condition is quite similar to that at 09 UT in Figure 16.

B. POLAR MAGNETIC SUBSTORM

The auroral substorm is associated with intense geomagnetic disturbances. Like auroral displays during auroral substorms, the onset of polar magnetic substorms is a universal time dependent phenomenon, but characteristics of the magnetic disturbance fields depend strongly on local time. Such geomagnetic disturbances are called the polar magnetic substorms. Recent progress in this study has made it possible to deduce a three-dimensional current system which may cause the polar magnetic substorm.

In order to study polar magnetic substorms, two completely independent approaches have been made in the past. The first approach has been to obtain simultaneous geomagnetic disturbance *vectors* at a number of stations and to infer a current system which could give rise to such a magnetic disturbance. This method of analysis is here called the *SD analysis*. In general, the current is assumed to be located on a spherical shell (the ionosphere). The result thus obtained does not necessarily mean, however, that such a current system exists in the ionosphere, but is simply a means of expressing geomagnetic disturbances. For this reason, it is called the *equivalent current system* and should not be confused with the actual current system. The other approach is mainly concerned with the *range* of geomagnetic disturbances (generally, the H component), rather than the disturbance vectors, and also with their daily or seasonal characteristics. Here this approach is called the *spiral analysis* and was initiated by STAGG (1926, 1935a, b, c) and later extended by NIKOLSKY (1947), MAYAUD (1956), BURDO (1959), and others; see also a review paper by HOPE (1961).

1. *SD Analysis*

CHAPMAN (1918, 1935), based on Moos' method, established the so-called SD current system. His analysis is equivalent to a Fourier analysis of the D field *along the circles of constant dipole latitude,*

$$D = c_0 + \sum_n c_n \sin(n\phi + \varepsilon_n), \tag{1}$$

where ϕ denotes dp longitude and ε_n the phase angle. The first term indicates the *Dst* component and the second term, the *DS* component. Chapman first obtained the

SD variation which is an average of the *DS* variation over each of the first two storm days. The current system associated with the *SD* variation is shown in Figure 19. It represents the statistical daily pattern of the magnetic variations in terms of the equivalent current system.

Later, VESTINE (1940), SILSBEE and VESTINE (1942), and FUKUSHIMA (1953) obtained the equivalent current systems or the distribution of the disturbance vectors and confirmed Chapman's result. As we shall see later, this analysis placed great emphasis

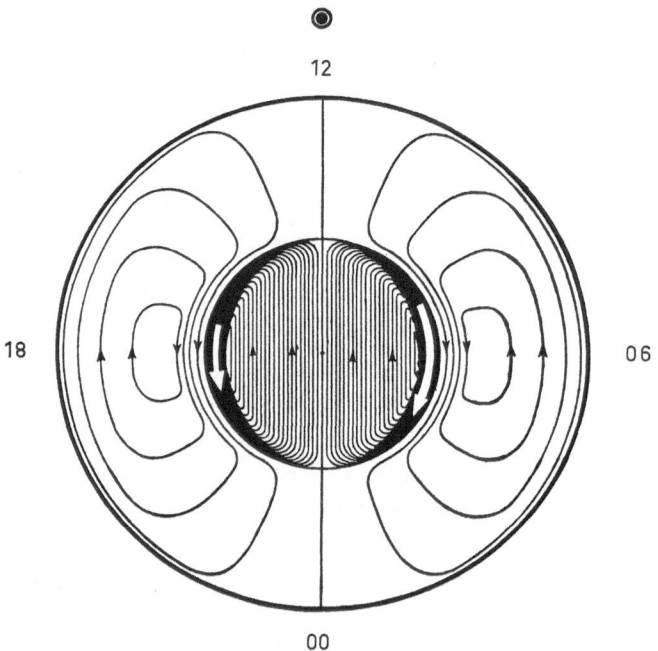

Fig. 19.   *SD* current system.

on the geomagnetic latitude circle of 67°, namely the auroral zone, which is now found to have only a statistical meaning. Therefore, it has failed to reveal the eccentric nature of geomagnetic phenomena in the polar region, namely that the major geophysical phenomena occur along the auroral oval and not along the auroral zone. This failure is not, of course, entirely due to incorrectness of the analysis, but is due in part to the difficulty in analyzing an eccentric oval system in a system based on dipolar coordinates, particularly when a close network of observations over the polar cap is not available.

In the *SD* analysis, high harmonic terms in (1) have been called *Di* where the suffix *i* signifies an 'irregular' component (CHAPMAN, 1951). Thus, (1) may be rewritten as

$$D \simeq Dst + DS_1 + DS_2 + Di,$$

where DS_1 denotes the diurnal component of DS, and DS_2 the semi-diurnal component. In general $DS_1 \gg DS_2$. Now, in Figure 20 let us compare $(Dst + DS_1)$ with typical daily magnetic records from College (dp lat 64.7°). According to SUGIURA and CHAPMAN (1960), D (in units of γ) is given by

$$D(H) \simeq -60 + 182 \sin(\phi + 206°)$$

along the auroral zone for a great storm. Comparing the actual trace H and $D(H)$ at the top, it is immediately clear that although DS_1 indicates the correct sign of changes,

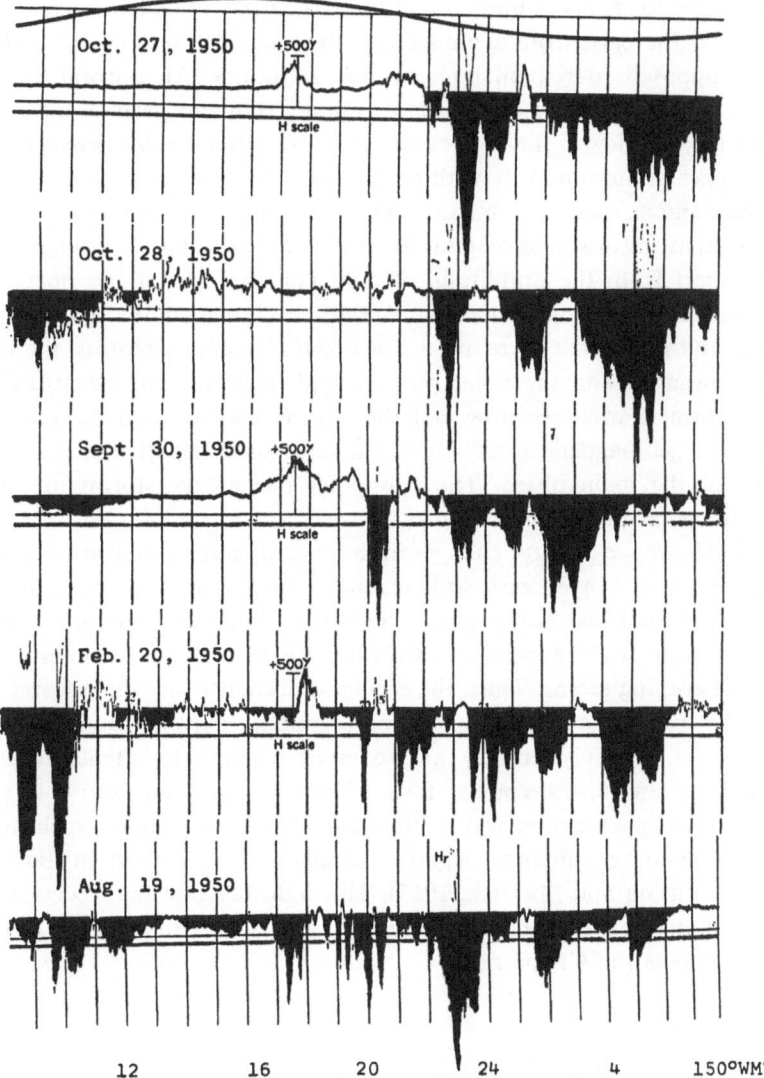

Fig. 20. Several examples of the daily magnetic records from College. The top curve shows the SD variation for a great storm, obtained by SUGIURA and CHAPMAN (1960).

the actual trace consists of short-lived impulses whose magnitude is far greater than the amplitude of $DS_1(H)$; negative impulsive changes, the so-called 'negative bays' are shaded to emphasize their impulsive nature. Therefore, Di is not a small and irregular fluctuation superposed on DS, but is the essential feature of geomagnetic disturbances in the auroral zone. Indeed, those impulsive changes are one of the major features of the *polar magnetic substorm*, namely

$$D \simeq \Sigma \, Di \simeq \Sigma \, \text{(the polar magnetic substorm)}.$$

Here we must realize an important change in our concept of polar magnetic disturbances. The SD analysis implies that the SD current system is fixed with respect to the sun, and the earth rotates under this fixed current system. The SD current intensity is supposed to remain constant for 24 hours. An auroral zone station observes the daily variation because it moves under different parts of the SD current system. Our present view is that a current system which has *some resemblance* to the SD system appears intermittently with the lifetime of order $1 \sim 3$ hours, almost in an impulsive way and as successive bursts, and the earth rotates under such a situation. We note also that the absolute magnitude of the current is much stronger than what we would expect from the amplitude of SD. This concept of the polar magnetic substorm is similar to what BIRKELAND (1908–1913) called the polar elementary storm. Birkeland noted that there are at least four elementary storms, the equatorial positive and negative and the polar positive and negative. The equatorial positive elementary storm is what we now call the storm sudden commencement and the initial phase of geomagnetic storms, and the equatorial negative elementary storm is identified as the main phase. The two polar elementary storms are the major features of the polar magnetic substorm.

In Figure 19, we see positive changes in the evening hours; they are called positive bays. This is expressed by an eastward current in the SD (equivalent) current system. When the SD current was constructed, the auroral zone was thought to be the instantaneous region of the aurora and thus the region of the most intense magnetic activity in the evening sector. Thus, the eastward current along the auroral zone was thought to be the strongest one generating secondary currents both poleward and equatorward of it. It is only recently that we have recognized an intense negative bay a little poleward of an auroral zone station. Figure 21 gives an example to show that when positive changes are observed at typical auroral zone stations, such as College, Healy, and Anchorage, intense negative changes are recorded at Barter Island (dp lat 70°) (WESCOTT and MATHER, 1965). This indicates that the SD current system does not accurately express the distribution of the disturbances. This problem will be discussed in detail in Chapter 3.

2. *Spiral Analysis*

Typically, the range of geomagnetic disturbances in high latitudes ($\geqslant$ dp lat 60°) shows three peaks during the course of a day (Figure 22). For example, at Wrangel Island (dp lat 64.7°) the first peak appears in the afternoon sector (~ 17 local time),

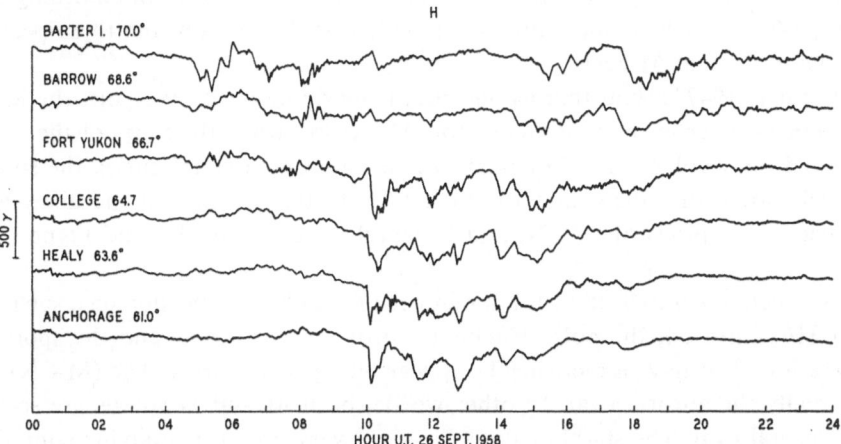

Fig. 21. Simultaneous magnetic records (September 26, 1958) obtained from the IGY North-South stations in Alaska (Wescott, E. M. and K. B. Mather: *J. Geophys. Res.* **70**, 29, 1965).

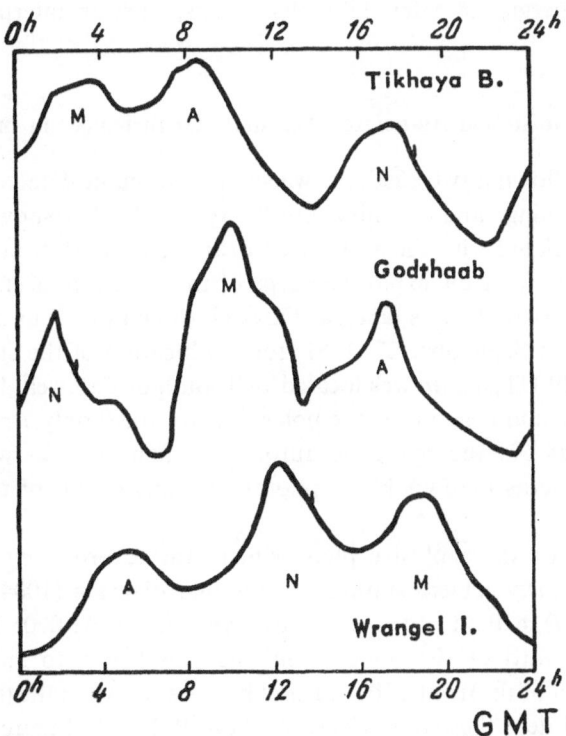

Fig. 22. Daily variation of the range of geomagnetic disturbances at three stations, Tikhaya Bay, Godthaab and Wrangel Island (Nikolsky, A. P.: *Problem Severa* **1**, 116, 1958).

the second in the midnight (00 LT) sector, and the third peak in the morning sector
(07 LT); Wrangel Island local time = UT − 12. For this reason, the three peaks are
called the A, N, and M peaks.

NIKOLSKY (1947) found that as one goes from dp lat 60° to the pole, the M peak
tends to occur later in the morning sector. Therefore, when the times of the M peak
are plotted on a polar map, they tend to line up along a spiral curve, the so-called
'M spiral'. MAYAUD (1956) and BURDO (1959) further elucidated the analysis and
confirmed three spirals, the A, N, and M spirals; see Section 1.4 and Figures 6a, 7
and 9.

As we mentioned in Section 1.4, a number of geophysical phenomena, such as the
Hα (or Hβ) emission, the polar blackout, auroral radar echos, and the sporadic E
layer are found to have a tendency to appear along such spirals. The (M + N) spiral
is essentially the auroral oval. In other words, both M and N spirals are segments
of the auroral oval. The spiral analysis has been very useful in studying complicated
features of geomagnetic disturbances in the polar cap, and has indeed revealed the
segments of the auroral oval.

Unfortunately, however, the analysis has relied mainly on the range, and the
results have been interpreted in terms of the precipitation pattern of energetic particles,
namely the so-called 'Störmer spiral'. Since most visible auroras are known to be caused
by electrons of energies of a few kilovolts or less, such an interpretation is a most
unlikely one.

2.2. Auroral Substorm and Associated Magnetic Disturbances in the Midnight Sector

In this and the following two sections, we shall describe in detail auroral displays in
the midnight, evening, and morning hours. We shall also show the geomagnetic
disturbances associated with them under or near the auroral displays.

A typical example of the explosive phase of the substorm is shown in Figure 23.
The series of photographs was taken at Farewell (dp lat 61.5°) in the midnight sector
(10 UT = 00 LT) on September 22, 1957; for the location of the station, see maps in
Appendix. At 1029 UT, an arc was located well south of Farewell. It suddenly became
bright at 1030 UT and began to move poleward. By 1037, only 7 min after the onset,
the entire sky was covered by active auroras. In fact, auroras were so bright and
active, all-sky cameras were unable to photograph any details of the structure of the
auroras.

Figure 24 shows the explosive phase which was recorded by the IGY network
of the Alaskan all-sky camera stations. In the first diagram (1004 UT [0004 LT] on
February 12, 1957), a faint arc is seen near Anchorage. At 0005 LT, a new arc was
formed just poleward of the pre-existing one and began to move poleward. The
successive location of the front of the auroral bulge is shown in the following diagrams.
The front arrived near the Arctic Coast at about 0018 LT. Figure 24 also shows the
equivalent current vectors at a number of Alaskan magnetic stations. The direction
of the vectors indicates the direction of the equivalent current, and the length is

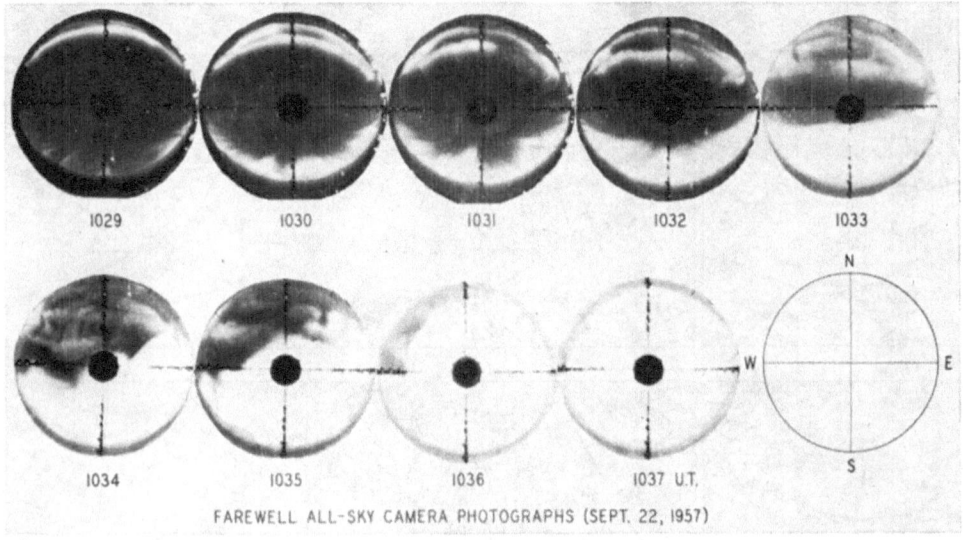

1029   1030   1031   1032   1033

1034   1035   1036   1037 U.T.

FAREWELL ALL-SKY CAMERA PHOTOGRAPHS (SEPT. 22, 1957)

Fig. 23. Typical example of the explosive phase of the auroral substorm observed in the midnight sector.

proportional to the magnitude of the total horizontal magnetic vectors; the length corresponding to a 100γ variation is shown in the first diagram. The directions of the actual horizontal magnetic disturbance vectors can be obtained by rotating the current vectors 90° counter-clockwise. The growth of the magnetic disturbance is seen first at Anchorage at 0008 LT and then at College, Healy, Big Delta, and Northway (at 0009 LT).

The highest latitude attained by the auroral bulge depends on the intensity of the substorm and also on the initial latitude where the poleward expansion originates. It is not uncommon to observe the bulge reaching as high as dp lat 80°. Figures 25a and b (in negative) show the expanding auroral bulge observed at Mould Bay (dp lat 79.1°). On December 2, 1965, the bulge appeared only near the horizon at about 0950 UT. In another instance, during an intense substorm which occurred at about 0920 UT on November 5, 1965, the bulge reached the zenith of Mould Bay. Figures 25a and b also show the magnetic changes (the X-component) associated with the above auroral activity. In the December 2, 1965 event, only a slight negative change was seen, while the November 5, 1965 event was associated with an intense negative bay which is superposed on a gradual positive change. It indicates that the westward current moved with the front of the auroral bulge as far as (or near) Mould Bay. For details of the poleward expanding bulge, see AKASOFU et al. (1966a).

2.3. Auroral Substorm and Associated Magnetic Disturbances in the Evening Sector

In the mid-evening hours (20~21 LT), the distance between the center-line of the oval and typical auroral zone stations, such as College and Kiruna, is of order 500 km.

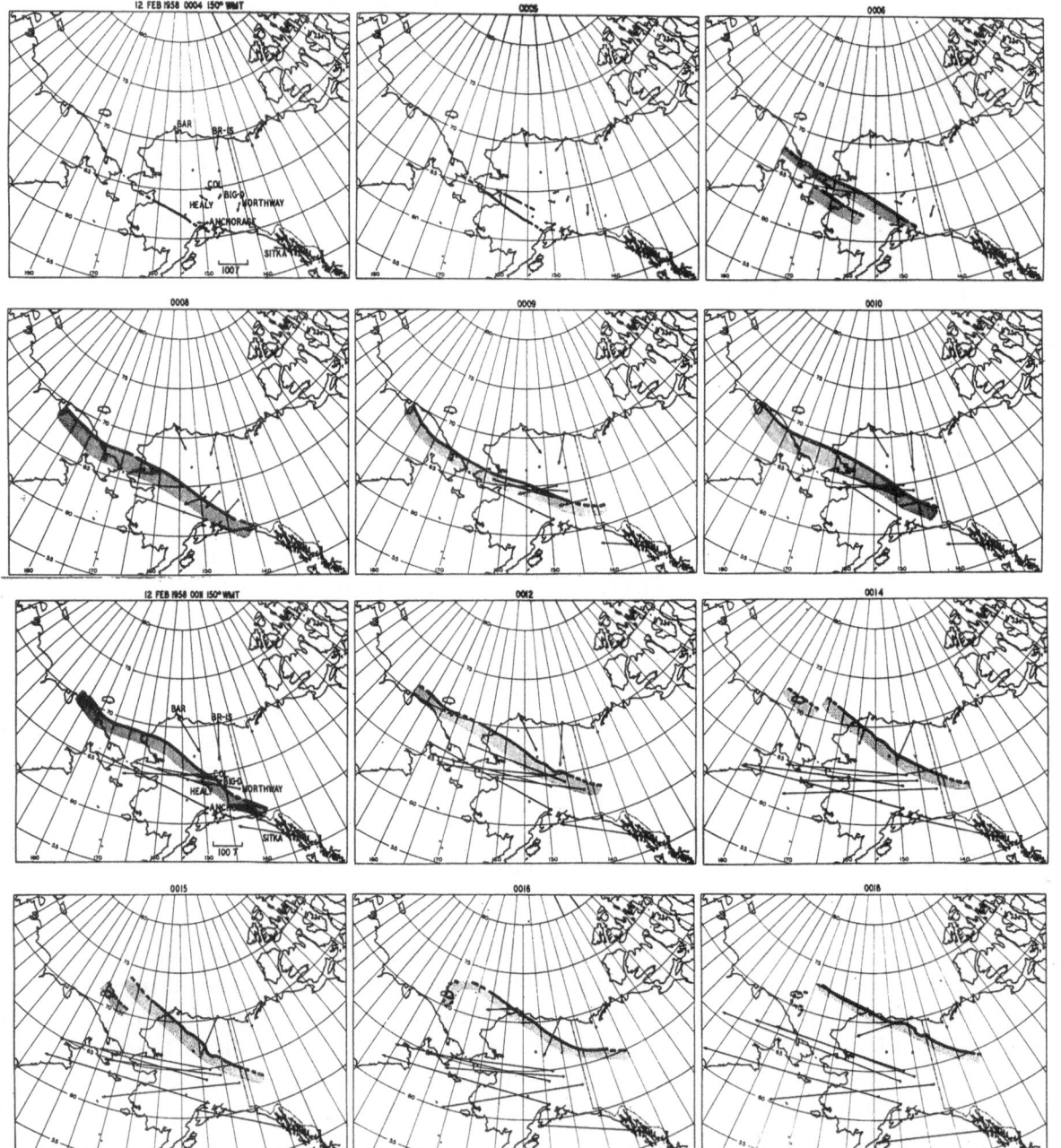

Fig. 24.   Poleward motion of the front of the auroral bulge. The simultaneous equivalent current vectors at Alaskan magnetic stations are also shown.

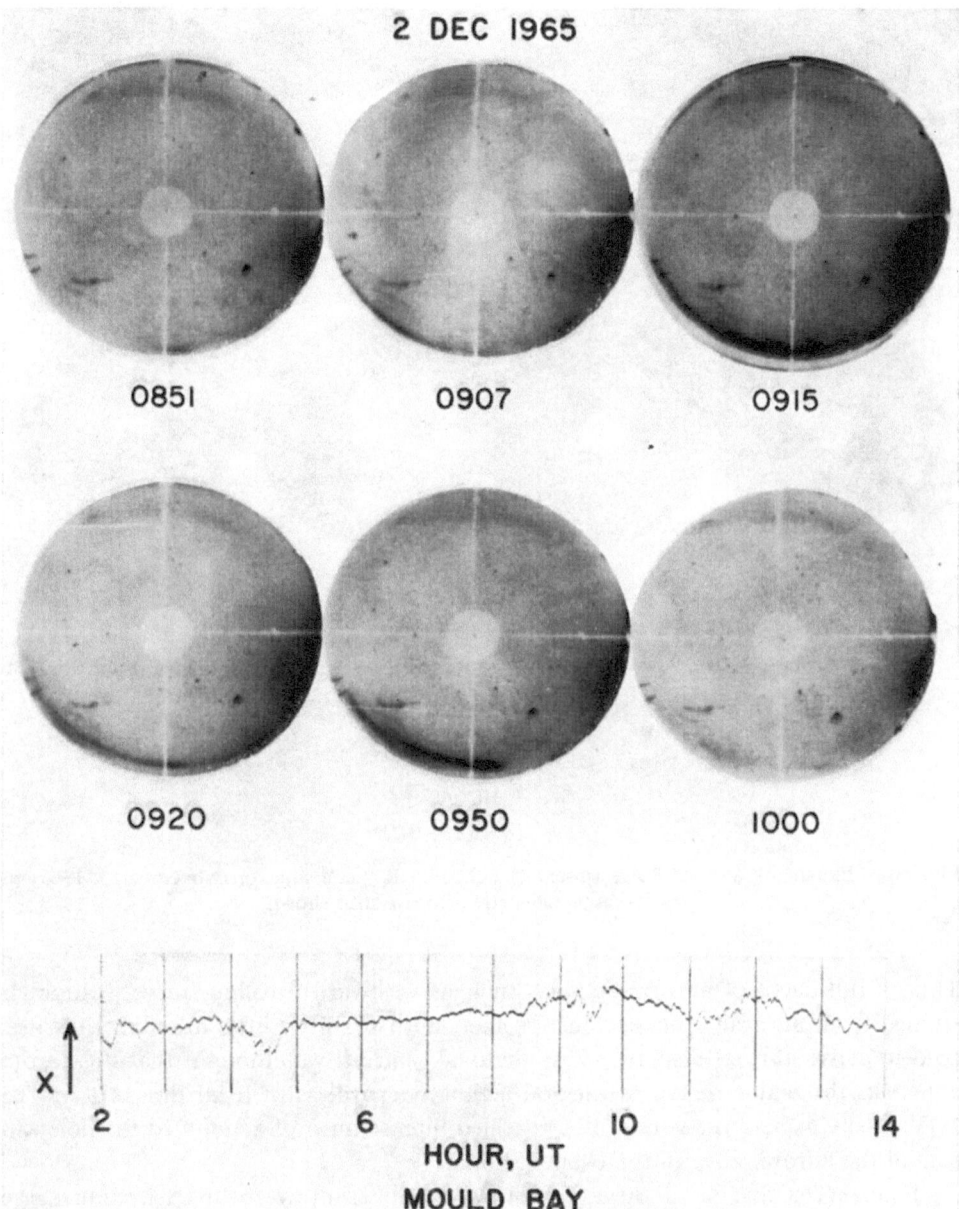

Fig. 25a.   Expanding auroral bulge observed at Mould Bay, Canada, on December 2, 1965; the simultaneous magnetic record is also shown.

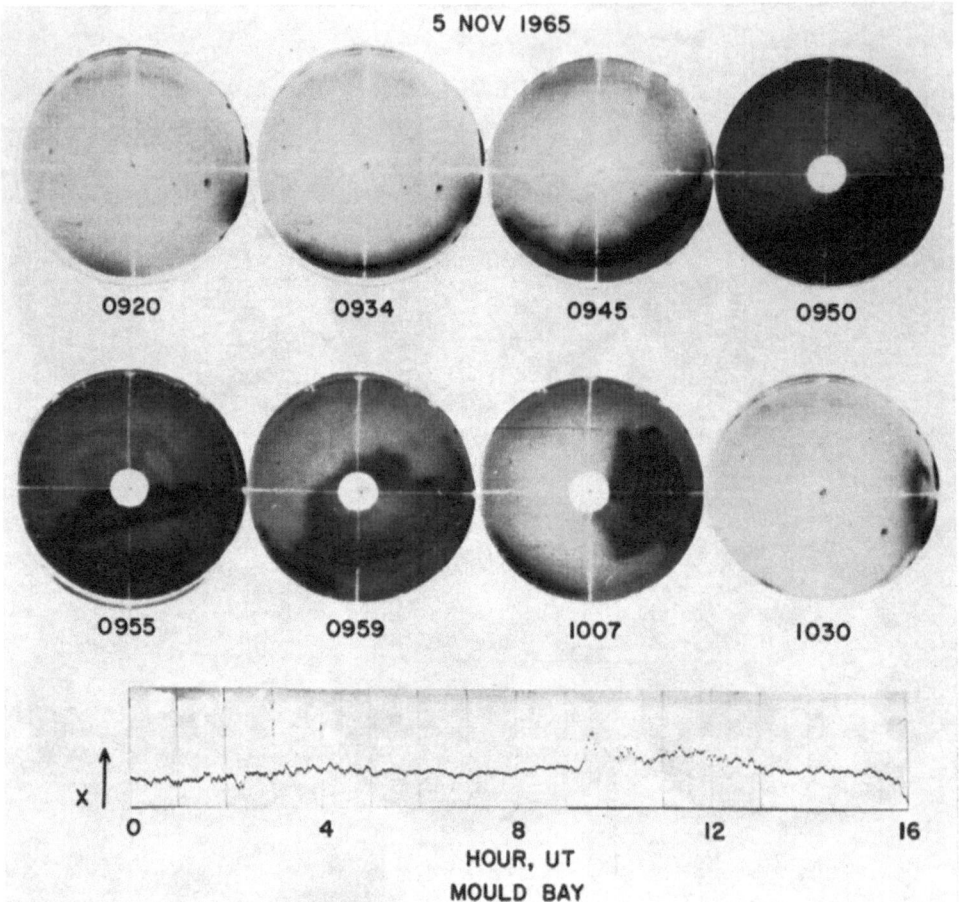

Fig. 25b. Expanding auroral bulge observed at Mould Bay, Canada, on November 5, 1965; the simultaneous magnetic record is also shown.

Thus, a full detail of auroral activity, such as westward traveling surges, cannot be studied from auroral zone stations. In fact, until recently, little attention has been paid to active auroral displays along the oval, since it was thought that the auroral zone was the major region of auroral activity, regardless of local times. It was the IGY all-sky camera network which revealed intense auroral activity to the poleward side of the auroral zone in the evening hours.

Figures 26a and b (in negative) show examples of westward traveling surges recorded at Fort Yukon, located about 200 km North of College. In both figures, the photographs in the first row show changes of auroras for every 30 or 60 min intervals

Fig. 26a, b. Examples to show that a positive magnetic bay is recorded at College (typical auroral zone station), when an intense westward traveling surge passed the northern sky of Fort Yukon which is located about 200 km dipole North of College.

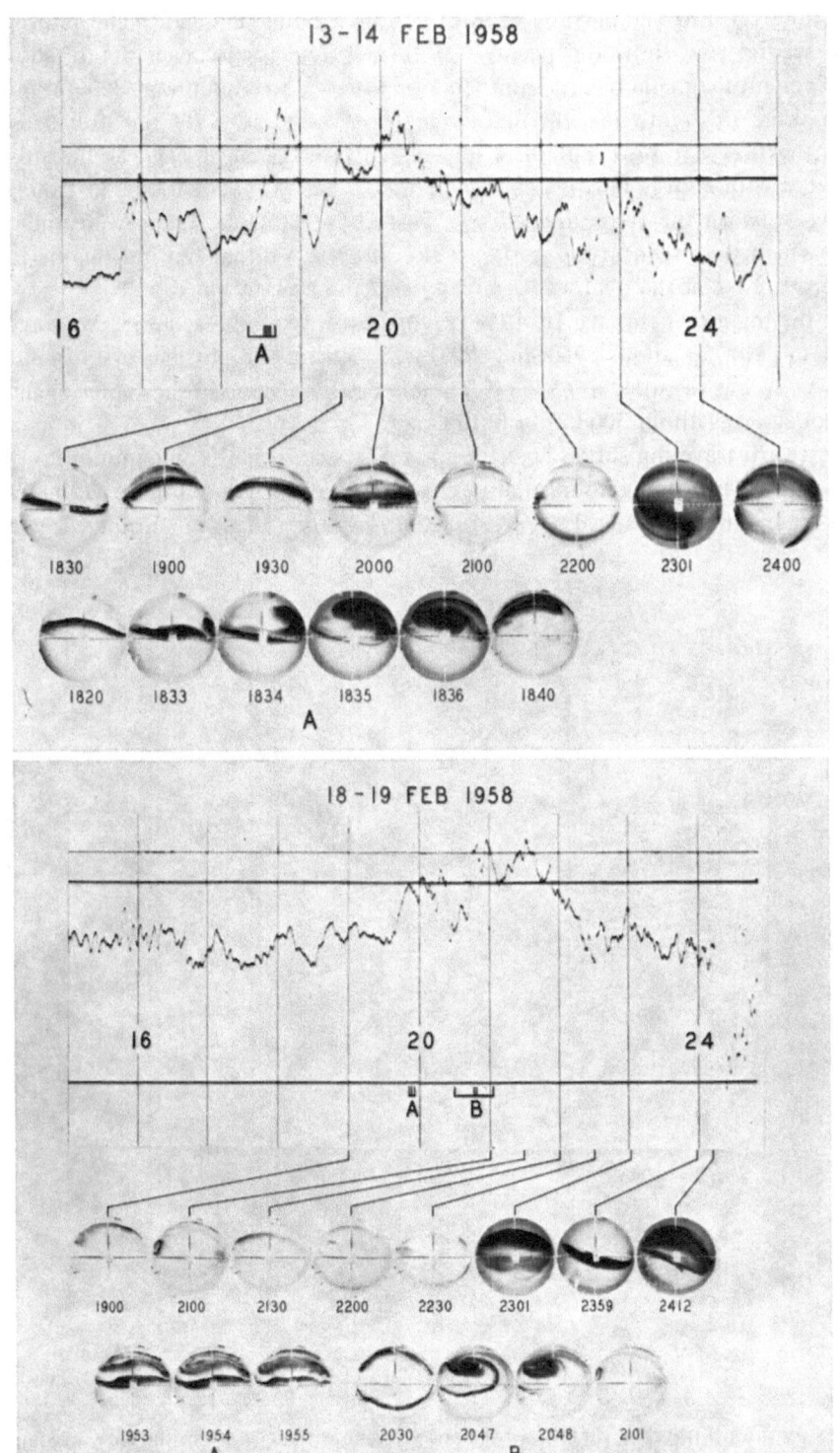

(depending on how significantly auroral conditions has changed). The photographs in the second row show the passage of surges. The top of each figure shows the *H* component magnetic record from College which is located about 200 km South of Fort Yukon. In Figure 26a, the first surge appeared at 1833 LT and traveled across the northern sky of Fort Yukon in less than 10 min. The passage of the surge was associated with a large positive change or the so-called 'positive bay' at College; the distance between the surge and College was about 300 km. There were then a few surges which were seen in the northern sky of Fort Yukon. During this period, the southern border of the oval was gradually shifting equatorward.

On the night of February 18, 1958 (Figure 26b), two intense surges were recorded at Fort Yukon, at about 1950 and 2080 LT. During the passage of both surges, a positive bay was recorded at College. The disturbance between the center of the surge and College was about 300 km in both cases.

Westward traveling surges travel a great distance. It is not uncommon to observe them in the early evening hours at about dp lat 70° and beyond. Figure 27 (in negative) shows an example of a surge recorded at Pyramida (dp lat 74.5°) in the early evening

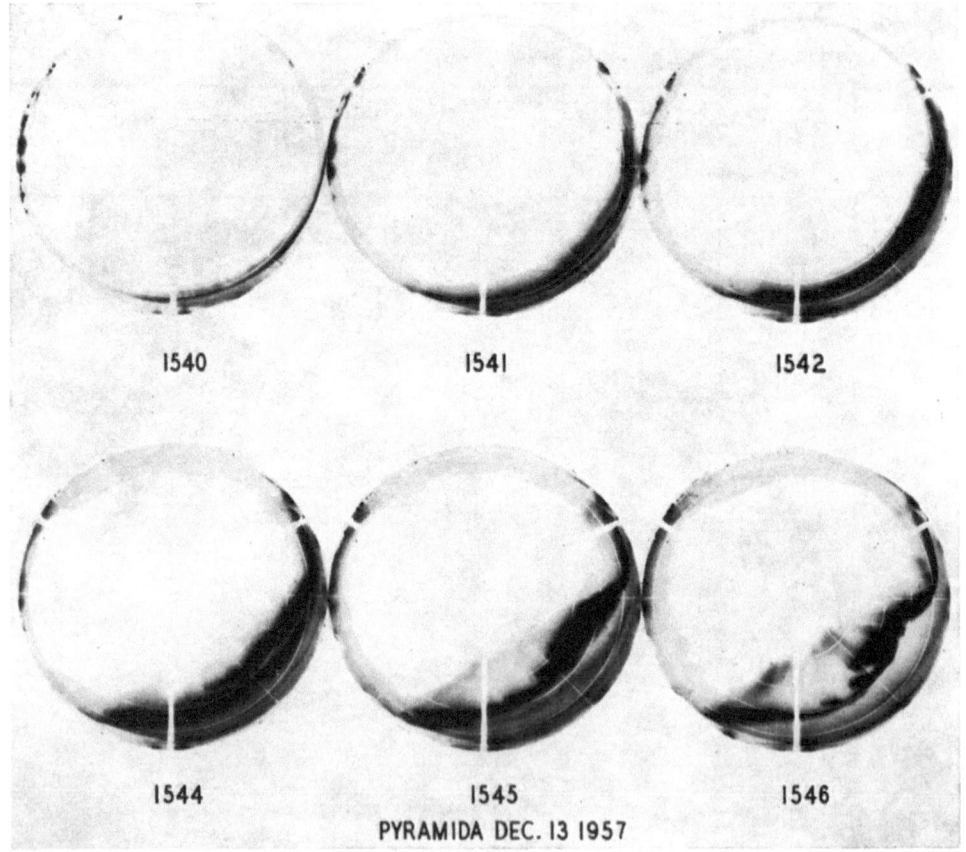

Fig. 27.   Westward traveling surge recorded at Pyramida (dp lat 74.5°) in the early evening hours.

hours. The surge produced an intense negative bay at the nearby station, Murchison Bay (dp lat 75.2°).

In Figures 26a, b, we noted that a positive bay was recorded at College when a westward traveling surge was passing well poleward of the station. This is the simplest situation. Characteristics of the magnetic variations associated with surges are, in general, extremely complicated. However, after examining a great number of individual events, it is possible to summarize the result as follows. Figure 28 shows three successive locations of a surge ($t=1$, 2, and 3) and a typical magnetic variation (the H component) at the North-South chain of magnetic stations (A, B, C, D, E); AKASOFU and MENG (1967a, b, c, d).

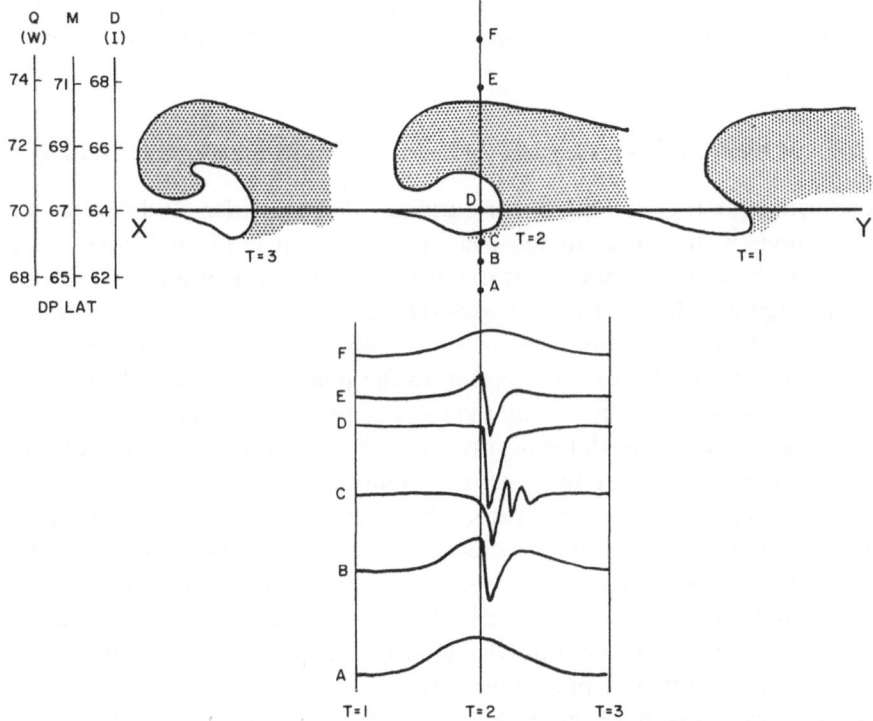

Fig. 28.   Magnetic variations (the H component) in the vicinity of a westward traveling surge.

Station A, which is located well equatorward of the surge, records only simple positive bays; in general, the change starts at least a few minutes earlier than at the other stations. At Station B, the corresponding variation is similar to that at A during an early phase of the event. However, the positive change is sharply terminated by the onset of a negative bay when the surge approaches the station. At Station C, the corresponding changes are irregular, but the first change is a negative one, followed by a positive change. Station D observes a sharply defined intense negative bay. The

variations at E and F are similar to B and A, respectively. The location of the line
xy, which is the initial location of an arc and thus the center-line of the oval, depends
greatly on the degree of magnetic disturbance. The three latitude scales in the left-
hand side of the diagram indicate three different conditions (quiet, medium, and
disturbed) in the mid-evening hours. The scales can also be applied to three different
intensities of surges, weak, medium, and intense, respectively. Such an analysis stresses
the importance of the distance between the station and the oval, rather than the dipole
latitude of the station. During a disturbed period, the center-line of the oval may
descend lower than 64°; thus Station D represents a typical auroral zone station, like
College, where an anomalously early appearance of active auroras and a negative
bay can be observed in the evening hours. Figure 28 will be used extensively in later
chapters. For details of the westward traveling surge and associated magnetic disturbances, see AKASOFU *et al.* (1965), AKASOFU *et al.* (1966b), and AKASOFU and
MENG (1967a, b, c, d).

2.4. Auroral Substorm and Associated Magnetic Disturbances in the Morning Sector

Auroral activity in the morning sector during a substorm differs at different locations
of the stations with respect to the oval. If the station is located poleward of the
oval before the onset of the substorm, the poleward expansion or the eastward
traveling surge may be observed. In general, however, the eastward surge is not as
well defined as the westward surge in the evening sector. In most events, a slight
equatorward motion of arcs occurs prior to the poleward motion. Within the oval,
eastward drifting irregular bands are the most common feature. At a station near the
equatorward edge of the oval, the disintegration of arcs and the eastward drift motion
of the resulting 'patches' are the most common features. Again, there occurs first a
slight equatorward motion of arcs before the disintegration starts. The disintegration
begins first at the equatorward boundary of the oval in the midnight sector and
spreads rapidly eastward and poleward. The resulting patches drift eastward with a
speed of order 200–300 m/sec. Figure 29 (in negative) shows an example of the
disintegration of arcs, observed at Meanook on February 21, 1958; it began about
20 min after the onset of a polar magnetic substorm in Alaska. The disintegration
of arcs is associated with a negative bay; the onset of the negative bay is less sharply
defined in the morning hours than that of the negative bay in the midnight sector,
and its magnitude is also less. For details of auroras in the morning sector, see
AKASOFU *et al.* (1966c).

SANDFORD (1964, 1967) discovered an extensive glow of λ 3914 and λ 5577 along
the auroral zone in the morning sector and called it the 'mantle aurora'. He noted
also that the mantle aurora was observed in the absence of discrete visible auroras.
On the other hand, DAVIS and DEWITT (1963) noted that extremely densely packed
patches often covered the entire morning sky over Byrd (dp lat $-71°$). Therefore, it
is quite likely that the 'mantle aurora' often consists of such patches. Indeed, the
intensity of the mantle aurora varies as a function of K_p.

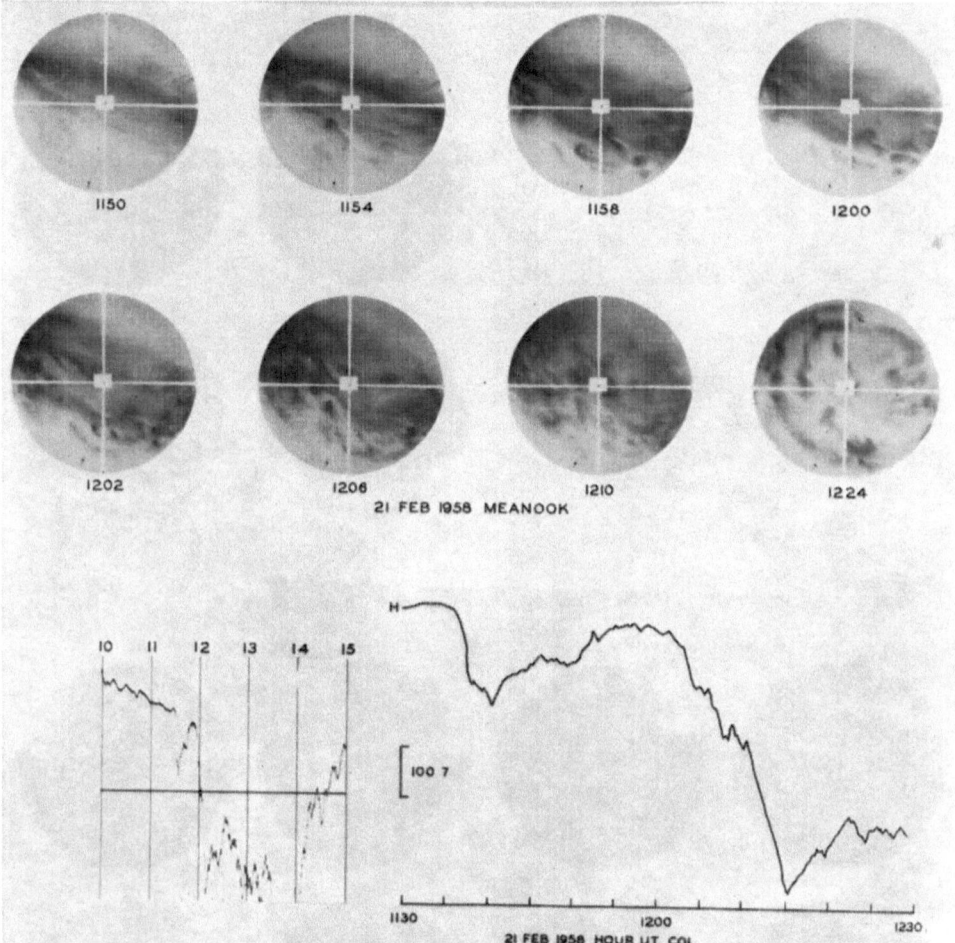

Fig. 29. Example of the disintegration of arcs in the morning sector.

2.5. Auroral Substorms in the Geomagnetically Conjugate Areas

DEWITT (1962) showed that auroras seen from Farewell (dp lat 61.4°) and Campbell Island (dp lat −57.3°), conjugate pairs, have similar forms and motions. Further, the conjugate auroras undergo similar variations in brightness and break-ups simultaneously.

BELON et al. (1967) made a very accurate examination of the conjugacy of auroral displays in the conjugate areas by using two jet aircraft. They showed that the displays (such as brightness, folds, and loops) are strikingly similar in the conjugate areas (Figure 30). By using the latest model of the geomagnetic main field, they also demonstrated that northern and southern auroras projected along geomagnetic field lines to the Northern Hemisphere are strikingly similar in their forms.

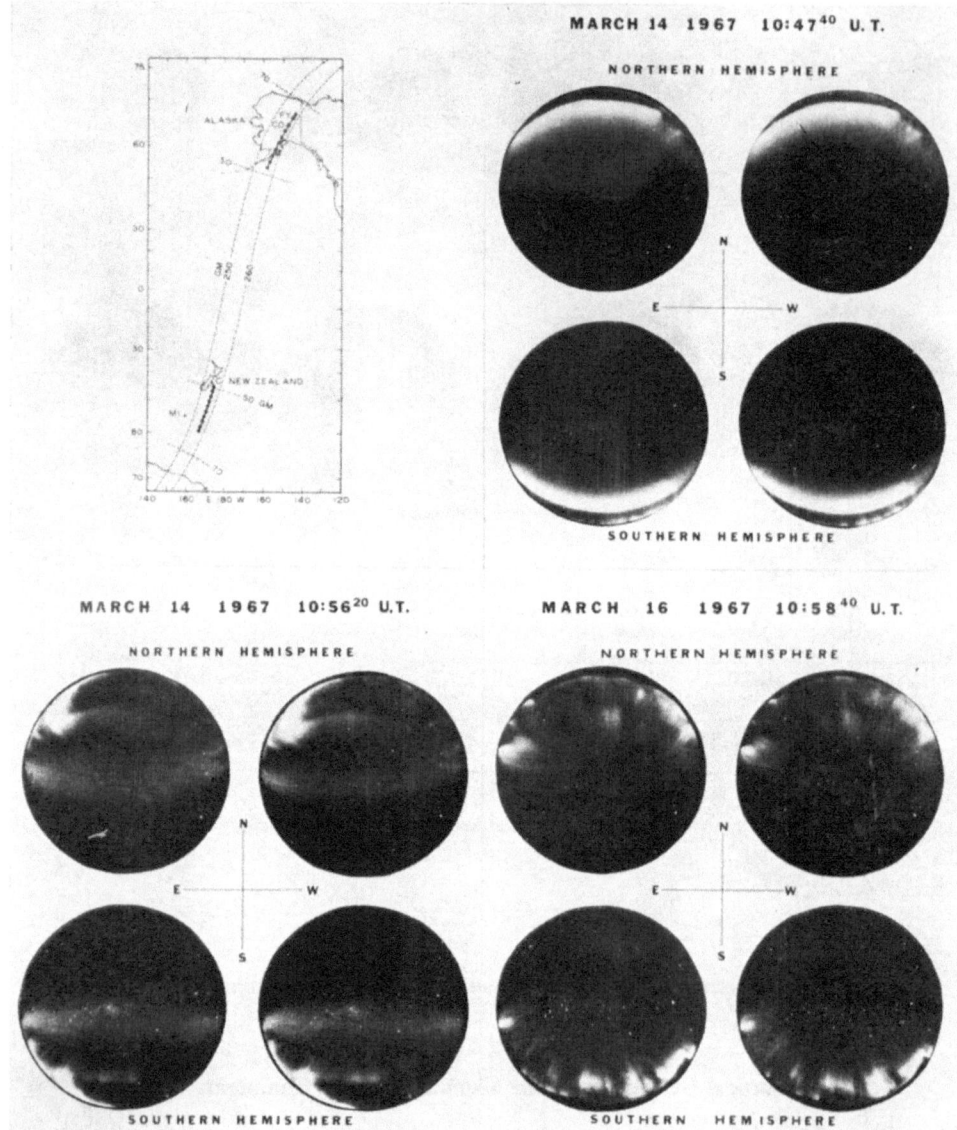

Fig. 30.   All-sky photographs taken at geomagnetically conjugate points (Belon, A. E., K. B. Mather, and N. W. Glass: *Antarctic J. of the U.S.*, 1967).

References

GENERAL

AKASOFU, S.-I.: 1965, 'Dynamic morphology of auroras', *Space Sci. Rev.* **4**, 498–540.

AKASOFU, S.-I.: 1966, 'Electrodynamics of the magnetosphere: Geomagnetic storms', *Space Sci. Rev.* **6**, 21–143.

REFERRED TO IN TEXT

AKASOFU, S.-I.: 1964, 'The development of the auroral substorm', *Planetary Space Sci.* **12**, 273–282.

AKASOFU, S.-I. and MENG, C.-I.: 1967a, 'The abnormally early appearance of active auroras', *J. Atmospheric Terrest. Phys.* **29**, 601-602.

AKASOFU, S.-I. and MENG, C.-I.: 1967b, 'The abnormally early appearance of the eastward motion of auroras in the evening', *J. Atmospheric Terrest. Phys.* **29**, 1029–1031.

AKASOFU, S.-I. and MENG, C.-I.: 1967c, 'Intense negative bays inside the auroral zone I. The evening sector', *J. Atmospheric Terrest. Phys.* **29**, 965–973.

AKASOFU, S.-I. and MENG, C.-I.: 1967d, 'Auroral activity in the evening sector', *J. Atmospheric Terrest. Phys.* **29**, 1015–1018.

AKASOFU, S.-I., KIMBALL, D. S., and MENG, C.-I.: 1965, 'Dynamics of the aurora. II. Westward traveling surges', *J. Atmospheric Terrest. Phys.* **27**, 173–187.

AKASOFU, S.-I., KIMBALL, D. S., and MENG, C.-I.: 1966a, 'Dynamics of the aurora. V. Poleward motions', *J. Atmospheric Terrest. Phys.* **28**, 497–503.

AKASOFU, S.-I., MENG, C.-I., and KIMBALL, D. S.: 1966b, 'Dynamics of the aurora. IV. Polar magnetic substorms and westward traveling surges', *J. Atmospheric Terrest. Phys.* **28**, 489–496.

AKASOFU, S.-I., MENG, C.-I. and KIMBALL, D. S.: 1966c, 'Dynamics of the aurora. VI. Formation of patches and their eastward motion', *J. Atmospheric Terrest. Phys.* **28**, 505–511.

BELON, A. E., MATHER, K. B., and GLASS, N. W.: 1967, 'The conjugacy of visual aurorae', *Antarctic J. U.S.* **2**, 124–127.

BIRKELAND, K.: 1908–1913, *The Norwegian aurora polaris expedition, 1902–1903*, Vol. I, Sections 1 and 2, H. Aschehoug & Co., Christiania.

BURDO, O. A.: 1959, 'Re certain laws of magnetic disturbance in the high latitudes', in *Physics of solar corpuscular streams and their action upon the upper atmosphere of the earth*, U.S.S.R. Academy of Sciences Press, Moscow, 1957, pp. 159–166, DRB Translation T321R.

CHAPMAN, S.: 1918, 'An outline of a theory of magnetic storms', *Proc. Roy. Soc.* **A95**, 61-83.

CHAPMAN, S.: 1935, 'The electric current-systems of magnetic storms', *Terr. Magn.* **40**, 349–370.

CHAPMAN, S.: 1951, *The earth's magnetism*, rev. 2nd ed., Methuen and Co., London.

DAVIS, T. N. and DEWITT, R. N.: 1963, 'Twenty-four-hour observations of aurora at the southern auroral zone', *J. Geophys. Res.* **68**, 6237–6241.

DEWITT, R. N.: 1962, 'The occurrence of aurora in geomagnetically conjugate areas', *J. Geophys. Res.* **67**, 1347–1352.

FUKUSHIMA, N.: 1953, 'Polar magnetic storms and geomagnetic bays', *J. Fac. Sci. Tokyo Univ.* **8**, 293–412.

HARANG, L.: 1951, *The aurorae*, Wiley, New York.

HEPPNER, J. P.: 1954, 'Time sequences and spatial relations in auroral activity during magnetic bays at College, Alaska', *J. Geophys. Res.* **59**, 329–338.

HOPE, E. R.: 1961, 'Low-latitude and high-latitude geomagnetic agitation', *J. Geophys. Res.* **66**, 747–776.

KHOROSHEVA, O. V.: 1961, 'The space and time-distribution of auroras and their relationship with high-latitude geomagnetic disturbances', *Geomagnetizm i Aeronomiya* **1**, 615–621.

MAYAUD, P. N.: 1956, 'Activité magnétique dans les régions polaires', *Ann. Geophys.* **12**, 84–101.

NIKOLSKY, A. P.: 1947, 'Dual laws of the course of magnetic disturbances and the nature of mean regular variations', *Terr. Magn. Atmos. Elect.* **52**, 147–173.

NIKOLSKY, A. P.: 1958, 'Magnetic disturbance in the circumpolar region of the Arctic', *Problem Severa* **1**, 116–132.

SANDFORD, B. P.: 1964, 'Aurora and airglow intensity variations with time and magnetic activity at southern high latitudes', *J. Atmospheric Terrest. Phys.* **26**, 749–769.

SANDFORD, B. P.: 1967, 'High latitude night-sky emissions', in *Aurora and Airglow* (ed. by B. M. McCormac), Reinhold Pub. Co., New York, pp. 443-452.

SILSBEE, H. C. and VESTINE, E. H.: 1942, 'Geomagnetic bays, their frequency and current-systems', *Terr. Magn.* **47**, 195-208.

STAGG, J. M.: 1926, 'Hourly character-figures of magnetic disturbance at Kew Observatory Richmond', 1913–1923, Meteorol. Office (London), *Geophys. Mem.* **32**.

STAGG, J. M.: 1935a, 'Numerical character-figures of magnetic disturbance in relation to geomagnetic latitude', *Terr. Magn. Atmos. Elect.* **40**, 255–262.

STAGG, J. M.: 1935b, 'The diurnal variation of magnetic disturbance in high latitudes', *Proc. Roy. Soc. London* **A149**, 298–311.

STAGG, J. M.: 1935c, 'Aspects of the current system producing magnetic disturbance', *Proc. Roy. Soc. London* **A152**, 277–298.

SUGIURA, M. and CHAPMAN, S.: 1960, 'The average morphology of geomagnetic storms with sudden commencement', *Abhandl. Akad. Wiss. Göttingen*, Sonderheft nr. 4.

VESTINE, E. H.: 1940, 'Disturbance field of magnetic storms', Trans. Wash. Assem; 1939 Publ. IATME, Bull. No. 11, 360–381.

WESCOTT, E. M. and MATHER, K. B.: 1965, 'Magnetic conjugacy from $L = 6$ to $L = 1.4$. 1. Auroral zone: conjugate area, seasonal variations and magnetic coherence', *J. Geophys. Res.* **70**, 29–42.

CHAPTER 3

POLAR MAGNETIC SUBSTORM

3.1. Introduction

In Chapter 2, it was shown that the auroral substorm is associated with a particular type of geomagnetic disturbance, called the polar magnetic substorm. We have also examined the characteristics of magnetic disturbances associated with the major features of auroral displays in Sections 2.2, 2.3 and 2.4.

In Section 2.1, it was also shown how the magnetic disturbance fields associated with polar magnetic substorms have been analyzed in the past. It was noted that the *SD* current system should be revised in at least two ways. First of all, the concept of the *SD* current system does not express the impulsive and intermittent nature of the polar magnetic substorm. Secondly, the *SD* (equivalent) current system does not accurately express the distribution of the magnetic disturbance fields during polar magnetic substorms. Further, we shall see in this chapter that it is not correct to assume that the entire *SD* current system is located in the ionosphere.

However, it is not the purpose of this monograph to review the historical progress of the studies of polar magnetic substorms. We simply note here that despite many past studies of polar magnetic substorms (cf. BIRKELAND, 1913; SILSBEE and VESTINE, 1942; FUKUSHIMA, 1953; KOKUBUN, 1965; MAYSURADZE, 1965; AKASOFU et al., 1965; ROSTOKER, 1966; HEPPNER, 1967; SCRASE, 1967) two major problems remain.

The first problem is to obtain the accurate changing distribution of the magnetic disturbance vectors over the entire earth during the lifetime of a substorm. There has been much discussion as to the flow pattern of the polar electrojet which causes polar magnetic substorms. There is no doubt that the polar electrojet flows along the auroral oval, which is the instantaneous location of the aurora (FELDSTEIN, 1963; AKASOFU, CHAPMAN and MENG, 1965). In Section 2.3 we noted that the auroral electrojet extends westward along the auroral oval and that the surge is the leading edge of the jet. Figures 31a, b and c show the *equivalent* current systems derived by SILSBEE and VESTINE (1942), AKASOFU et al. (1965) and FELDSTEIN (1966), respectively. Figure 31a represents a classical example of the equivalent current system which is similar to the *SD* current system. It consists of two polar jets which flow along the auroral zone, one flows westward and the other eastward; the latter is significantly weaker than the former. Figures 31b and c indicate that the auroral electrojet flows westward all around the auroral oval. However, there is an important difference between them in the nature of their eastward currents in the afternoon and evening hours. In Figure 31b, it is proposed that the eastward current is a leakage current from the

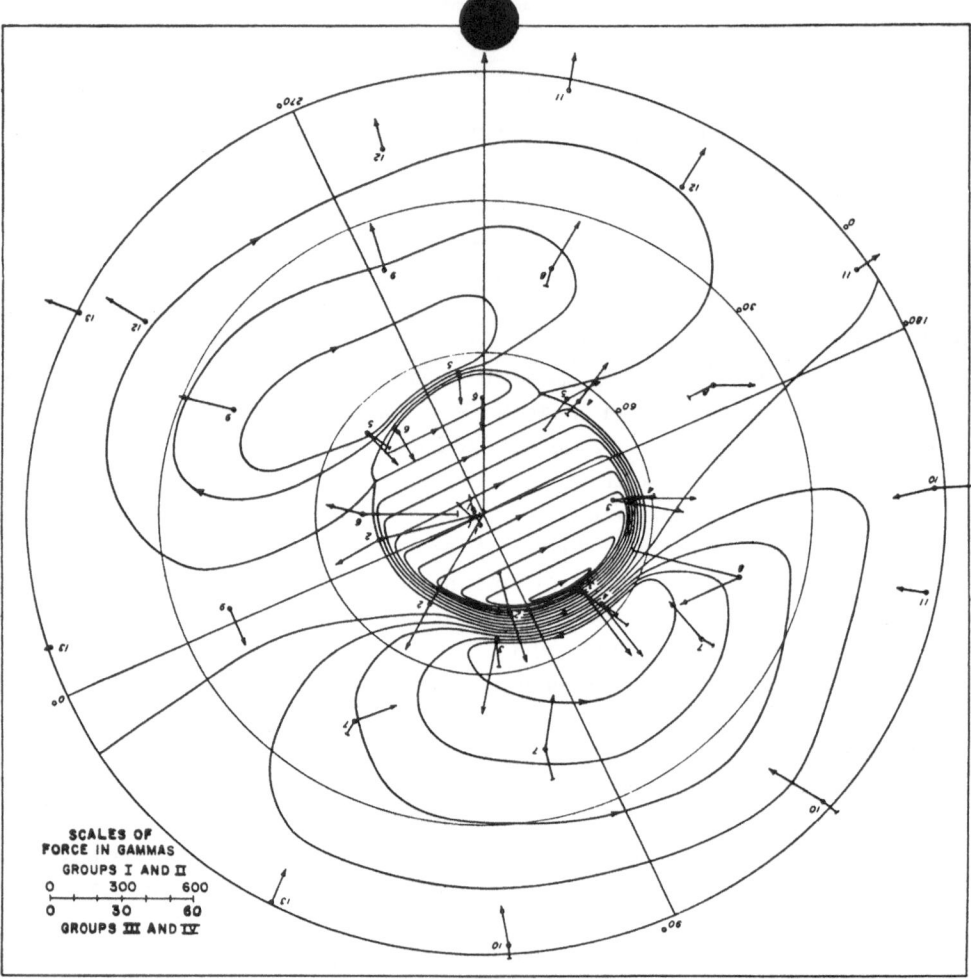

Fig. 31a.   Equivalent current system for an average polar magnetic substorm (Silsbee, H. C. and
E. H. Vestine: *Terr. Magn.* **47**, 195, 1942).

auroral electrojet, but in Figure 31c it is suggested to be an independent electrojet.
In the spiral analysis (Section 2.1.2) the distribution of particular types of disturbances
is expressed by the A spiral.

Although there has been much discussion as to the nature of this eastward current,
the present network of observatories is not close enough to accurately determine the
distribution of the disturbance fields over the entire polar region, so that it is not
possible to determine an accurate equivalent current system. The difficulty involved
in this work may be seen in the following example. Figure 32 shows simultaneous
H or X records of a polar magnetic substorm that occurred between 1800 and 2100
UT on December 16, 1957, from a number of stations in the northern polar region.

In the midnight sector (Dixon (dp lat 63.0°) and Tixie Bay (dp lat 60.4°)) it showed as a negative bay of order 500 γ, most intense at about 1830 UT. In the early morning sector it showed as a less intense negative bay in the auroral zone at Cape Wellen (dp lat 61.8°), College (dp lat 64.7°), Barrow (dp lat 68.5°), and Meanook (dp lat

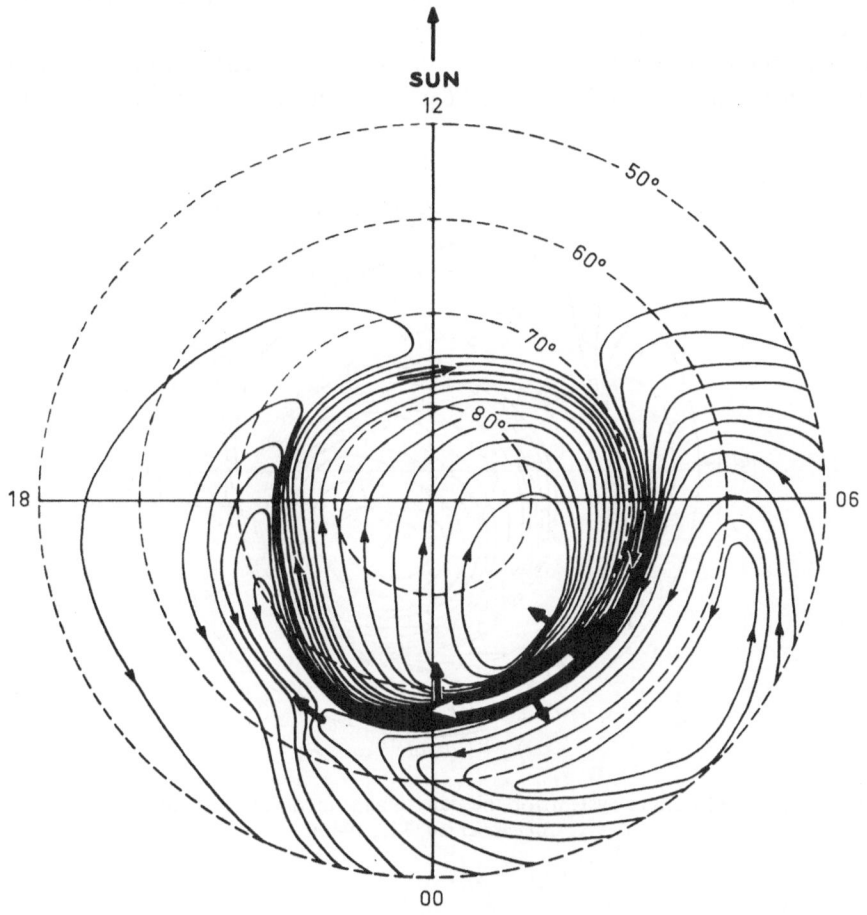

Fig. 31b. Equivalent current system for an intense polar magnetic substorm (Akasofu, S.-I., S. Chapman, and C.-I. Meng: *J. Atmospheric Terrest. Phys.* **27**, 1275, 1965).

61.8°). In the afternoon sector of the auroral zone, it showed as a positive bay of order 100 γ at Reykjavik (dp lat 70.2°), and a combination of a positive and negative bay at Kiruna (65.3°); such changes occur when a surge travels near the poleward horizon of the station (see Figure 28). At Murchison Bay (dp lat 75.2°), there was an intense negative bay of order 300 γ, indicating that the northern boundary of the surge passed there. There was also a delay of almost 30 min in the epoch of its maxi-

mum intensity there, as compared with that in the midnight sector. In the noon sector
of the auroral zone (e.g., at Churchill (dp lat 68.7°)) there was very little systematic
variation. However, at Baker Lake (dp lat 73.8°) the disturbance was strikingly similar
to that at Murchison Bay and Tixie Bay. This is an important indication that a part
of the westward electrojet which caused the negative bay at Tixie Bay extended that
far. It is quite obvious, however, that it is not possible to construct a unique distribution of the equivalent current system on the basis of Figure 32.

The second problem is to infer a three-dimensional current system in the magnetosphere, rather than the two-dimensional equivalent current system, from the distribu-

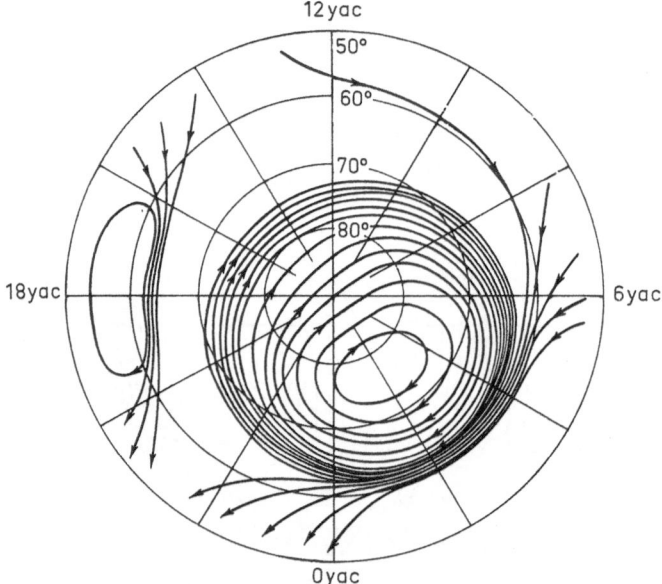

Fig. 31c.   Equivalent current system for the daily magnetic variation (Feldstein, Y. I.: *Planetary
Space Sci.* **14**, 121, 1966).

tion of the disturbance vectors. In order to construct a three dimensional model, it is
necessary first to examine critically the conventional practice of obtaining an equivalent current system and to make use of satellite observations of the magnetic field.
For example, a striking resemblance of simultaneous records taken from the synchronous satellite at the geocentric distance of 6.6 earth radii and at Honolulu strongly
suggests that low latitude magnetic disturbances associated with polar magnetic substorms are of extra-ionospheric origin (CUMMINGS and COLEMAN, 1968; for details see
Section 9.2).

In this chapter we discuss a few isolated and intense polar magnetic substorms
and examine their growth and decay in polar, middle and low latitudes.

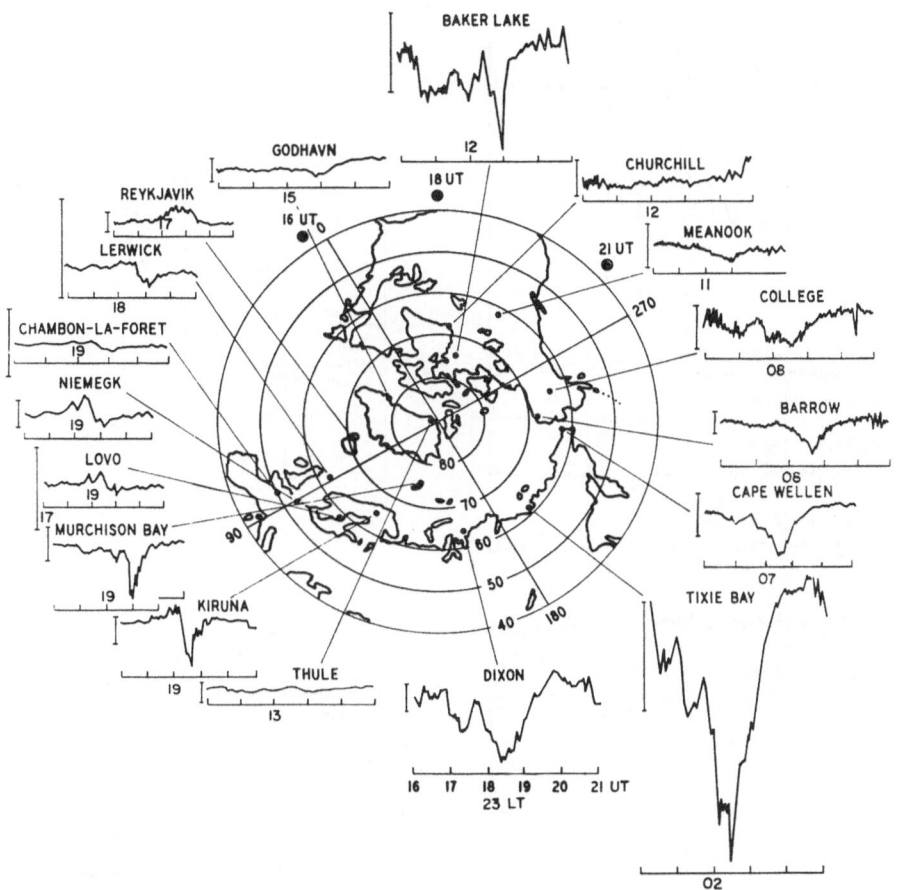

Fig. 32. Collection of magnetic records (H or X component) from a number of polar magnetic stations in the Northern Hemisphere (Akasofu, S.-I., S. Chapman, and C.-I. Meng: *J. Atmospheric Terrest. Phys.* **27**, 1275, 1965).

3.2. Polar Magnetic Substorms on December 13, 1957

Figures 33a and b show a collection of magnetic records from a number of stations on December 13, 1957 (ZAYTSEV and FELDSTEIN, 1967). At least two intense substorms occurred on that day; their onset times were about 0730 UT and 1400 UT, respectively. They are most clearly recognizable in the College record as two negative bays; a similar negative bay was recorded at Churchill (dp lat 68.7°), Arktika-1 and Baker Lake (dp lat 73.8°); these stations were located in the oval.

The corresponding changes for the first storm at other stations are, however, very complicated. At the eastern tip of Siberia, Cape Wellen (dp lat 61.8°), a weak positive bay was followed by a more intense negative bay; the situation was quite similar to that of Station B in Figure 28. Tixie (dp lat 60.4°), which was located in the early evening sector at that time, recorded a positive bay (Station A in Figure 28); note

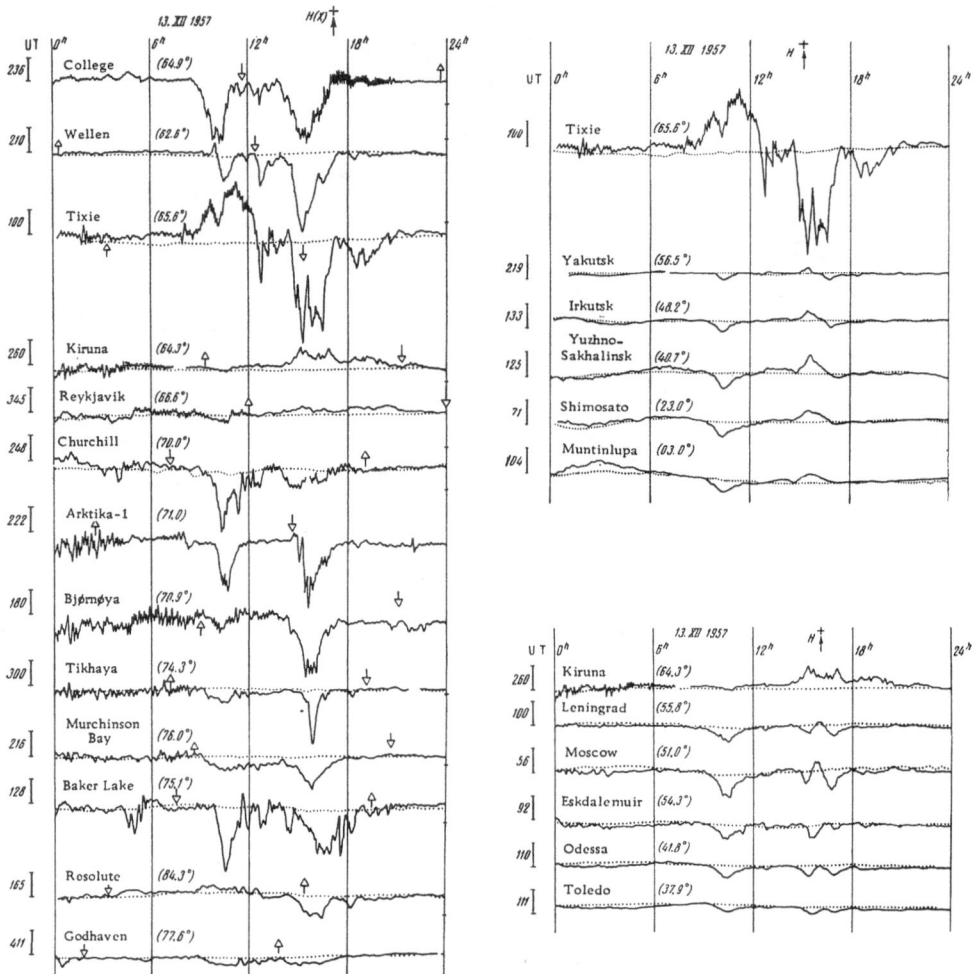

Fig. 33a,b.   Collection of magnetic records from a number of stations on December 13, 1957
(Zaytsev, A. N. and Y. I. Feldstein: *Geomagnetizm i Aeronomiya* 7, 449, 1967).

however that a weak negative bay was superposed on the positive bay at about the maximum epoch of the negative bay at College. There was only a small indication of the substorm at Kiruna (dp lat 65.3°), and Reykjavik (dp lat 70.2°). On the other hand, a definite negative bay ($\sim -150\ \gamma$) was observed at Tikhaya (dp lat 71.5°) and Murchison Bay (dp lat 75.2°) which were located several degrees North of Kiruna and Reykjavik. At stations below dp lat 60°, the first substorm was associated with a negative change. We shall discuss in detail this particular type of change in this chapter and in Section 9.2.

The second substorm was recorded as a negative bay at most of the stations, except Kiruna and Reykjavik; at that time these stations were in the afternoon sector where the eastward current flows. At stations below dp lat 60°, the second substorm was associated with either a positive change, or a negative change followed by a positive change.

ZAYTSEV and FELDSTEIN (1967) determined the distribution of the disturbance vectors in the polar region at four instants, 1030, 1125, 1250 and 1545 UT; Figures 34a–d. They also drew the approximate location of the westward current and the eastward current on the maps. They noted that at 1030 and 1545 UT the current system was similar to Figure 31c. During the period between the two substorms (1125 and 1250 UT), there was little systematic distribution of the vectors. We noted in Section 2.1 that when a polar substorm field is expressed in terms of the equivalent current, the current system which has a certain resemblance to the *SD* current, appears only intermittently; the daily variation of the disturbance fields in the polar

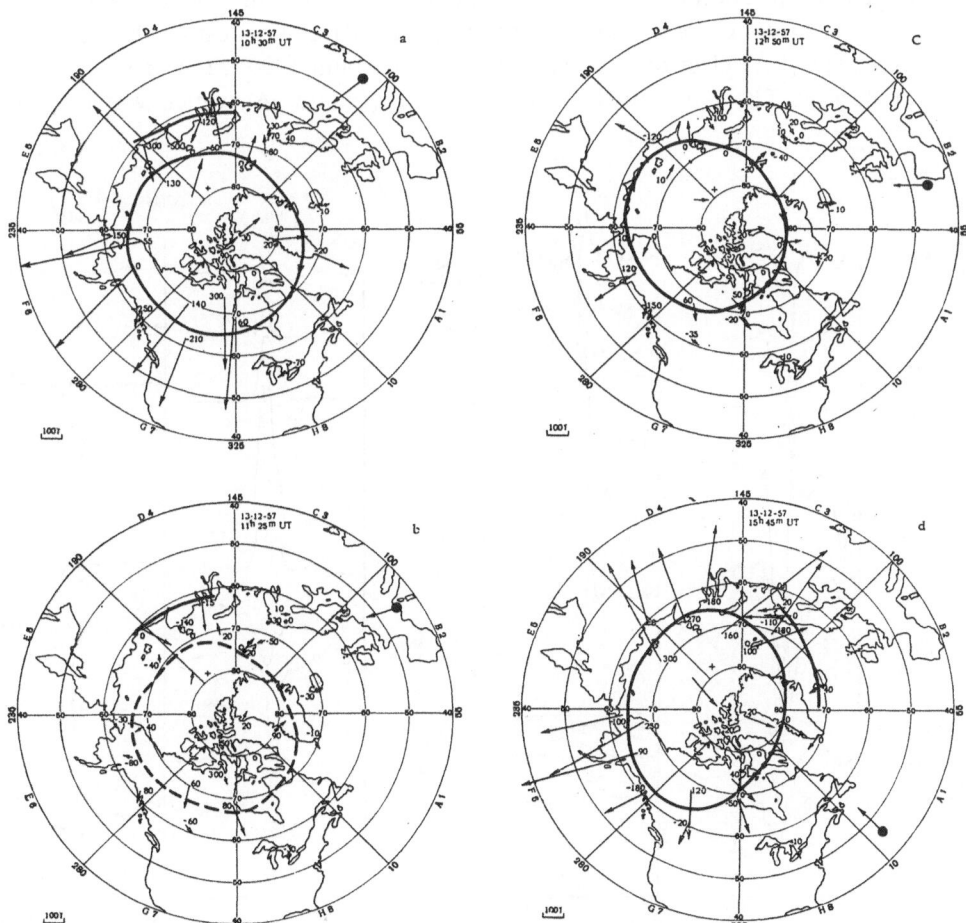

Figs. 34a–d.   Distribution of magnetic disturbance vectors at 1030, 1125, 1250 and 1545 UT on December 13, 1957 (Zaytsev, A. N., and Feldstein, Y.-I.: *Geomagnetizm i Aeronomiya* 1, 449, 1967).

region does not result from the fact that the earth rotates under a fixed current pattern of constant intensity. The study of ZAYTSEV and FELDSTEIN (1967) suggests that the current system shown in Figure 31c appears intermittently, with quiet periods between.

3.3. Polar Magnetic Substorm on December 16, 1964

Figures 35a and b show the *H* and *D* component magnetic records from stations in and near the auroral oval; Figure 36 shows the approximate location of the auroral oval at 1400 UT, together with the locations of the stations whose records are used in Figures 35a and b. The oval is the strip in which quiet arcs are found with a very high probability of occurrence; during auroral substorms, however, the width of the oval in the dark hemisphere greatly increases (it becomes much wider than is shown in Figure 36). The stations in Figure 35b are mostly in the sunlit hemisphere.

The record from College suggests that this particular polar substorm had a double structure. The polar electrojet appeared to grow at about 1200 UT near College, but not at Barrow. The Bar-I all-sky camera station (Figure 36), which was the only

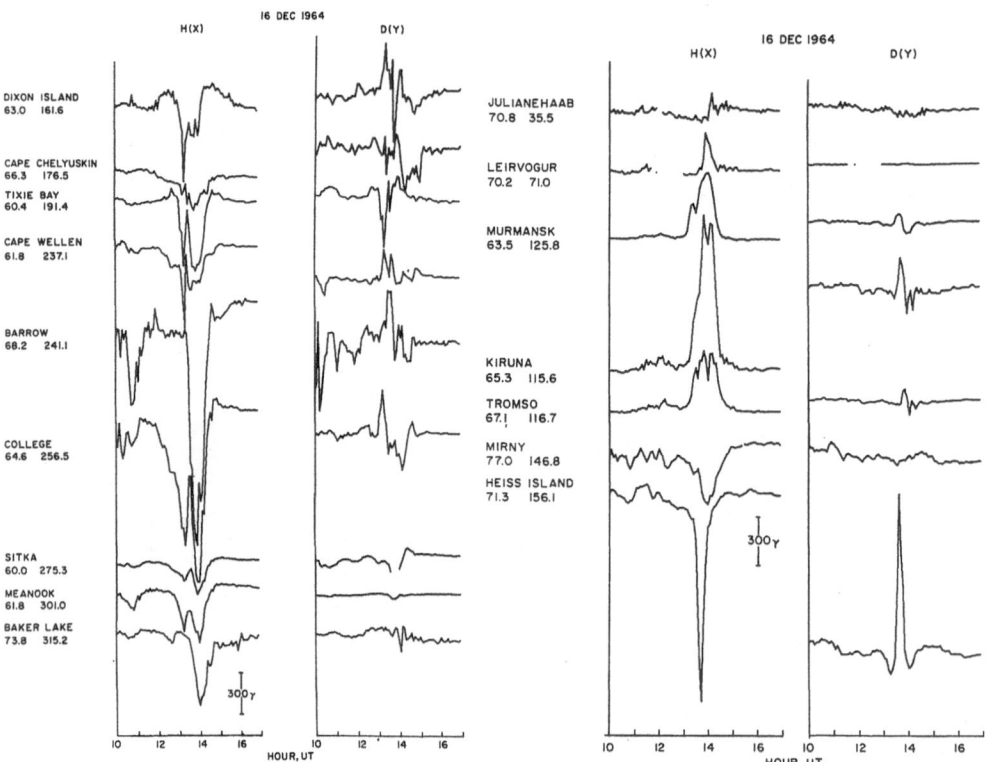

Figs. 35a, b.   *H* and *D* component magnetic records from stations in and near the auroral oval on December 16, 1964 (Akasofu, S.-I., and C.-I. Meng: *J. Geophys. Res.* **74**, 1969).

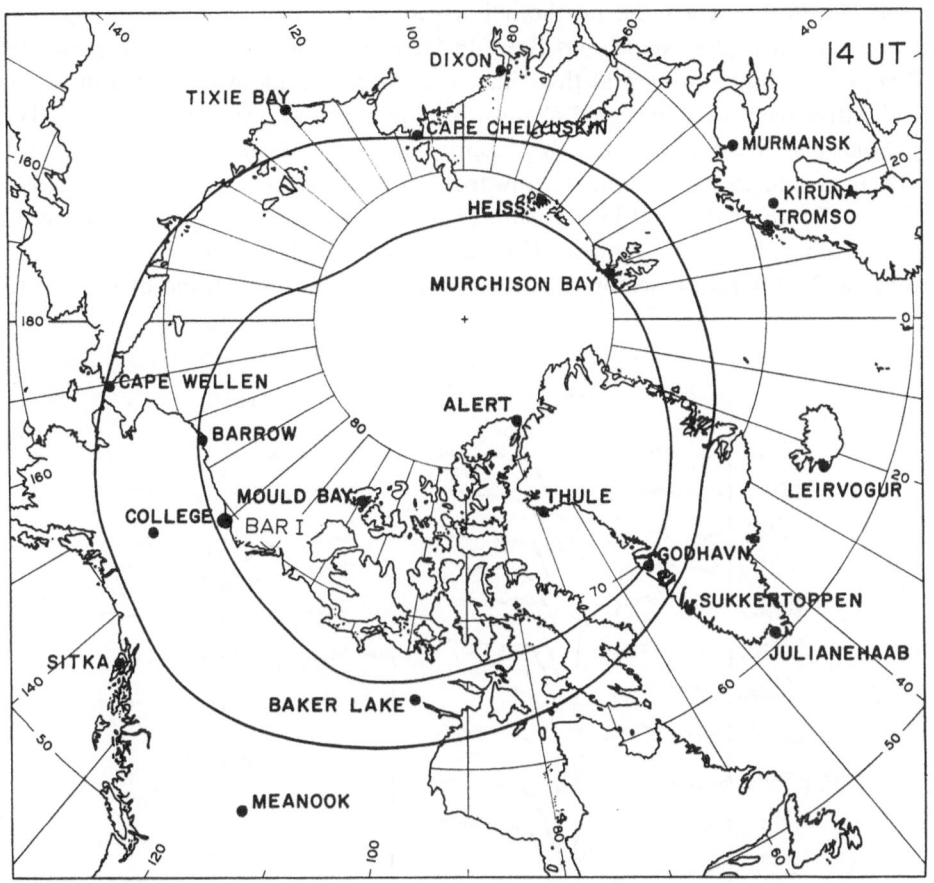

Fig. 36. Approximate location of the auroral oval and the location of the stations whose records are used in Figs. 35a, b (Akasofu, S.-I. and C.-I. Meng: *J. Geophys. Res.* **74**, 1969).

station with fair weather on that night in Alaska, recorded a sudden brightening of auroras very near the southern horizon between 1239 and 1245 UT. However, the auroral activity was limited near the southern horizon, so that the growth of the electrojet appeared to occur along dp lat 65° in Alaska. At about 1320 UT, the polar electrojet began to diminish, but about 10 min later (about 1330 UT) it increased suddenly all along the auroral oval (College, Cape Wellen, Heiss Island, Godhavn ($-150\ \gamma$), and Baker Lake). Unfortunately, the Murchison magnetic station was discontinued after the IGY, but its southern conjugate station, Mirny, recorded a negative bay; MENG (1968) showed that in December, the ratio of the magnitude of bays at Murchison to those at Mirny was 2.3. Thus we may infer that the intensity of the negative bay, if observed at Murchison, would be of order $-825\ \gamma$. The Bar-I all-sky films showed a violent poleward explosive motion of auroras, which began at 1327 UT near the southern horizon. The poleward expanding bulge passed the zenith at 1330 UT and reached the northern horizon at 1335 UT. The expanding bulge appeared to reach close to Mould Bay (dp lat 79.1°) since an intense negative bay (~300 γ) was observed there at that time; a similar example was studied in Section 2.2. Negative bays were observed also to the South of the oval (Dixon Island, Tixie Bay, Sitka and Meanook) in the *dark* sector. This suggests that the polar electrojet greatly increased in width, both northward and southward, as well as in intensity.

To the South of the oval in the *sunlit* hemisphere, positive bays were observed at several stations, including Tromsø, Kiruna, Murmansk, Leirvogur and Julianehaab, located in the afternoon sector of the auroral zone; the magnitude of the positive

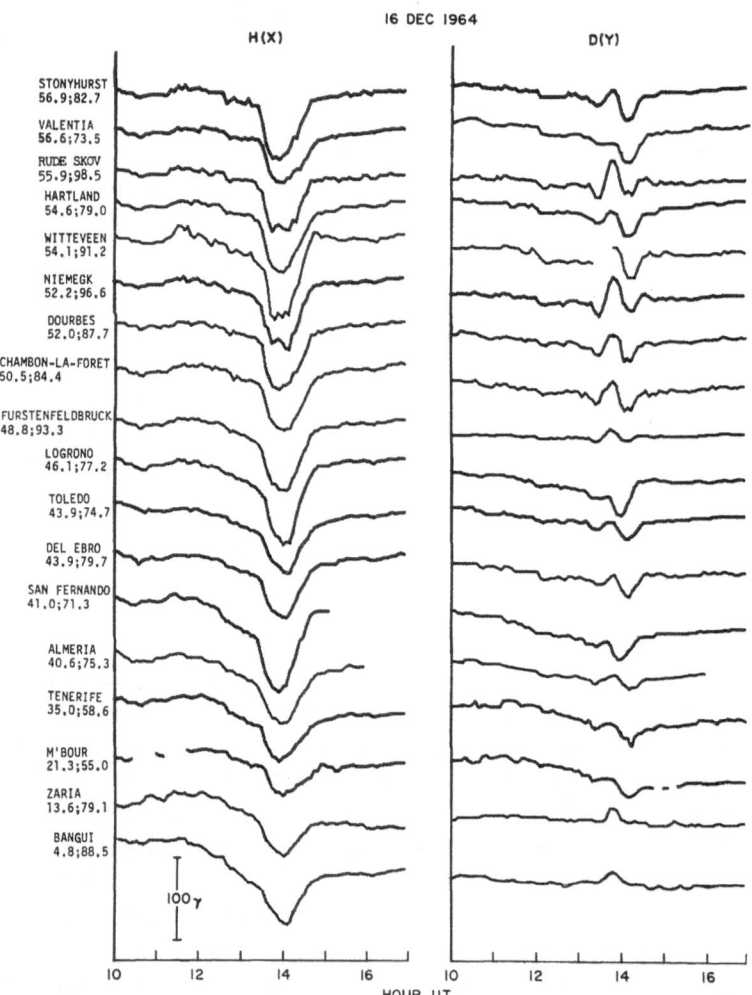

Figs. 37a–d.   *H* and *D* component magnetic records below dipole latitude 60° on December 16, 1964; (a) West Europe sector, (b) East Europe sector, (c) Pacific sector and (d) America sector (Akasofu, S.-I. and C.-I. Meng: *J. Geophys. Res.* **74**, 1969).

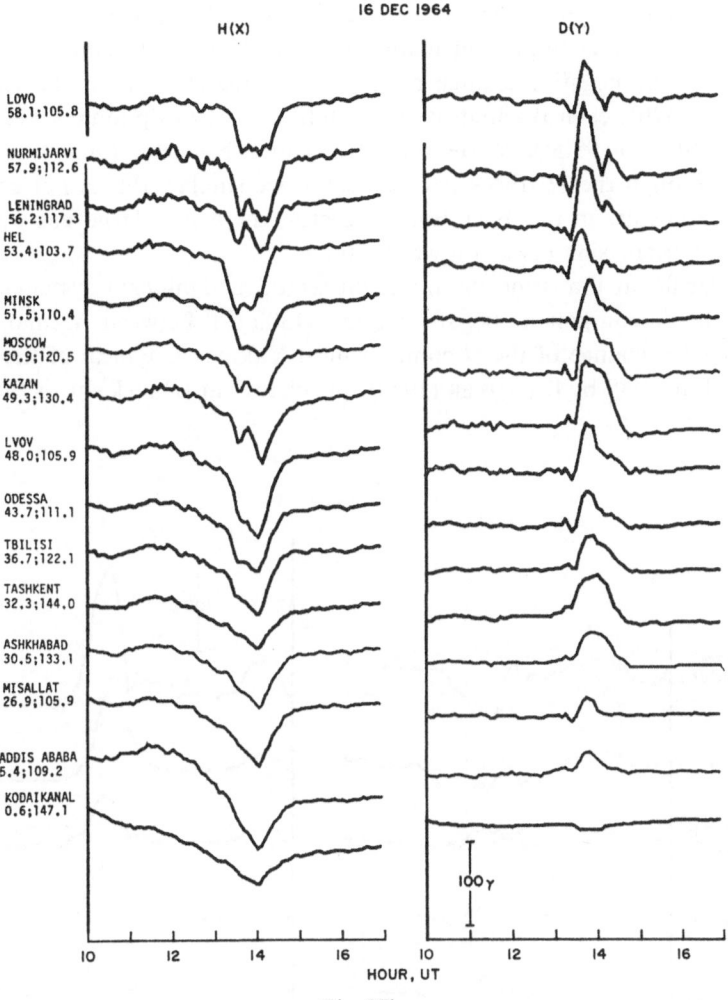

Fig. 37b.

bay decreased toward the noon sector. At Dixon and also at Tixie, a weak positive bay began when the negative bay was in progress at College, but it was suddenly indented by an intense negative bay at about the time of the sudden enhancement at College. Such complicated changes and the associated auroral activity were discussed in Section 2.3; see Figure 28. During these rather systematic changes of the *H* component, the *D* component changed irregularly at all stations, except Heiss Island.

Figures 37a–d show the magnetic variations in West Europe (dp long < 100°), East Europe (dp long > 100°), the Pacific and America, respectively. In each sector, the magnetic variations differ considerably from one sector to another; within each sector there is much similarity among the stations.

In the afternoon sector (Figures 37a and b) the most distinct feature is the onset of a negative change over the entire latitude span at 13 UT, when the polar electrojet was enhanced. Another important feature is a small positive change of the H component during the negative change at higher latitude stations in the East Europe sector (Figure 37b); even the high latitude stations east of dp long 90° (Figure 37a), Rude Skov (dp long 98.5°), Witteveen (91.2°) and Niemegk (96.6°) show such a tendency. Although the H traces in each sector are similar, this is not so for the D traces. In particular, in the East Europe sector, there is a large eastward change systematically decreasing toward lower latitudes.

In the Pacific, at that time the midnight sector, a significant systematic positive change in the H component began at about 1300 UT between Memambetsu and Muntinlupa. The change of the H component at Yakutsk is typical of the 'transition type' studied in detail by ROSTOKER (1966); the change at Irkutsk shows the transition

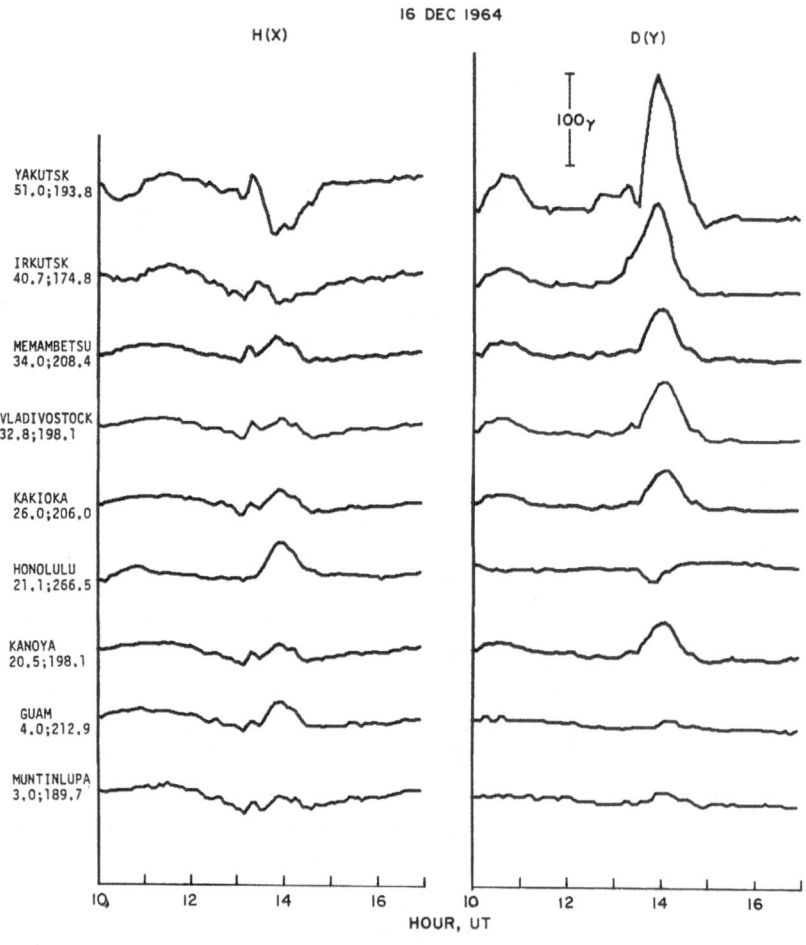

Fig. 37c.

between Yakutsk and Memambetsu. The D component shows a large positive change, diminishing toward lower latitudes; the only exception is Honolulu, which was located near the eastern edge of the Pacific sector; in terms of the D component change, Honolulu belongs to the American sector (Figure 37d).

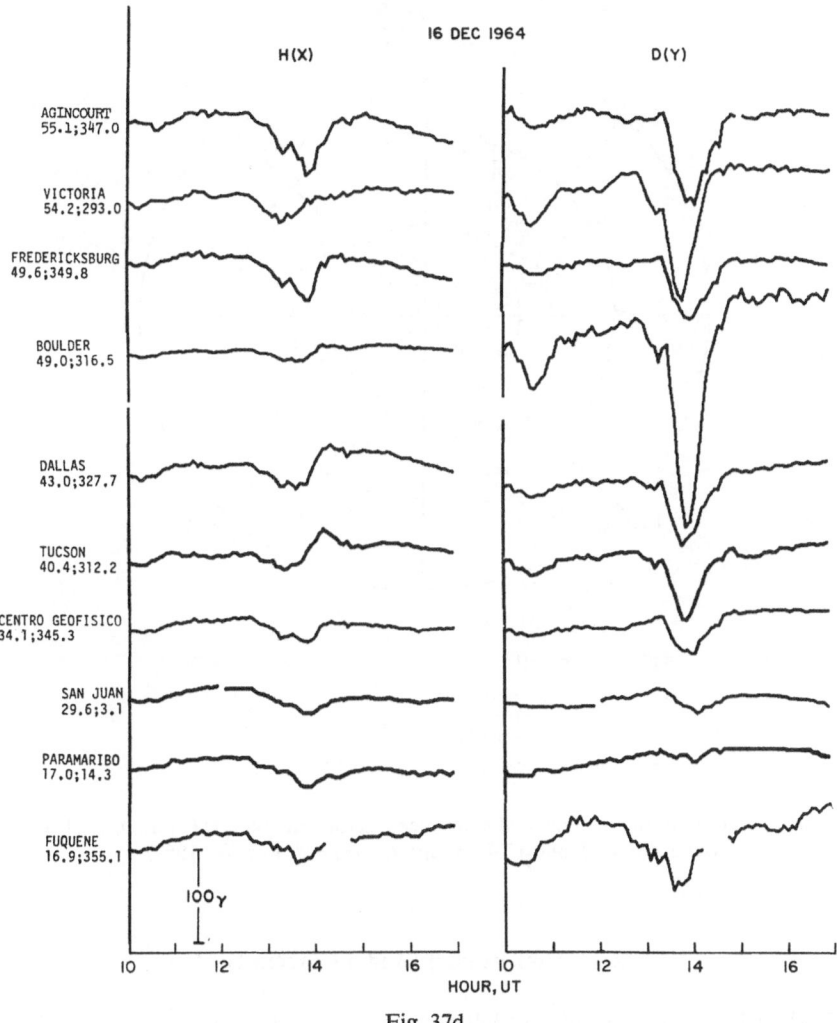

Fig. 37d.

In America, then in the morning sector, the H component variation was complicated, but in a certain systematic way. The change at Agincourt and Fredericksburg, similar to that at Meanook, suggests that the variation was first a negative one which was followed by a positive change. The magnitude of the positive change, however, varied with the latitude. The D component change was negative; the magnitude was the largest at Boulder and diminished toward higher and lower latitudes. Figure 38 shows the distribution of the magnetic disturbance vectors at 1400 UT, which was the maximum epoch of the substorm. For details of the time variation of the distribution of the vectors, see AKASOFU and MENG (1969).

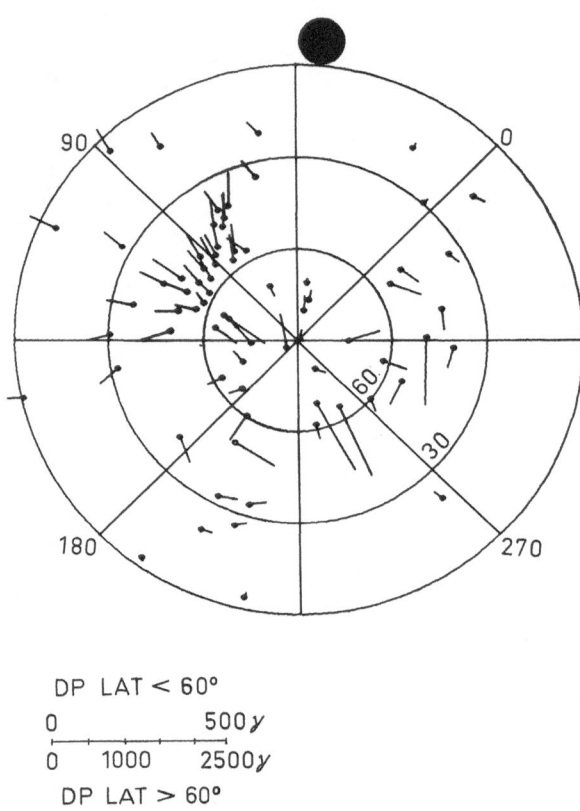

DP LAT < 60°

DP LAT > 60°

Fig. 38. Distribution of the magnetic disturbance vectors at 1400 UT on December 16, 1964 (Akasofu, S.-I. and C.-I. Meng: *J. Geophys. Res.* **74**, 1969).

3.4. Three-Dimensional Model Current System

In the past, it has been common practice, as a matter of convenience, to express the observed distribution of magnetic disturbance vectors in terms of the *equivalent current system* on a spherical shell concentric with and above the earth. It is a simple and mathematically correct means of representing the way in which the observed magnetic disturbance vectors are distributed over the earth.

Here, as an example, we examine the equivalent current diagram constructed by SILSBEE and VESTINE (1942) in Figure 31a. In the diagram, which corresponds to the average for many bays, it is assumed that the currents flow entirely on a spherical shell, that is div$J = 0$ on the shell. The westward nightside polar electrojet completes its circuit by return current mainly across the polar cap but partly also in lower latitudes. The lower latitude return currents flow eastward in the dark sector and westward in the daylight sector, causing positive and negative 'bays', respectively.

The equivalent current system for the substorms should, however, not be confused with the actual current system. In the past, the ionosphere has tacitly been regarded as this particular spherical shell, and thus the equivalent current system has been interpreted as a purely ionospheric current system.

For this reason observations additional to those made at the earth's surface are of vital importance for our study. There are now at least two types of observations which should be carefully taken into account in this respect. They are: (1) magnetic observations by a synchronous satellite (cf., ATS) at a geocentric distance of 6.6 earth radii, which determine whether low latitude bays (both positive and negative) during polar magnetic substorms are caused by extra-ionospheric currents (CUMMINGS and COLEMAN, 1968), and (2) magnetic observations by polar orbiting satellites which test whether an electric current flows into or out from the auroral oval (ZMUDA et al., 1966; ZMUDA et al., 1967).

CUMMINGS and COLEMAN (1968) showed that a striking similarity exists between the magnetic records from the synchronous satellite over the Pacific and those from Honolulu (for details, see Section 9.2). Thus, the equivalent current system cannot be interpreted as an ionospheric current system. The ionospheric component seems to have made only a minor contribution there. If the bays are entirely due to an ionospheric current, the field direction at the two points should be opposite. Based on the study of the asymmetric development of the main phase and also of the normality of the SD variation at Huancayo, the same conclusion was made by AKASOFU and CHAPMAN (1964, 1967) who concluded that the SD variation in the middle-low latitude belt cannot be due to an ionospheric current. An extensive study of the field of the low latitude negative bays by using a North-South chain of stations also confirmed this conclusion (AKASOFU and MENG, 1968). Keeping this important evidence in mind, we shall now examine Figure 38.

In Figure 38 the magnetic vectors are directed westward in the morning sector, and eastward in the late evening sector. Their magnitude decreases rapidly toward lower latitudes; this is also seen in Figures 37b, c, d. In Figure 31a, these directions of the vectors may be ascribed to ionospheric currents converging poleward and diverging equatorward. Since the low latitude positive and negative bays are mostly due to extra-ionospheric currents, a significant part of these East-West vectors is also likely to be produced by extra-ionospheric currents.

Therefore, we examine here a model current system in which the current flows into the auroral ionosphere along the field lines in the morning sector; it then flows along the auroral zone, westward in the dark sector and eastward in the daylight sector; finally it flows out along the field lines in the evening sector; the current completes its circuit by flowing in the equatorial plane, eastward in the dark sector and westward in the daylight sector. A dipole field configuration is assumed, and the currents are assumed to flow along the field lines that cross the equatorial plane at a circle of radius 6.55 times that of the earth. These lines intersect the earth's surface at latitude 67°.

Such a model is similar, in principle, to what BIRKELAND (1913) and ALFVÉN (1939, 1940, 1950) proposed for the storm current system. They proposed that polar magnetic substorms are caused by electric currents that flow into the auroral iono- sphere along the field lines in the daylight sector, and then after flowing along the auroral zone flow out along the field lines in the dark sector. Birkeland thought that the current was provided by a beam of electrons coming directly from the sun. Alfvén proposed that the current was provided by space charges generated at the nearly circular boundary of what he called the 'forbidden region'. Here, without considering for a moment the nature and cause of the current, we examine whether or not such a current system can reasonably reproduce the observed distribution of the magnetic vectors. Our model is a slightly modified version of KIRKPATRICK's model (1952); his model is reproduced here as Figure 39. The only major difference is that we rotate his current system clockwise by 90°, with respect to the sun-earth

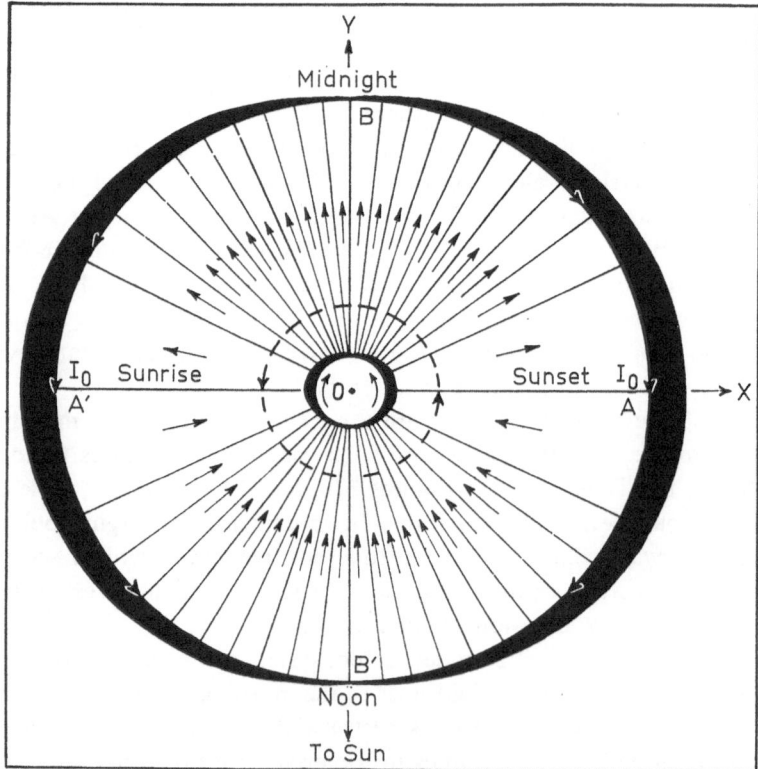

Fig. 39.   Kirkpatrick's model three-dimensional current system (Kirkpatrick, C. B.: *J. Geophys. Res.* **57**, 511, 1952).

line, so that his noon meridian is the 06 local time meridian in our model. Thus, taking the coordinates $(x, y, z$ and $r, \theta, \phi)$ with the Oy axis directed toward the sun, the current system consists of:

(i) a sheet current of intensity $J_0 \sin(\phi + 3\pi/2)$ per radian of ϕ flowing along the line of force $(r = 6.55a \sin^2 \theta)$; where a denotes the earth's radius.

(ii) a line current of intensity $J_0 \sin \phi$ flowing in the auroral circle $(\theta \simeq 23°)$ which is located at an altitude of 100 km, and

(iii) a line current of intensity $2J_0 \sin(\phi + \pi)$ flowing along the equatorial circle of radius 6.55a, where J_0 is taken to be 10^6 amperes.

Figures 40a and b show the calculated latitudinal distribution of the resulting H component along the noon-midnight meridian and of the E component along the dawn-dusk meridian. They also show the magnitude of the observed H and E components on these meridians at 1400 UT. The model calculation fairly well reproduces the observed distribution for both components below dp lat 55°.

Kirkpatrick's model current may be visualized by deforming the (ionospheric) spherical surface for the equivalent current system in Figure 31a into a surface generated by rotating a dipole field line of equatorial crossing distance 6.55 earth radii around the dipole axis.

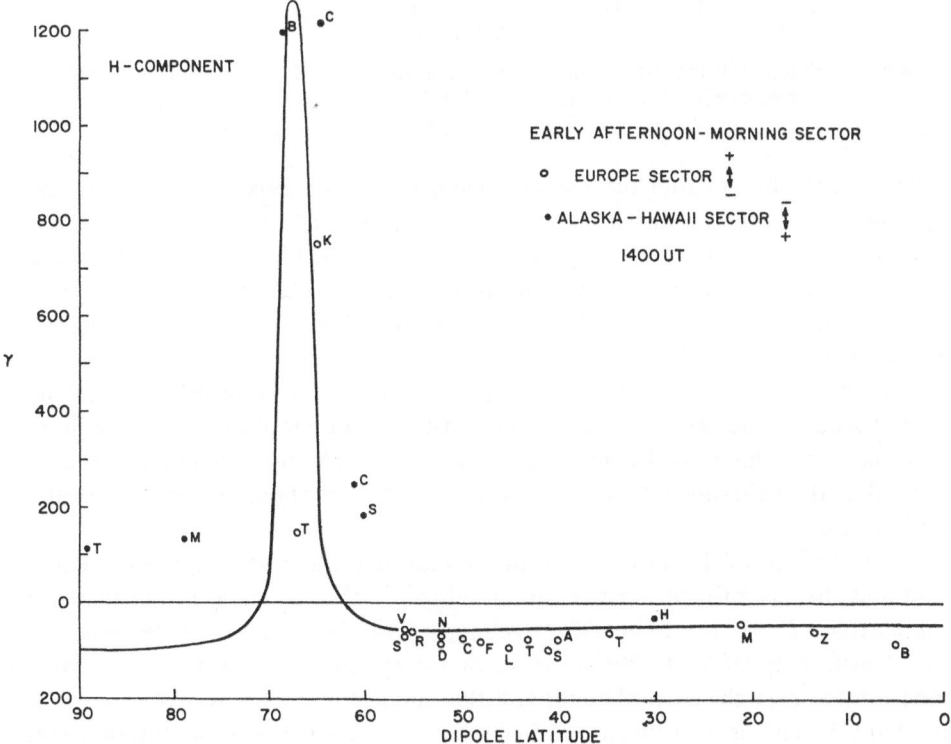

Fig. 40a. Latitudinal distribution of the H component for the model current system, together with the observation (Akasofu, S.-I. and C.-I. Meng: *J. Geophys. Res.* **74**, 1969).

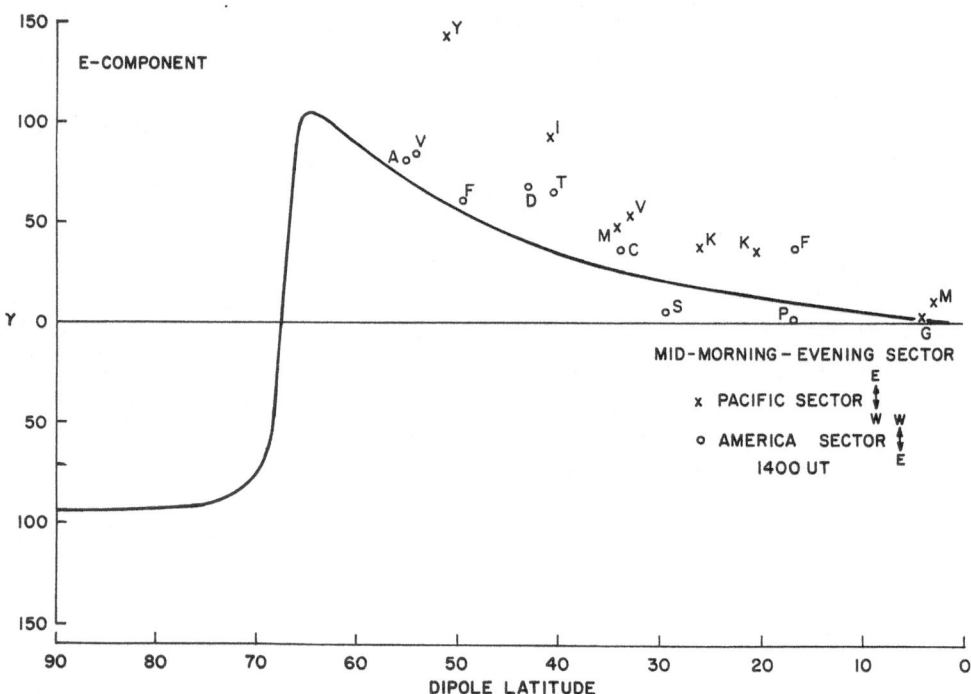

Fig. 40b. Latitudinal distribution of the *E* component for the model current system, together with the observations (Akasofu, S.-I., and C.-I. Meng: *J. Geophys. Res.* **74**, 1969).

It is quite obvious that the model current system adopted here is too simple to reproduce the observed distribution of magnetic vectors above dp lat 55°. The discrepancy between the observation and the model calculation may be caused by inadequacy of the model in the vicinity of the auroral oval. In order to account for the distribution of magnetic vectors in the polar regions, the model current system should be modified in the following ways:

(a) The westward auroral electrojet flows along the auroral oval, not along the auroral zone. Therefore, the extra-ionospheric current should flow into or flow out from the oval. This is well confirmed by the polar orbiting satellites which demonstrate that the field-aligned currents are observed only above the auroral oval, and nowhere else.

(b) The width of the auroral oval in the midnight sector during a polar magnetic (and thus the auroral) substorm is too broad to be represented, as in our model, by a line current. There, the current flows along a wide belt, perhaps extending from a little North of dp lat 60° to 80° at the maximum epoch of the substorm, though more concentrated near the auroral zone; at Sitka, the *Z* component was of order $-300\ \gamma$ at 1400 UT. This is well illustrated in Figure 40a; see the Alaska-Hawaii sector. It was noted earlier that active auroras moved poleward from the auroral zone to about dp lat 80°. Thus, the polar cap is greatly influenced by the continuously varying concentrated auroral current. This variability makes it difficult to use the magnetic field distribution in the polar cap as the basis for determining the appropriate model current system for the substorm. VESTINE and CHAPMAN (1938) and KIRKPATRICK (1952) normalized the distribution by using the observed value at the dipole pole and demonstrated that the Birkeland-Alfvén model cannot well reproduce the distribution in lower latitudes.

(c) Since the electrojet is likely to flow along active auroras (namely in the extra-conductive strip in the ionosphere), the Cowling conductivity plays an important role for the ionospheric part of the current (the Cowling current); BOSTRÖM (1964). Further, since the whole ionosphere has an anisotropic conductivity, the Hall current can also be generated by introducing space charges from outside the ionosphere, as implied in our model.

ATKINSON (1967) demonstrated by an analog computer model calculation that the combination of a field-aligned current system (similar to that used in this paper) and the Hall (ionospheric) current may well reproduce the extremely complicated distribution of the magnetic vectors in and around the regions of active auroras; in his model calculation, ATKINSON (1967) took into account the distribution of active auroras, such as the poleward expanding bulge and the westward traveling surge. His diagram of the distribution of magnetic disturbance vectors is reproduced here in Figure 41.

Three-dimensional models, such as those discussed above, should be examined by using simultaneous records from a close network of observatories and also by *in situ*

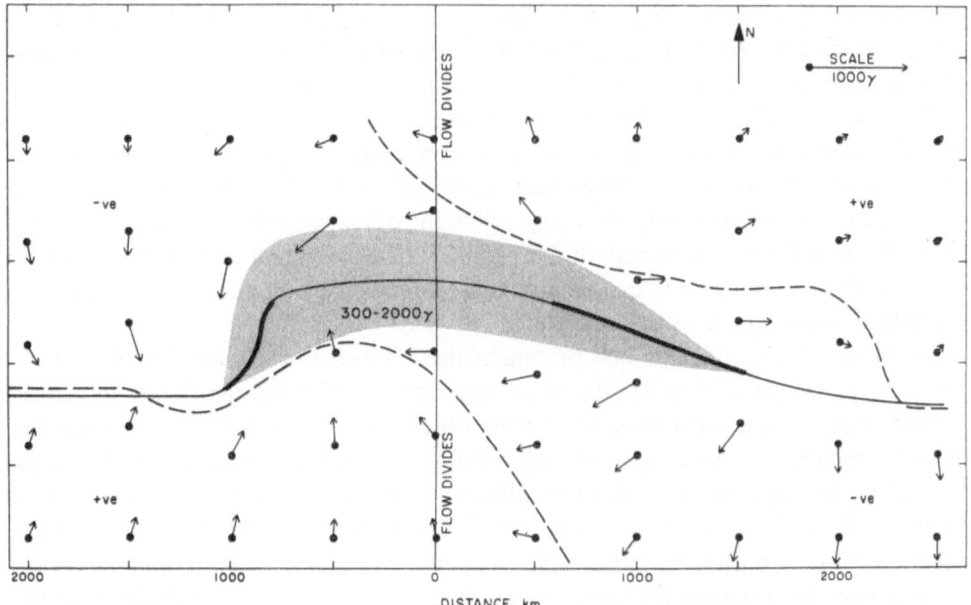

Fig. 41.   Distribution of the disturbance vectors in and near the auroral bulge, determined by an analog computer (Atkinson, G.: *J. Geophys. Res.* **72**, 6063, 1967).

observations by rocket or ion-cloud experiments. It is of great importance to examine whether an intense ionospheric (Hall) current across the polar cap exists, or whether magnetic field variations well inside the polar cap are mainly caused by the field aligned current. An ideal location and time will be at about dp lat 75° in the midnight sector, when the expanding auroral bulge or the westward traveling surge is seen equatorward of the station. CAHILL (1959) made a rocket observation in the polar cap and detected an ionospheric current during substorms, suggesting that there is an intense ionospheric current across the polar cap.

The correct distribution of the Hall current will help us in determining the electrostatic field in the ionosphere, which can then be projected into the magnetosphere, in the way discussed by TAYLOR and HONES (1965); see also Section 10.3.

3.5. Polar Magnetic Substorms in Geomagnetically Conjugate Areas

It has been realized that polar magnetic substorms at geomagnetically conjugate points show a striking similarity (NAGATA and KOKUBUN, 1960; WESCOTT, 1962; BRYUNELLI, 1962; ONDOH and MAEDA, 1962/3; BOYD, 1963; BOBROV, 1963; YUDOVICH, 1963; WESCOTT and MATHER, 1965a, b, c, d). This suggests that the polar electrojets in geomagnetically conjugate areas are driven by a common electric field, suggesting that the auroral field lines, are, as a first approximation, equipotential. Figures 42a, b show some examples of the H component magnetic records from College (dp lat 64.7°) and Macquarie Island (dp lat −61.1°), (MENG and AKASOFU, 1968). Although details are somewhat different, the major changes are very similar, except for the magnitude of positive bays, which are greater at the summer station than at the winter station (WESCOTT and MATHER, 1965a; MENG and AKASOFU, 1968). This evidence suggests that positive bays are fundamentally different from negative bays.

WESCOTT and MATHER (1965d) showed that the correlation becomes increasingly poorer when one examines higher latitude conjugate pairs inside the auroral oval. By using the Shephard Bay-Scott Base conjugate pair, they showed that in some cases one station records a positive bay, while the other records a negative bay. MENG and AKASOFU (1968) examined this problem further in detail and showed that this breakdown of conjugacy is caused by the asymmetry of the geomagnetic field configuration with respect to the equator.

In Chapter 2.3, we examined the magnetic variations in the vicinity of westward traveling surges. In Figure 28, if a polar cap station is far enough from the surge, it records a positive bay (Station F). On the other hand, Station E is close enough to be influenced by an intense (westward) auroral electrojet and records a negative bay. Therefore, if the conjugate stations observe bays of opposite signs, the distance between the station and the surges should be different. Figure 43 shows this situation schematically. In both seasons (northern winter and summer), the H component variations across the surge are the same. MENG and AKASOFU (1968) noted that there is a strong tendency for a higher occurrence of indented negative bays in the Northern Hemisphere than in the Southern Hemisphere during the northern winter months

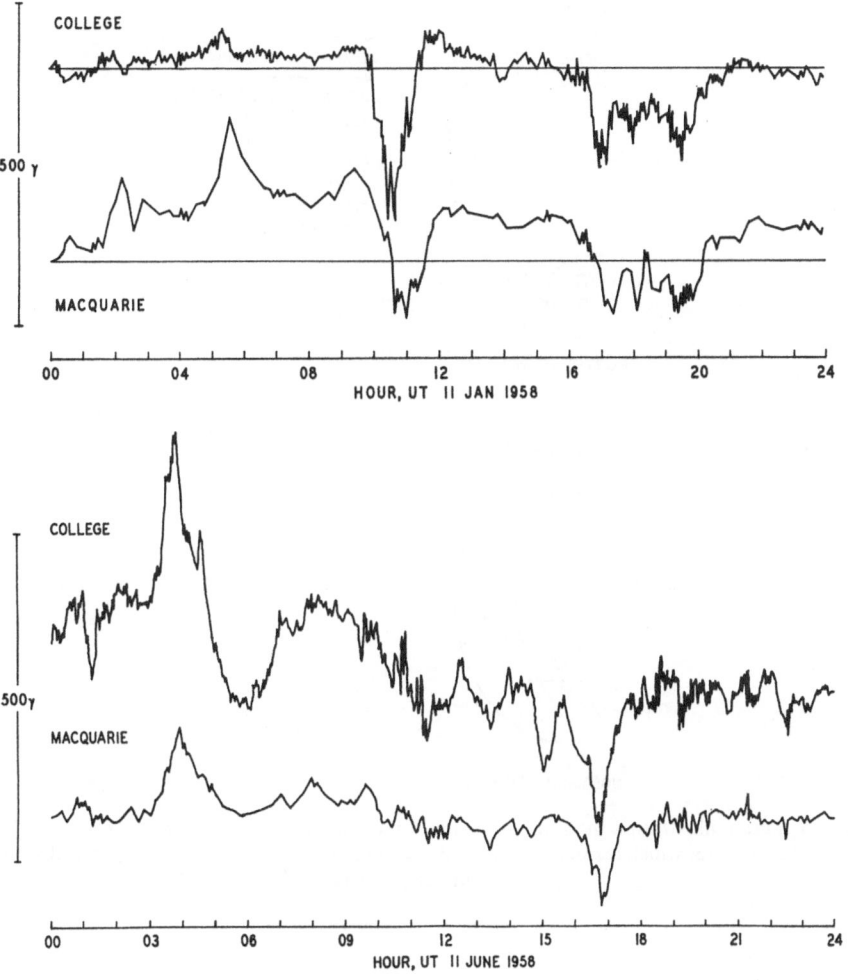

Figs. 42a,b.   *H* component magnetic records from the geomagnetically conjugate areas, College and Macquarie Island (Meng, C.-I., and S.-I. Akasofu: *Radio Science* **3**, 751, 1968).

(and also for a higher occurrence of simple positive bays in the Southern Hemisphere than in the Northern Hemisphere). This evidence suggests that conjugate areas are displaced by about 200 ~ 500 km. Therefore, the poleward boundary of the surge (and thus of the precipitation of auroral electrons) is closer to the northern conjugate station than to the southern conjugate station. Figure 44 illustrates a suggested model for the northern summer months. In this model, the conjugate station of Station A is Station B, instead of B′ which is determined on the basis of the spherical harmonic coefficients of the main field. Auroral electrons reach higher latitudes in the Southern Hemisphere (the local winter polar cap) than in the Northern Hemisphere (the local summer polar cap).

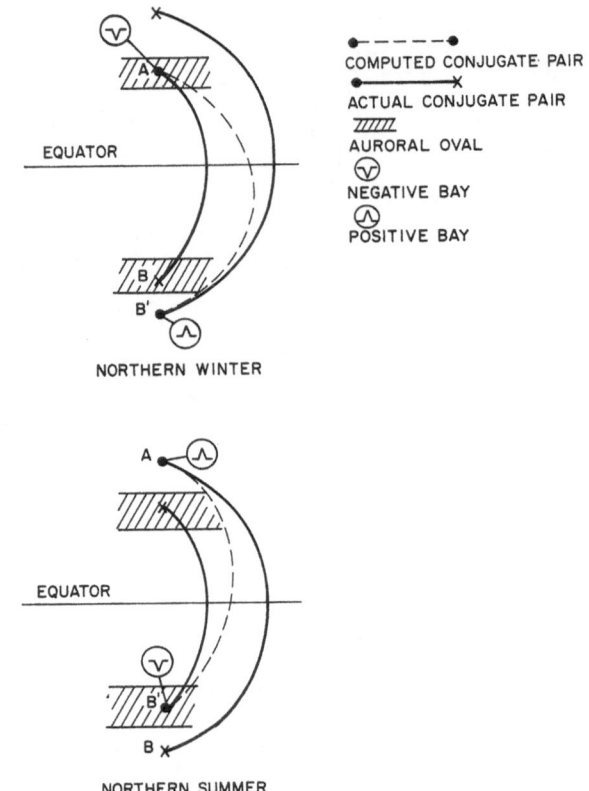

Fig. 43.   Diagram illustrating the shift of the pattern of the magnetic variations in the Northern and Southern Hemisphere, which results in a poor correlation at the computed geomagnetically conjugate areas in the polar cap.

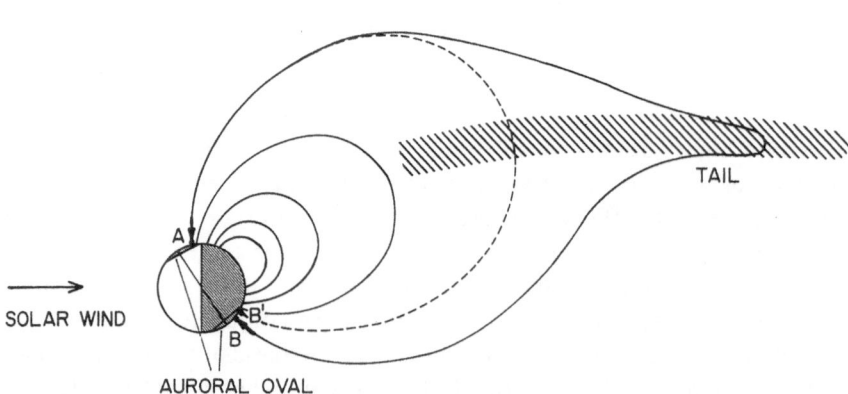

Fig. 44.   Schematic drawing showing the asymmetry of the magnetosphere which results in the shift of the conjugate areas.

References

GENERAL

CHAPMAN, S. and BARTELS, J.: 1940, *Geomagnetism*, The Clarendon Press, London.
MATSUSHITA, S. and CAMPBELL, W. H. (eds.): 1967, *Physics of geomagnetic phenomena*, Academic Press, New York.

REFERRED TO IN TEXT

AKASOFU, S.-I. and CHAPMAN, S.: 1964, 'On the asymmetric development of magnetic storm fields in low and middle latitudes', *Planetary Space Sci.* **12**, 607–626.
AKASOFU, S.-I. and CHAPMAN, S.: 1967, 'The normality of the *SD* variation at Huancayo and the asymmetry of the main phase of geomagnetic storms', *Planetary Space Sci.* **15**, 205–207.
AKASOFU, S.-I. and MENG, C.-I.: 1968, 'Low latitude negative bays', *J. Atmospheric Terrest. Phys.* **30**, 227–241.
AKASOFU, S.-I. and MENG, C.-I.: 1969, 'A study of polar magnetic substorms', *J. Geophys. Res.* **74**.
AKASOFU, S.-I., CHAPMAN, S., and MENG, C.-I.: 1965, 'The polar electrojet', *J. Atmospheric Terrest. Phys.* **27**, 1275–1305.
ALFVÉN, H.: 1939, 'Theory of magnetic storms', I, *Kungl. Sv. Vef.-Akademiens Handl.* **18**, No. 3.
ALFVÉN, H.: 1940, 'Theory of magnetic storms', II, III, *Kungl. Sv. Vef.-Akademiens Handl.* **18**, 1940.
ALFVÉN, H.: 1950, *Cosmical electrodynamics*, Clarendon Press, Oxford, England.
ATKINSON, G.: 1967, 'The current system of geomagnetic bays', *J. Geophys. Res.* **72**, 6063–6067.
BIRKELAND, K.: 1913, *The Norwegian aurora polaris expedition 1902–1903*, Vol. I, Section 2, H. Aschehoug, Christiania.
BOBROV, M. S.: 1963, 'Magnetic disturbances at conjugate points as a source of data on the upper atmosphere and solar corpuscular radiation', *Geomagnetizm i Aeronomiya* **3**, 436–442.
BOSTRÖM, R.: 1964, 'A model of the auroral electrojets', *J. Geophys. Res.* **69**, 4983–4999.
BOYD, G. M.: 1963, 'The conjugacy of magnetic disturbance variations', *J. Geophys. Res.* **68**, 1011–1013.
BRYUNELLI, B. Ye.: 1962, 'Variations of the magnetic field at conjugate points', *Geomagnetizm i Aeronomiya* **2**, 772–779.
CAHILL, L. J. Jr.: 1959, 'Detection of an electrical current in the ionosphere above Greenland', *J. Geophys. Res.* **64**, 1377–1380.
CUMMINGS, W. D. and COLEMAN Jr., P. J.: 1968, 'Simultaneous magnetic field variations at the earth's surface and at the synchronous equatorial distance', *Radio Sci.* **3**, 758–761.
FELDSTEIN, Y. I.: 1963, 'Some problems concerning the morphology of auroras and magnetic disturbances at high latitudes', *Geomagnetizm i Aeronomiya* **3**, 183–192.
FELDSTEIN, Y. I.: 1966, 'Peculiarities in the auroral distribution and magnetic disturbance distribution in high latitudes caused by the asymmetrical form of the magnetosphere', *Planetary Space Sci.* **14**, 121–130.
FUKUSHIMA, N.: 1953, 'Polar magnetic storms and geomagnetic bays', *J. Fac. Sci. Tokyo Univ.* **8**, 293–412.
HEPPNER, J. P.: 1967, 'High latitude magnetic disturbances', *Aurora and Airglow* (ed. by B. M. McCormac), Reinhold, New York, pp. 75–92.
KIRKPATRICK, C. B.: 1952, 'On current systems proposed for S_D in the theory of magnetic storms', *J. Geophys. Res.* **57**, 511–526.
KOKUBUN, S.: 1965, 'Dynamic behaviour and north-south conjugacy of geomagnetic bays', *Rep. Ionos. Sp. Res. Japan* **19**, 177–200.
MAYSURADZE, P. A.: 1965, 'Dynamics of the field of magnetic variations at high latitudes', *Geomagnetizm i Aeronomiya* **5**, 841-845.
MENG, C.-I.: 1968, *Polar auroral and magnetic substorms: their morphology and relation to the ring current*, Ph.D. Dissertation, Geophysical Institute, University of Alaska.
MENG, C.-I. and AKASOFU, S.-I.: 1968, 'Polar magnetic substorms in the conjugate areas', *Radio Sci.* **3**, 751–757.
NAGATA, T. and KOKUBUN, S.: 1960, 'On the earth storms. IV. Polar magnetic storms, with special reference to relation between geomagnetic disturbances in the northern and southern auroral zones', *Rep. Ionos. Sp. Res. Japan* **14**, 273–290.

ONDOH, T. and MAEDA, H.: 1962–1963, 'Geomagnetic-storm correlation between the Northern and Southern Hemispheres', *J. Geomag. Geoelec.* **14**, 22–32.

ROSTOKER, G.: 1966, 'Midlatitude transition bays and their relation to the spatial movement of overhead current systems', *J. Geophys. Res.* **71**, 79–95.

SCRASE, F. J.: 1967, 'The electric current associated with polar magnetic sub-storms', *J. Atmospheric Terrest. Phys.* **20**, 567–579.

SILSBEE, H. C. and VESTINE, E. H.: 1942, 'Geomagnetic bays, their frequency and current-systems', *Terr. Magn. Atmos. Elect.* **47**, 195–208.

TAYLOR, H. E. and HONES, E. W. Jr.: 1965, 'Adiabatic motion of auroral particles in a model of the electric and magnetic fields surrounding the earth', *J. Geophys. Res.* **70**, 3605–3628.

VESTINE, E. H. and CHAPMAN, S.: 1938, 'The electric current-system of geomagnetic disturbance', *Terr. Mag.* **43**, 351–382.

WESCOTT, E.: 1962, 'Magnetic activity during periods of auroras at geomagnetically conjugate points', *J. Geophys. Res.* **67**, 1353–1355.

WESCOTT, E. M. and MATHER, K. B.: 1965a, 'Magnetic conjugacy from $L = 6$ to $L = 1.4$. 1. Auroral zone: Conjugate area, seasonal variations, and magnetic coherence', *J. Geophys. Res.* **70**, 29–42.

WESCOTT, E. M. and MATHER, K. B.: 1965b, 'Magnetic conjugacy from $L = 6$ to $L = 1.4$. 2. Mid-latitude conjugacy', *J. Geophys. Res.* **70**, 43–48.

WESCOTT, E. M. and MATHER, K. B.: 1965c, 'Magnetic conjugacy from $L = 6$ to $L = 1.4$. 3. Low latitude conjugacy', *J. Geophys. Res.* **70**, 49–52.

WESCOTT, E. M. and MATHER, K. B.: 1965d, 'Magnetic conjugacy at very high latitude; Shepherd Bay – Scott Base relationship', *Planetary Space Sci.* **13**, 303–324.

YUDOVICH, L. A.: 1963, 'Magnetic activity at conjugate points', *Geomagnetizm i Aeronomiya* **3**, 583–586.

ZAYTSEV, A. N. and FELDSTEIN, Y. I.: 1967, 'Polar disturbances and current system according to data obtained in the winter season of the IGY', *Geomagnetizm i Aeronomiya* **7**, 449–454.

ZMUDA, A. J., HEURING, F. T., and MARTIN, J. H.: 1967, 'Dayside magnetic disturbances at 1100 kilometers in the auroral oval', *J. Geophys. Res.* **72**, 1115–1117.

ZMUDA, A. J., MARTIN, J. H. and HEURING, F. T.: 1966, 'Transverse magnetic disturbances at 1100 kilometers in the auroral region', *J. Geophys. Res.* **71**, 5033–5045.

CHAPTER 4

IONOSPHERIC SUBSTORM

4.1. Introduction

The ionosphere reacts violently to the magnetospheric substorm and exhibits various stormy features. In this monograph, we shall be mainly concerned with the heavy ionization in the lower ionosphere and the redistribution of the ionization during magnetospheric substorms.

The heavy ionization in the lower ionosphere provides important information on the precipitation of electrons which have appreciably higher energies than those which produce visible auroras. For this reason, this aspect of the ionospheric substorm is closely associated with the X-ray substorm which will be discussed in the next chapter. The most convenient device with which to examine this anomalous ionization is the riometer, which measures the heavy absorption of cosmic radio noise which passes through this ionization before reaching the earth. The anomalous ionization can be observed by other means; radar has been used extensively to study radio auroras; the anomalous ionization causes 'disturbances' in the propagation condition of short waves or VHF waves.

The absorption coefficient κ is proportional to

$$\frac{n_e}{v^2 + (\omega \pm \omega_B)^2}$$

where n_e denotes the number density of electrons, v the electron collision frequency, ω_B the gyro-frequency of the electron due to the component of the earth's field along the direction of propagation (cf. RATCLIFFE, 1959) and ω the observing frequency; the plus sign for an ordinary wave and the minus sign for an extraordinary wave. Figure 45 shows the height-integrated absorption for a vertically incident plane ordinary wave of frequency 30 Mc/sec for a model atmosphere when energetic electrons which have an energy spectrum of type

$$J(> E) = J_0 e^{-E/E_0}$$

over College, Alaska (BAILEY, 1968), where J denotes the flux of electrons having energy greater than E, J_0 the flux above zero energy and E_0 the so-called e-folding energy. Figure 45 is produced for the day condition and Figure 46 for the night condition; $h(N_{max})$ denotes the height at which the number density of electrons produced by the precipitation of energetic electrons peaks. The numbers on each line indicate the critical frequency of the anomalous ionization (or fEs).

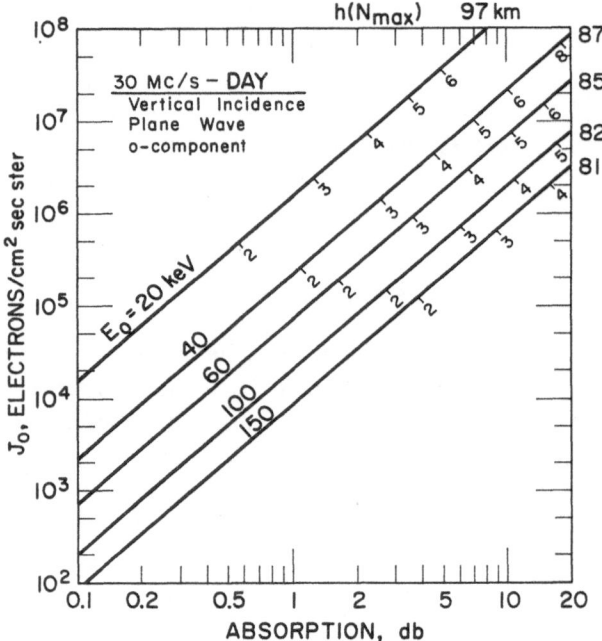

Fig. 45. Integrated absorption of cosmic radio noise when electron precipitation (characterized by J_0 and E_0) occurs in the daytime ionosphere over College, Alaska (Bailey, D. K.: *Rev. Geophys.* **6**, 289, 1968).

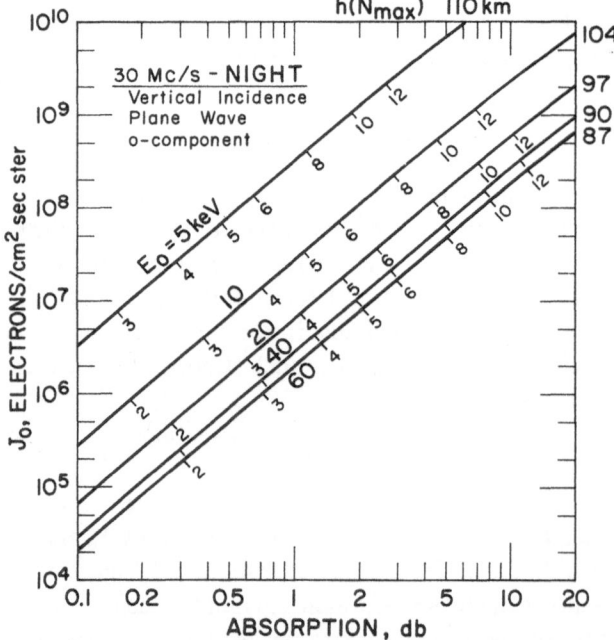

Fig. 46. Integrated absorption of cosmic radio noise when electron precipitation (characterized by J_0 and E_0) occurs in the nighttime ionosphere over College, Alaska (Bailey, D. K.: *Rev. Geophys.* **6**, 289, 1968).

4.2. Typical Daily Variation of the Cosmic Radio Wave Absorption

In this section we examine typical daily riometer (27.6 Mc/s) records from College, Alaska; Figure 47a shows the records from five successive days during a highly disturbed period during the IGY, and Figure 47b from five selected days during moderately disturbed periods during the IQSY. In Figure 47a, we can see a heavy

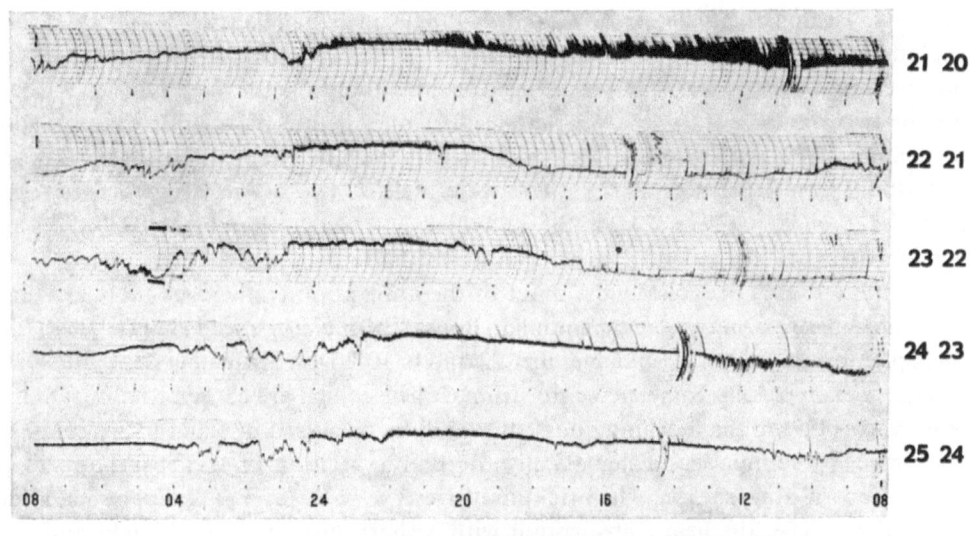

COLLEGE RIOMETER   SEPT 1957

Fig. 47a.   Several daily riometer records (IGY) from College, Alaska; time in LT.

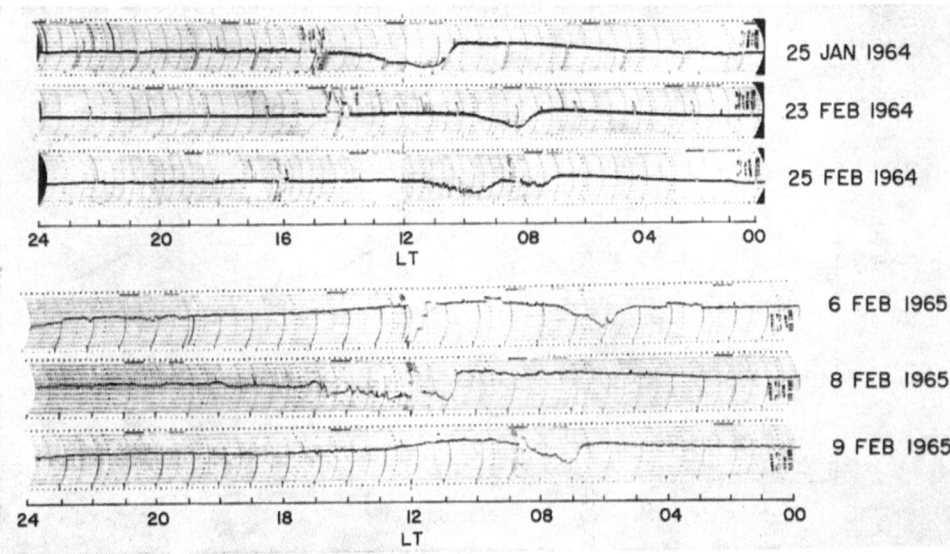

Fig. 47b.   Several daily riometer records (IQSY) from College, Alaska.

and almost continuous absorption of cosmic radio waves during the daylight hours, particularly on September 22 and 23. This continuous absorption diminishes gradually after 16 LT. The absorption during the period between 16 LT and 00 LT is characterized by brief impulsive type absorptions; typical examples are seen at about 2030 LT on September 21, 1900 LT, 2200 LT, 2300 LT on September 22, 2240 LT on September 23 and 1720 LT, 2200 LT on September 24. This period is suddenly terminated in the midnight hours by the onset of an intense absorption with varying intensity. Examples are found at about 0045 LT on September 21/22, 0050 LT on September 23/24 and 0010 LT on September 24/25. A careful inspection shows, however, that such absorption events last for only about one to three hours and that one or two similar events follow after the first one, before dawn. For example, on September 21/22, the first event (0045 LT) was followed by another event which began at 0340 LT; on September 22/23, there were at least two events after the first one (0050 LT); on September 23/24, the second event began sharply at 0230 LT; on September 24/25, the second event began at 0250 LT.

As the morning progresses, the onset of the absorption tends to be more gradual than those onsets occurring in the midnight hours. Examples are seen at about 0640 LT on September 21, 0600 LT on September 22 and 0810 LT on September 23. Figure 47b shows six examples of riometer records from College which are chosen to illustrate intense absorption in the morning hours; they will be discussed in detail in Section 4.6.

Figure 48 summarizes schematically the daily variation of the absorption at a typical auroral zone station. The brief impulsive type in the evening hours is hereafter called the E type, the heavy absorption with a sharp onset in the midnight and the early morning hours is called the N type, and the absorption with a gradually commencing type in the late morning hours is called the M type. This classification is based on characteristics of auroral displays associated with each type of absorption.

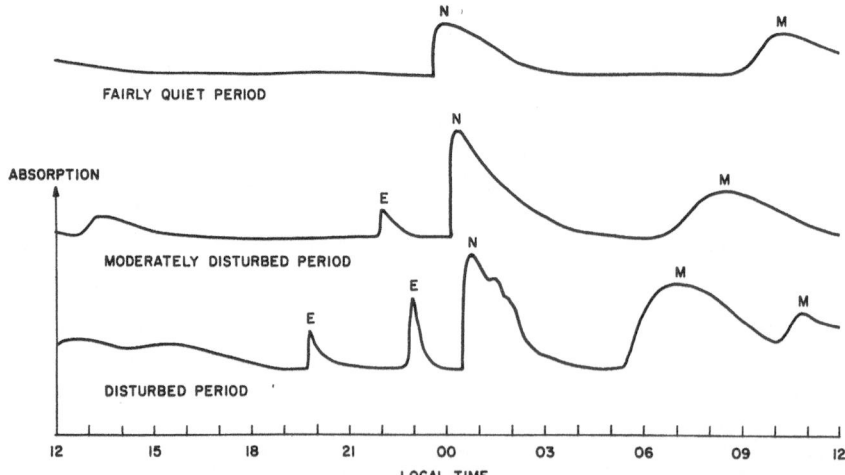

Fig. 48. Schematic diagram showing the daily variation of the cosmic radio noise absorption for different magnetic conditions.

In earlier studies, both the E and the N types were combined together and were called either the F (fast) type (PARTHASARATHY and BERKEY, 1965) or the night event (HARGREAVES and COWLEY, 1967). The M type has been referred to as the slowly varying absorption (SVA); when the M type occurs late in the morning hours, it has been called the daytime absorption. The three types will be discussed in detail in Sections 4.4, 4.5 and 4.6, respectively. In the following, we shall review first some of the statistical studies of absorption.

4.3. Statistical Daily Variation Pattern

The daily riometer records, such as those examined in Section 4.2, can be collected from a number of stations at different dp latitudes for a statistical analysis. It is clear

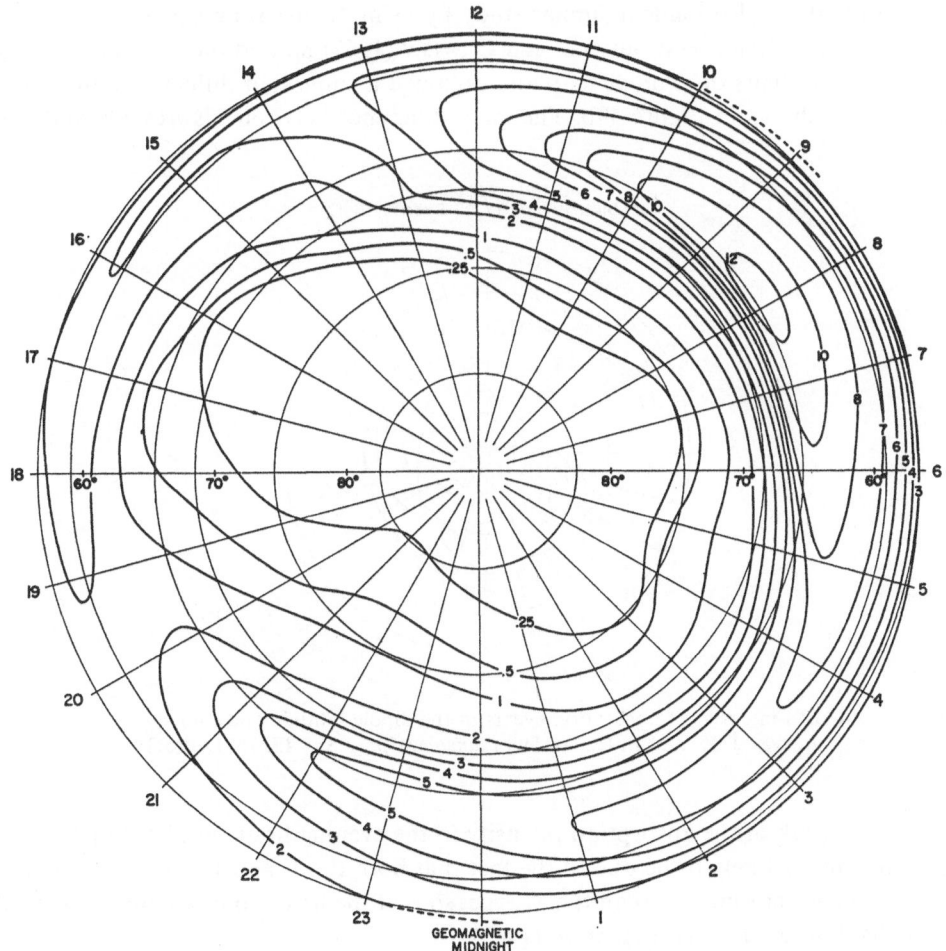

Fig. 49a. Percentage of the time that cosmic radio noise absorption of 1.0 dB or more occurred at 30 Mc/sec (Hartz, T. R., L. E. Montbriand, and E. V. Vogan: *Canadian J. Phys.* **41**, 581, 1963).

from the examples seen in Section 4.2. that there are at least three important parameters to be considered; they are the occurrence frequency, intensity and duration
(HARGREAVES and COWLEY, 1967). It is thus not surprising to have different results,
depending on how the three parameters are weighted in different statistical analyses.

Hartz *et al.* (1963) made an extensive analysis of riometer records from six
Canadian stations. They compiled the data by listing the occurrence and duration
of absorption of intensity greater than 1 dB in each half-hour of the day and obtained
the time percentage occurrence as a function of geomagnetic latitude and mean
geomagnetic time. Figure 49a shows their results.

In this statistical study, the heaviest absorption is encountered at about 08 LT at
dp lat 67°. Further, a station at this latitude encounters two peaks of the occurrence
of absorption in the course of a day, one in the early evening ($\sim$22 LT) and the
other at about 08 LT. A station at dp lat 60° encounters a single peak at about 08 LT.
DRIATSKIY (1966) also made a similar study by using Soviet riometer data.

On the other hand, HARGREAVES and COWLEY (1967) plotted the median intensity
of absorption events on a polar map and obtained a somewhat different result; their
polar plot is shown in Figure 49b. The main difference between Figures 49a and 49b

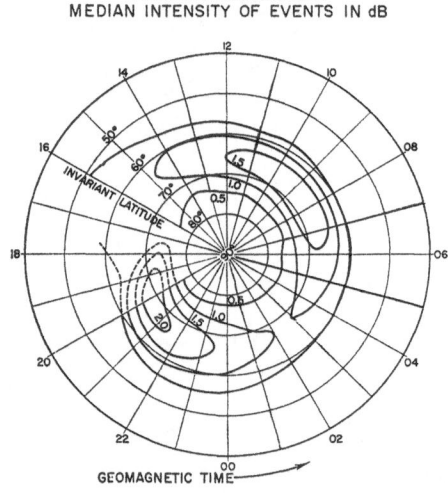

Fig. 49b.   Median intensity of absorption events in the dipole latitude and time coordinates (Hargreaves, J. K. and F. C. Cowley, *Planetary Space Sci.* **15**, 1571, 1967).

is a distinct peak of the absorption intensity in the evening hours at about dp lat 67°.
The difference is likely to be due to the fact that both the E and N types are intense,
but have a shorter duration than the M type, so that the first two types are suppressed
by the method used by HARTZ *et al.* (1963).

As noted by a number of workers (cf. BASLER, 1963), there is a considerable seasonal
variation of the intensity of the absorption. Figure 50 shows the average absorption

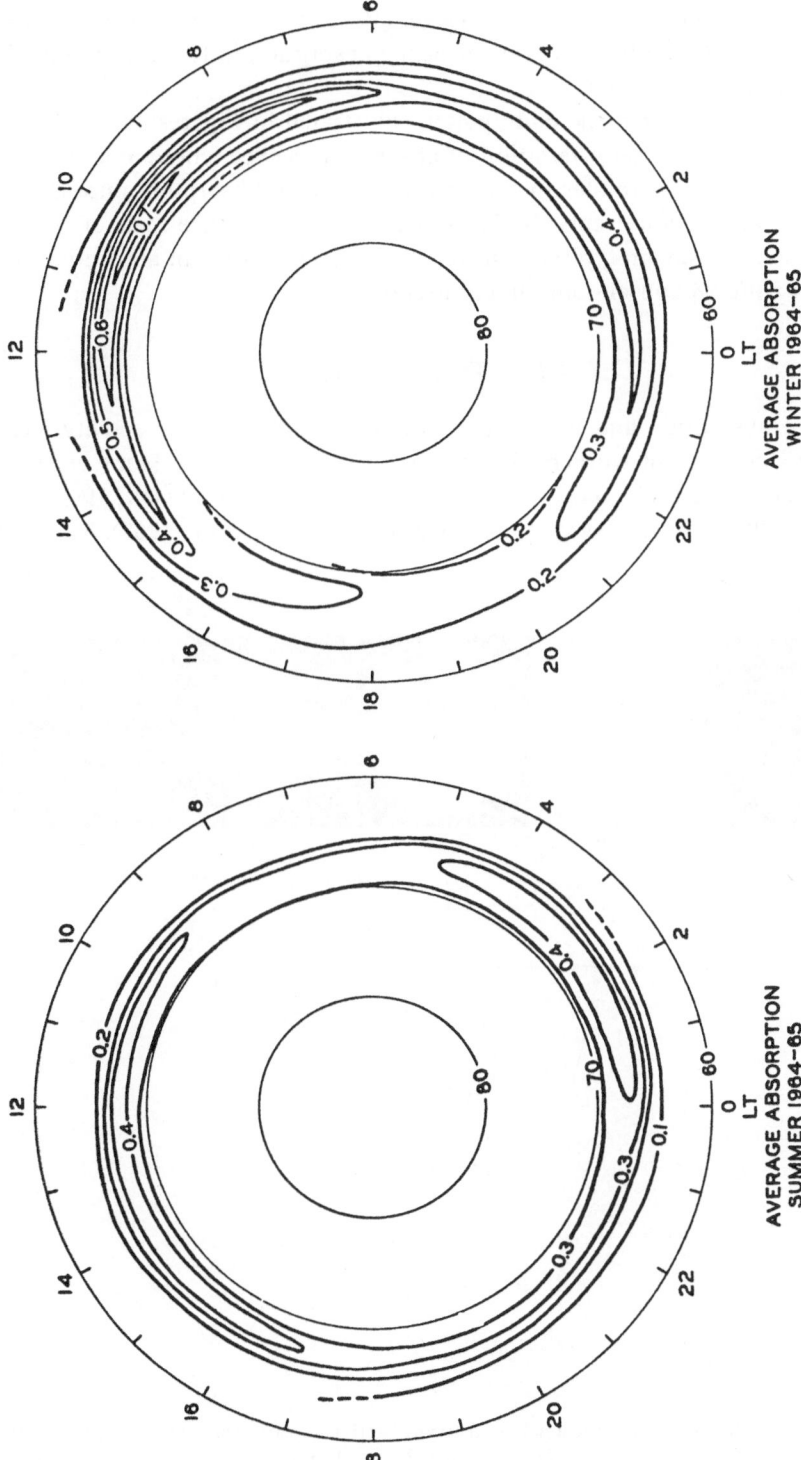

Fig. 50. Average cosmic radio noise absorption in the dipole latitude and time coordinates during the summer and winter months of 1964/65 (Hook, J. L.: *J. Atmospheric Terrest. Phys.* **30**, 1341, 1968).

for summer and winter, constructed by HOOK (1968), using Alaskan riometer records (dp lat 61° ~ 69°). It can be seen that the heavy absorption in the late morning hours is not clearly seen in summer.

It is quite obvious that the daily pattern illustrated in Figures 48, 49a, b and 50 does not mean that the daily variation of absorption is experienced at a point on the earth because the earth rotates under such a pattern of absorption once a day. On the contrary, absorption events (N, E, M types) occur in a time scale of order only a few hours, as we saw in typical riometer records. Therefore, an absorption pattern similar to Figure 48 appears only intermittently.

4.4. N Type Absorption

The N type absorption and its relation to auroral and magnetic activity has been studied by a number of workers (LITTLE and LEINBACH, 1958; HOLT et al., 1962; HOLT and OMHOLT, 1962; BASLER, 1962; ANSARI, 1964; GUSTAFSSON, 1964; EATHER and JACKA, 1966; PARTHASARATHY and BERKEY, 1965; BERKEY, 1968). The purpose

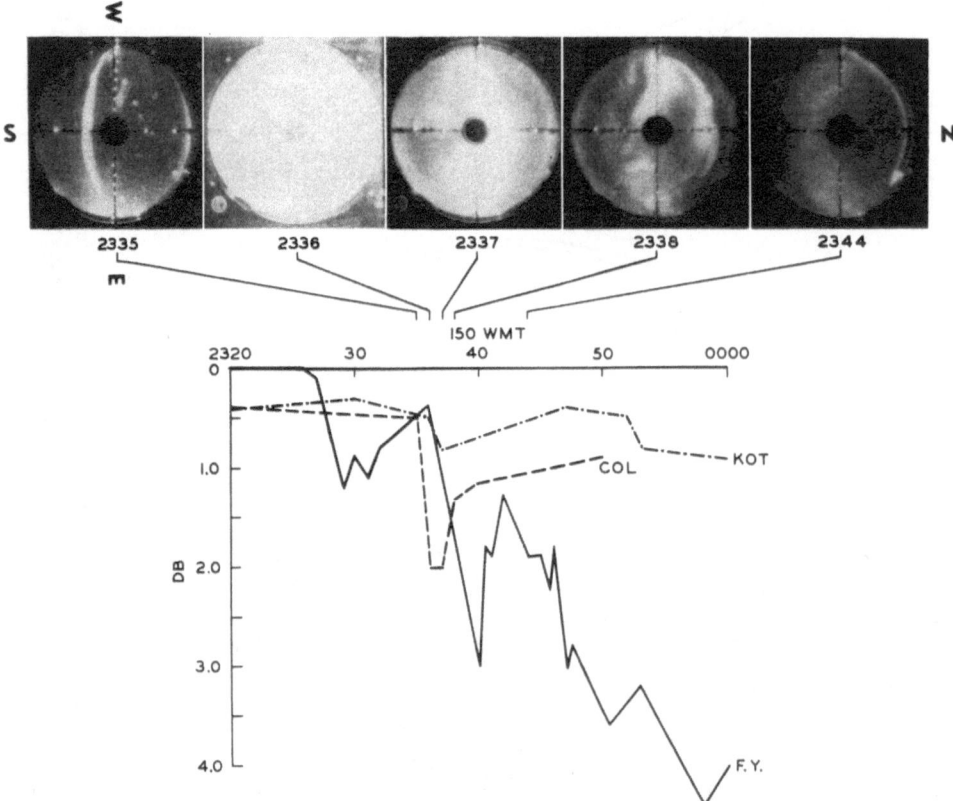

Fig. 51.   N type absorption and the associated auroral activity (Parthasarathy, R. and F. T. Berkey: *J. Geophys. Res.* **70**, 89, 1965).

of this section and the following two sections is to relate the three types of absorption to the pattern of auroral and polar magnetic substorms. This will enable us to construct a pattern of the growth and decay of the ionospheric substorm over the entire polar cap.

The N type absorption is associated with the poleward expansive motion of the aurora during an auroral substorm. Figure 51 shows an example which occurred on January 26, 1962 (PARTHASARATHY and BERKEY, 1965). An arc, which was located a little South of College at 2335 LT, became suddenly active at 2336 LT and moved rapidly poleward. The sudden and violent nature of the onset of the auroral substorm is well recognized by comparing the photographs taken at 2335 LT and 2336 LT. The cosmic radio noise absorption began first at College, but it developed most strongly at Fort Yukon; at Kotzebue, about 700 km West of College, there was only a slight absorption during the event.

Figure 52 shows another example of the close association between the poleward explosive motion of the aurora and the sharp onset of absorption. An arc, which

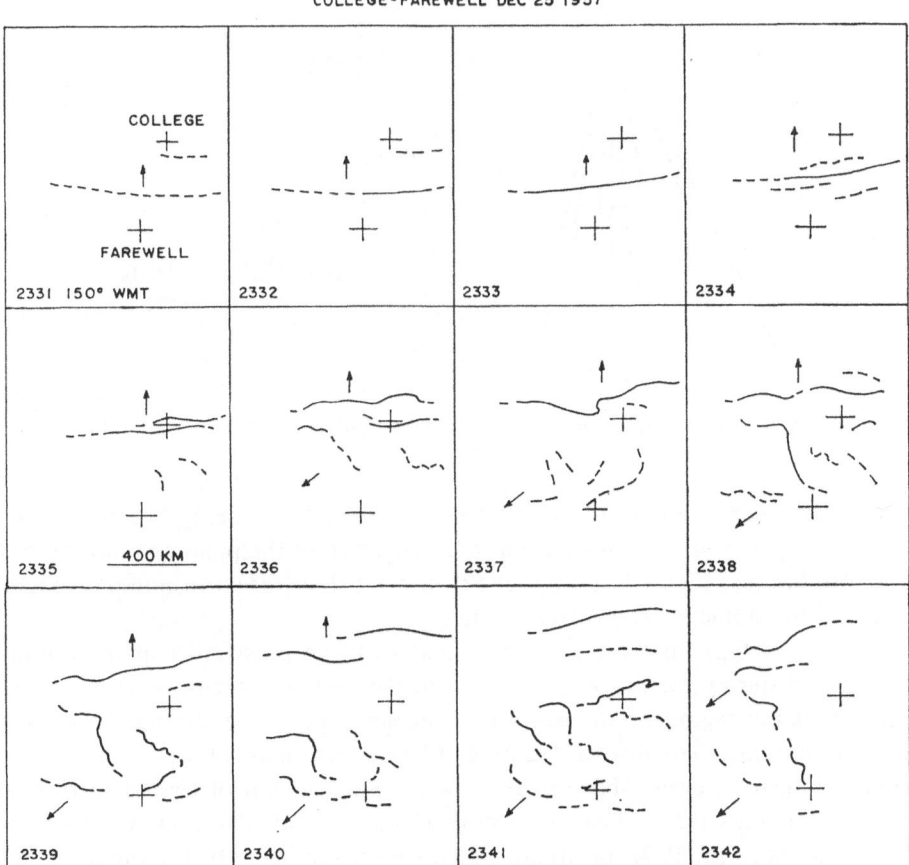

Fig. 52a,b, c.   N type absorption and the associated auroral and magnetic activities.

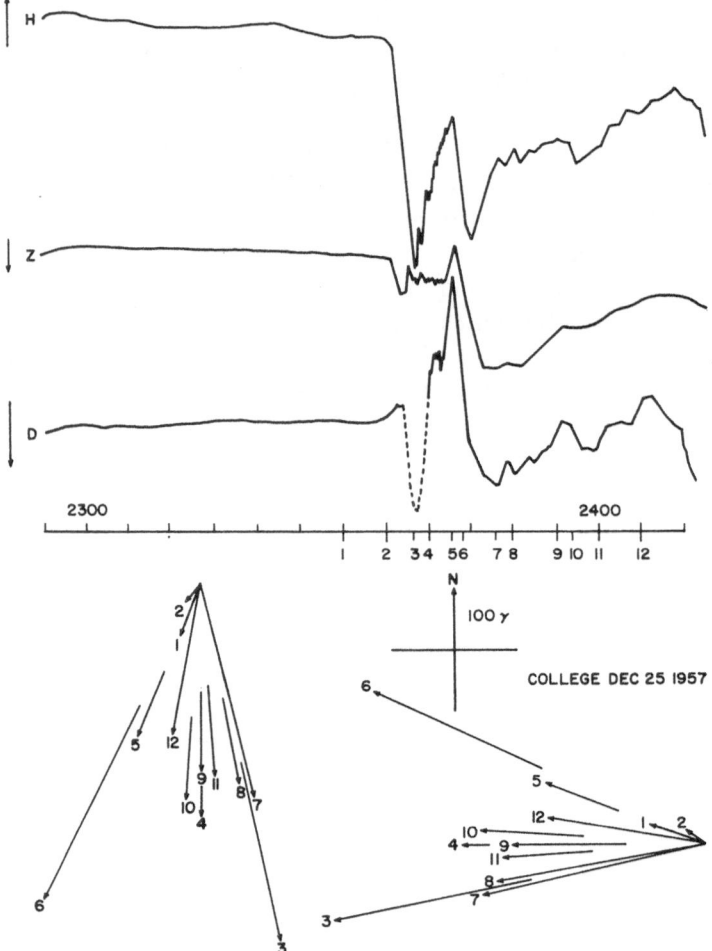

Fig. 52b.   Lower left, magnetic vectors; lower right, equivalent current vectors.

was located between Farewell and College at 2331 LT, suddenly became active at 2332 LT and poleward expansion followed. The speed of the poleward motion of the arc was 1070 m/sec; Figure 52a. Figures 52b and 52c show the corresponding magnetic record and the riometer record from College.

Since the N type absorption is associated with the poleward expansive motion of the aurora during the auroral substorm in the midnight sector, it is not difficult to infer that the region of the absorption expands poleward during the explosive phase and contracts equatorward during the recovery phase. Figure 53a shows an example of such an event observed by a North-South chain of Canadian riometers (JELLY and BRICE, 1967). The absorption began first at Val d'Or (dp lat 61°) at 0300 UT on May 15, 1964, but there was a systematic delay of the onset at higher latitude stations. The expanding bulge reached at least as high as dp lat 67° at about

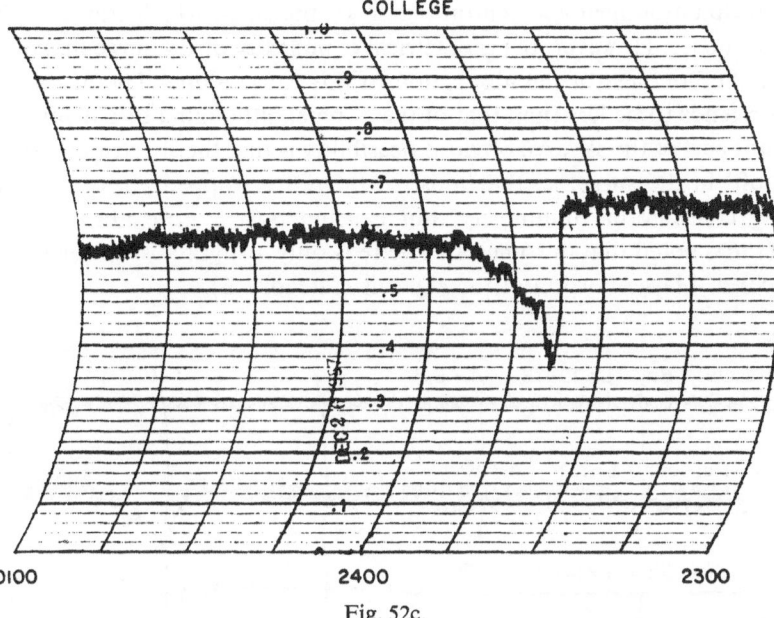

Fig. 52c.

0350 UT, 50 min after the onset. A number of examples were recently studied also by
JELLY (1968). A similar example was also studied by DRIATSKIY (1966) who used
riometer records from three stations, Dixon Island, Heiss Island and NP-10.

Another important feature in Figure 53a is a clear indication of the recovery
between dp lat 61° and 68° a little after 0340 UT, suggesting that the most intense
region of precipitation was near the front of the expanding bulge. Many more examples should be examined in a similar way to study the distribution of precipitation
of energetic electrons in the auroral bulge. Figure 53 shows a beautiful example of the
usefulness of a North-South chain of stations in studies of polar upper atmospheric
phenomena. Future progress in polar upper atmospheric research depends greatly
on such a systematic positioning of stations.

Figure 53b shows another example of the expansion of the region of absorption
observed by the Canadian North-South chain of riometers on September 26, 1963
(LIN *et al.*, 1968). The onset of the N event is clearly seen at about 0347 UT at Cape
Jones (dp lat 67°). The region of the absorption expanded progressively poleward and
reached Coral Harbour (dp lat 76°) a little before 0430 UT. Fortunately, the Alouette I
satellite passed over Canada during the substorm and observed an unusual flux of
energetic electrons (>40 keV), confirming the poleward expansion of the precipitation area.

4.5. E Type Absorption

The E type absorption is associated with a westward traveling surge. The brief duration of this type appears to be due partly to the fact that there is a strong concentra-

tion of precipitation near the central region of the surge which travels with a speed of about 1 km/sec. However, the geometrical complexity of the surge makes it difficult to ascribe the briefness of the absorption to this cause alone. The magnitude of the absorption tends to be less than that of the N type, although a bright surge can cover a significant part of the sky. An obvious possibility is that the spectrum of the electrons is very 'soft' in the surge, so that most of them are stopped above a 100 km level, generating considerable luminosity, but causing only a little absorption. In fact, BERKEY (1968) noted that in some cases there is only a slight absorption even if bright auroras cover a large portion of the sky (his Category B).

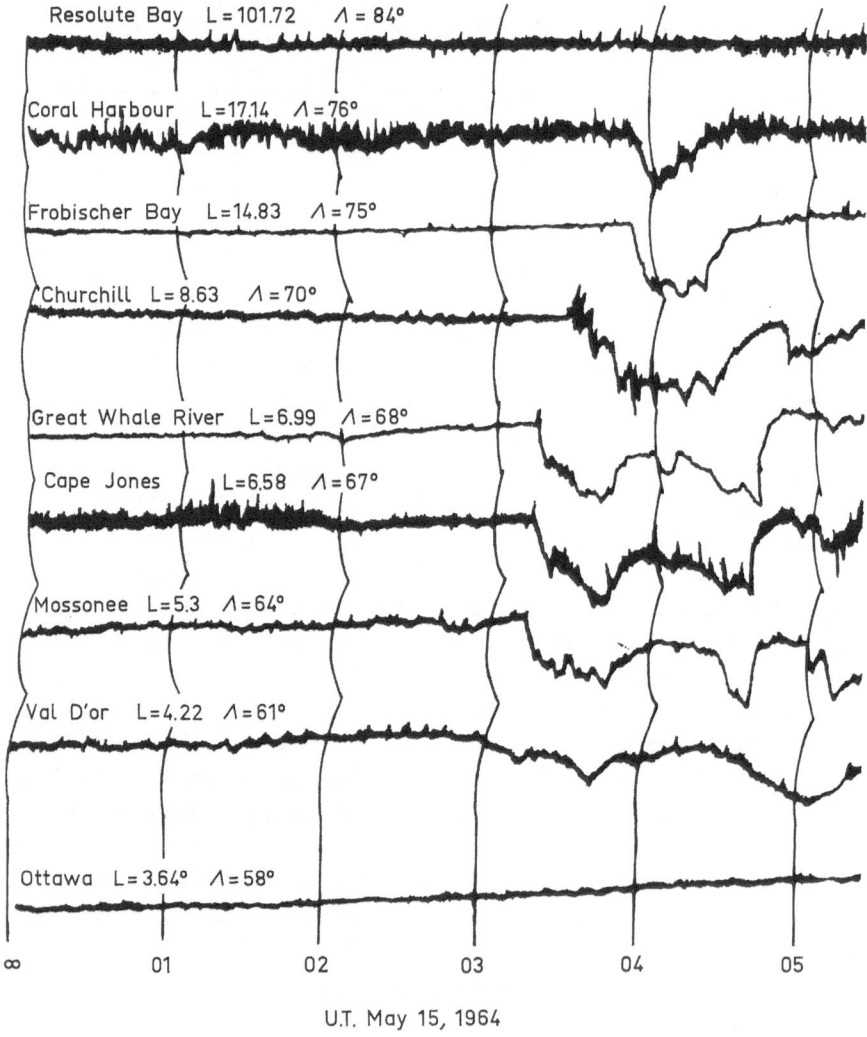

U.T. May 15, 1964

Fig. 53a.   Poleward expansion of the N type absorption recorded by the Canadian North-South chain of riometers (Jelly, D. and N. M. Brice: *J. Geophys. Res.* **72**, 5919, 1967).

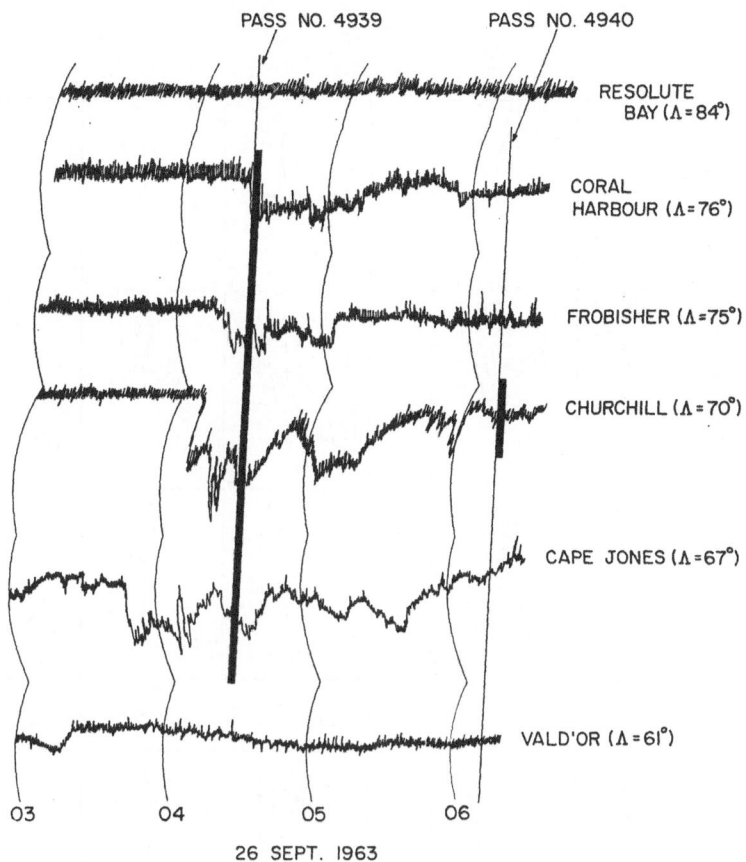

PASS NO. 4939   PASS NO. 4940

RESOLUTE
BAY (Λ = 84°)

CORAL
HARBOUR (Λ = 76°)

FROBISHER (Λ = 75°)

CHURCHILL (Λ = 70°)

CAPE JONES (Λ = 67°)

VALD'OR (Λ = 61°)

03   04   05   06

26 SEPT. 1963

Fig. 53b. Poleward expansion of the N type absorption recorded by the Canadian North-South chain of riometers; Alouette pass numbers are indicated (Lin, W. C., I. B. McDiarmid, and J. R. Burrows: *Canadian J. Phys.* **46**, 80, 1968.

Figure 54 shows an example of an intense westward traveling surge and the associated absorption at College and Barrow. At College, the E type absorption was recorded at about 2016 LT, a few minutes before the surge passed over Fort Yukon. The surge reached near Point Barrow at about 2027 LT, after traveling across Alaska. At Point Barrow, the E type absorption was recorded at about 2029 LT.

Figure 55 shows another intense surge which passed over Fort Yukon on February 11, 1958. The surge was associated with a large-scale counter-clockwise motion and was very bright. The E type absorption at Fort Yukon reached the peak value at about 2102 or 2103 LT. The auroral activity did not reach the zenith of College which is located about 200 km South of Fort Yukon, so that only a weak E type absorption was recorded there. Several examples of the E type absorption and the associated all-sky photographs can also be found in the paper by EATHER and JACKA (1966). Figure 56 shows one of their examples.

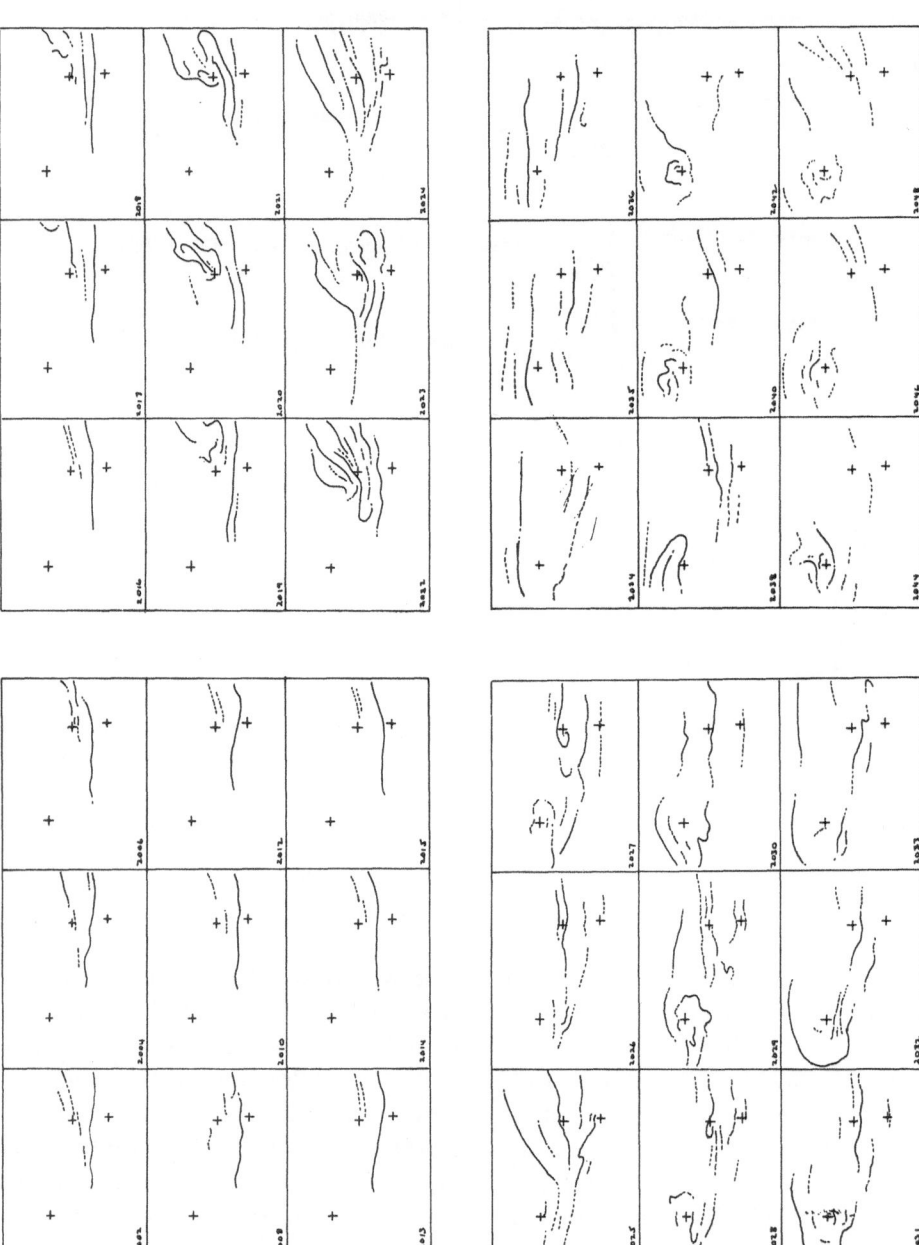

Fig. 54a,b.   Example of a westward traveling surge and the associated cosmic radio noise absorption at College and Barrow.

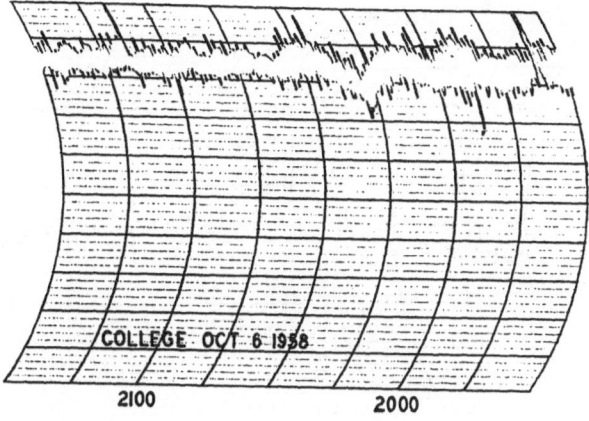

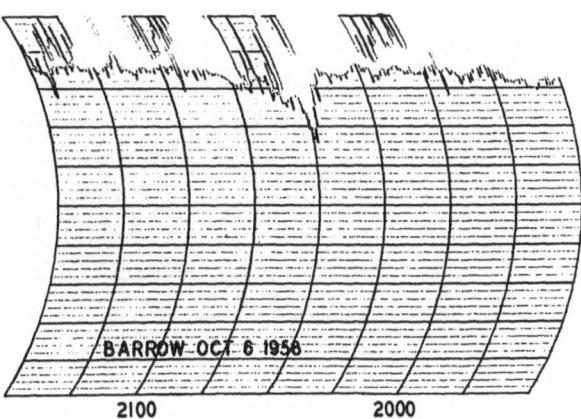

Fig. 54b.   Cosmic noise absorption at Barrow and College (in Figure 54a, the upper left and lower right, respectively).

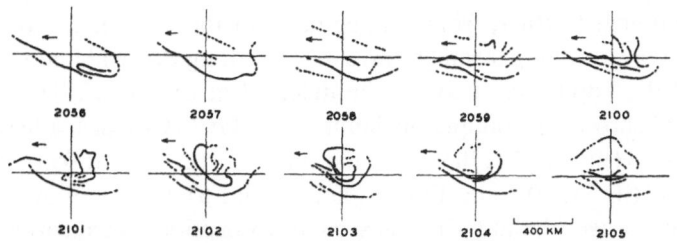

Fig. 55a, b.   Example of a westward traveling surge and the associated cosmic radio noise absorption at Fort Yukon and College; the center of the crosses is located over Fort Yukon.

FORT YUKON

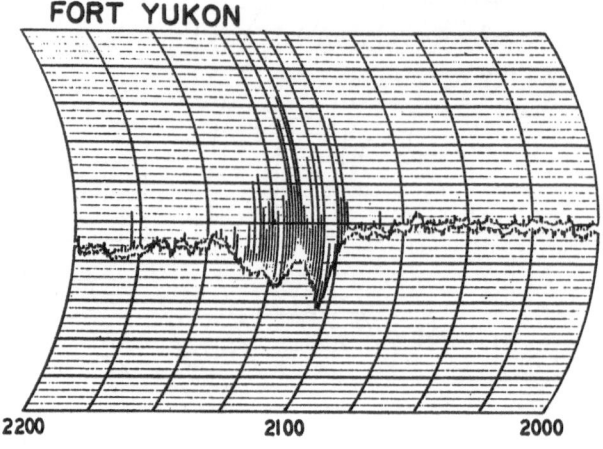

2200   2100   2000

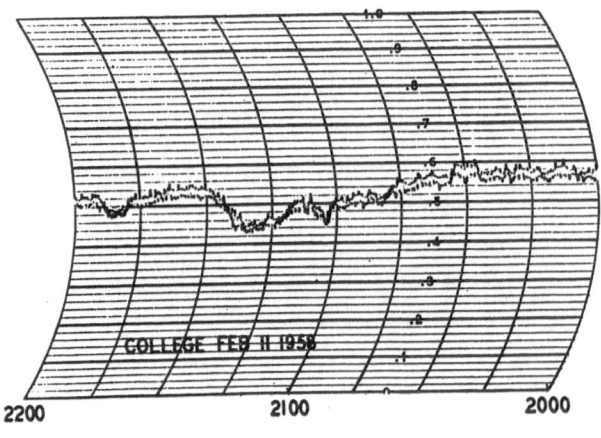

2200   2100   2000

Fig. 55b.

As mentioned in Section 2.3., the westward traveling surge travels along the pre-existing arc in the evening sector, namely along the auroral oval. Thus, the E type absorption is expected to be more common at dp lat 70° than at dp lat 65° in the early evening hours. Further, from the geometry of the surge it can be inferred that the absorption occurs along a rather narrow band along the oval and that absorption events recorded at dp lat 70° may not be detected at dp lat 65°. This is indeed the case. Figure 57 shows an example on January 21, 1967 (CHIVERS, 1967). There were two absorption events on that day. The first one was recorded at 0537 UT (1937 LT) at Barter Island (dp lat 70°) and Fort Yukon (dp lat 67°), but not at College (dp lat 64.7°) and Anchorage (dp lat 60°). The second absorption was recorded at 0930 UT (2330 LT) at the three high latitude stations, Barter Island, Fort Yukon and College, but did not extend to Anchorage. The situation will, however, be quite different

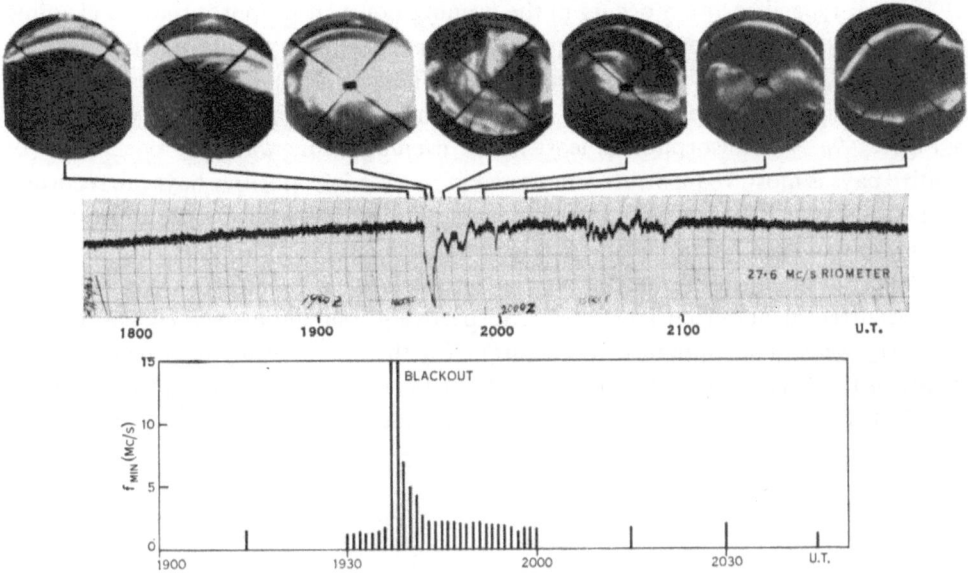

Fig. 56.   Example of a westward traveling surge and the associated cosmic radio noise at Mawson
(Eather, R. H. and F. Jacka: *Australian J. Phys.* **19**, 215, 1966).

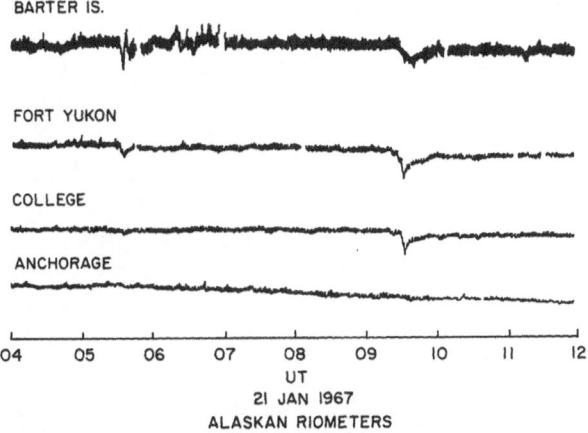

Fig. 57.   Cosmic radio noise absorption recorded by the Alaskan North-South chain of riometer
stations (Chivers, H. J. A.: Private communication, 1967).

during intense geomagnetic storms when the oval expands equatorward. If the oval
descends to as low as dp lat 65° in the early evening hours, both quiet auroral and
westward traveling surges appear abnormally early, resulting in an abnormally early
appearance of the E type events. On such an occasion, the N type also tends to appear
earlier than usual; this corresponds to the disturbed condition in Figure 28 in Section
2.3.

Since E type absorption occurs in the evening hours, it is appropriate to examine whether or not positive (magnetic) bays are associated with E type absorption; positive bays are most common in the evening hours. There is no or only slight absorption associated with positive bays. This is in agreement with the statistical study illustrated in Figure 49a. The absorption is least in the evening hours when the occurrence of positive bays is most frequent. As mentioned in Section 2.3., positive bays are recorded when the westward traveling surges are passing well poleward of an auroral zone station, so that such a poor correlation is not particularly surprising. Figure 58 shows the College magnetic and riometer records on January 23, 1958. Two intense positive bays occurred at about 06 and 09 UT, respectively, but the associated absorptions were very weak. The amount of absorptions for these two events may be compared to that of the N type which began at about 1210 UT and the two M type absorptions at 1430 UT and 1800 UT; although the magnitude of the negative bays associated

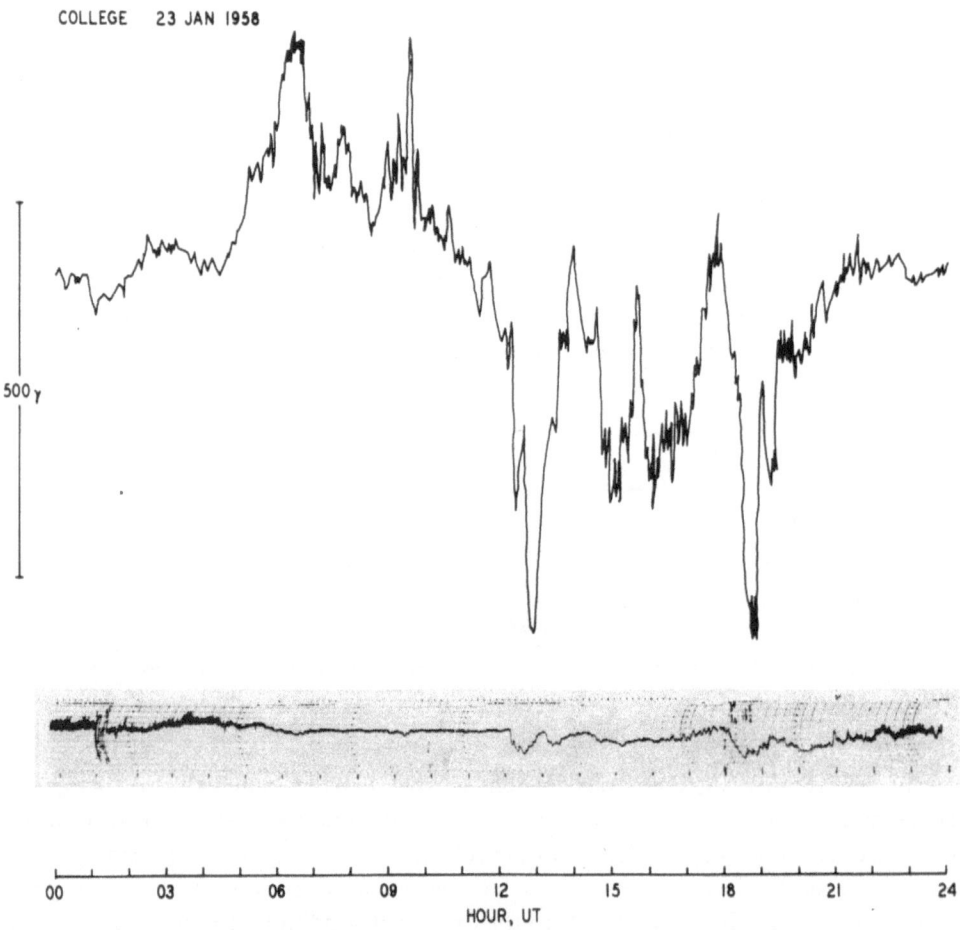

Fig. 58. Comparison of College magnetic and riometer records.

with them were only twice as great as that of the positive bays, the amount of the N and M type absorptions was considerably greater than those associated with the positive bays. This suggests an important difference in the nature of positive and negative bays (Section 3.1).

However, if a negative (magnetic) bay is superposed on a positive bay, an appreciable absorption occurs during that period (Section 2.3; Figure 28). The negative bay occurs when a westward traveling surge passes a little poleward of the station. Figure 59 shows such an example on June 24, 1958; a sudden superposition of an intense negative bay began at about 0650 UT. Prior to this event, a positive bay had been in progress, but an appreciable absorption began only at about 0650 UT.

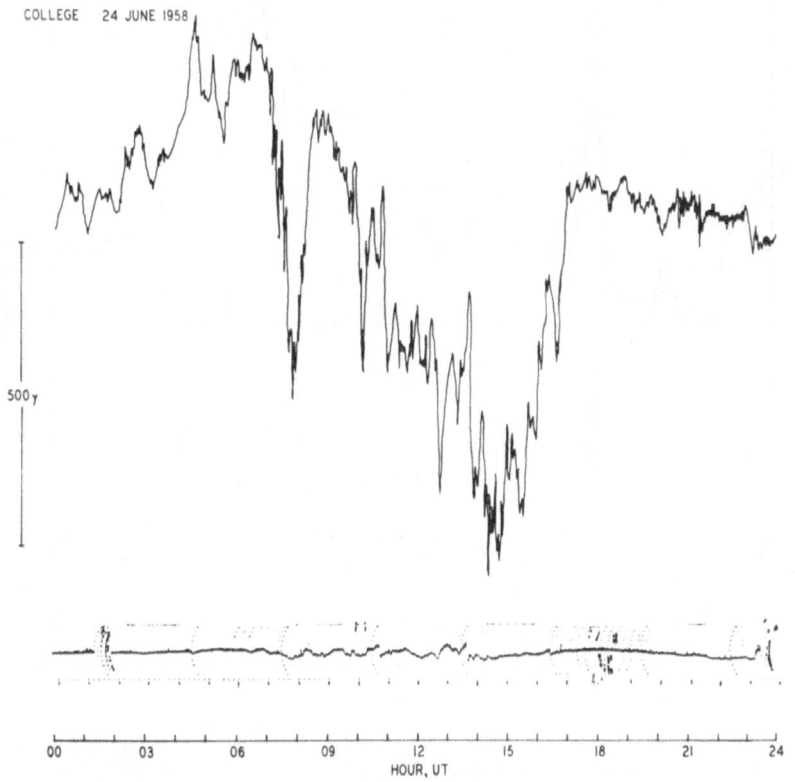

Fig. 59. Comparison of College magnetic and riometer records.

4.6. M Type Absorption

The association between the M type absorption and eastward drifting bands or patches has been noted by a number of workers (HEPPNER *et al.*, 1952; ANSARI, 1964; EATHER and JACKA, 1966; BERKEY, 1968). The M type absorption is characterized not only by a rather slow onset, but also by the lack of a simultaneous explosive increase of the brightness of the aurora. Figure 60 shows a good example in which the N type and the M type are well contrasted (ANSARI, 1964); the upper diagram shows the absorption and the intensity of λ 5577 by looking in the direction 12° South of zenith and the lower diagram 12° North of zenith. It shows two absorption events, beginning at 0025 LT and 0135 LT, respectively. The first event had a sharp onset, so that it is

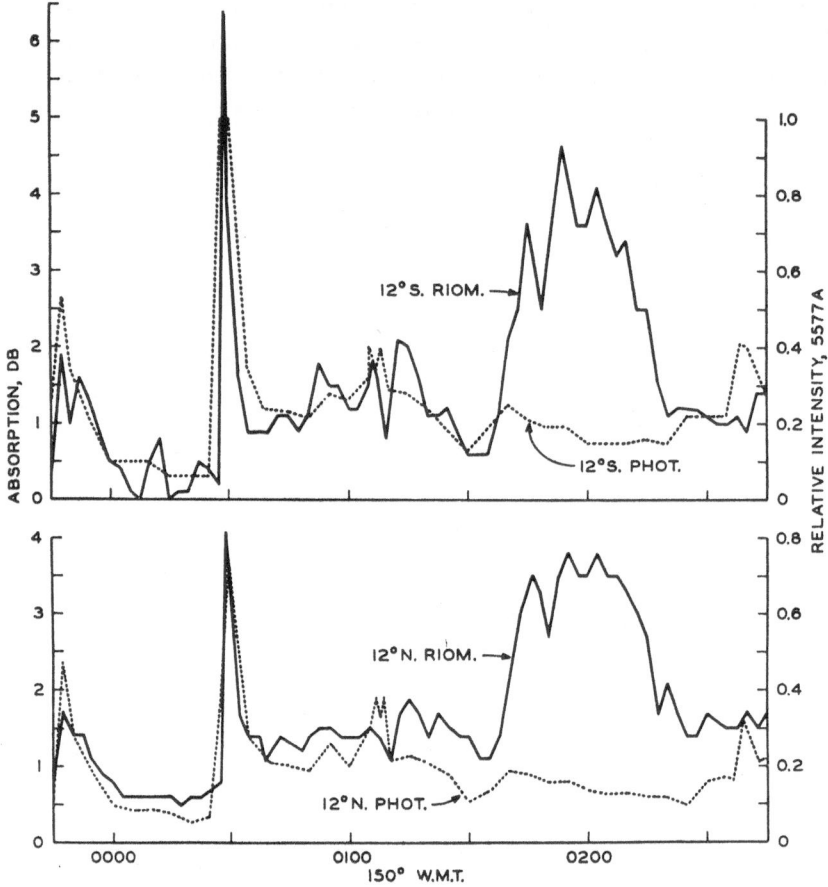

Fig. 60.   Comparison of the cosmic radio noise absorption and the λ 5577 intensity for both the N type and M type events (Ansari, Z. A.: *J. Geophys. Res.* **69**, 4493, 1964).

undoubtedly the N type event; the intensity of λ 5577 also had a sharp increase. On the other hand, the onset of the second event was slow; furthermore, there was only a slight indication of increase of λ 5577.

One obvious possibility which might account for the difference between the E or N type and the M type is that the M absorption is associated with a precipitation of 'harder' electrons than that for the E or N type (ANSARI, 1964). Since they penetrate deep into the atmosphere, the absorption will be greater for the same value of J_0, and the resulting luminosity will be less (because of a more efficient deactivation of excited oxygen atoms). For example, in Figure 45, for $J_0 = 10^5/cm^2$ sec ster, the amount of absorption is about 0.3 dB and 3 dB for $E_0 = 20$ keV and 100 keV, respectively. There seems to be some controversy on the spectral changes of the precipitating electrons. JOHANSEN (1965) did not observe a significant change in the ratio of luminosity to absorption and concluded that there was little change in the spectra. EATHER and JACKA (1966) originally proposed a contribution by protons, but recently withdrew their suggestion by supporting Ansari's conclusion. This problem will be discussed further in the next chapter.

As noted in Sections 4.3 and 4.4 the M type absorption occurs not only in the early morning hours (when simultaneous auroral activity can be examined), but also in the late morning hours, midday hours or even in the early afternoon hours. In fact, the statistical studies in Section 4.4 suggest that the M type absorption is most intense in the late morning hours and often extends as far as the noon sector. Since such late M type events cannot be examined in terms of simultaneous auroral activity and since they are not necessarily well correlated with magnetic activity, they have been somewhat separately treated from the early morning M type (BROWN, 1964; ANSARI, 1965; FEDYAKINA, 1963; BEWERSDORFF et al., 1967).

The lack of a good correlation of such late M events at an auroral zone station with magnetic activity is due partly to the fact that the auroral electrojet flows along the auroral oval (and thus an auroral zone station is located well equatorward of the oval in the late morning hours) and also due partly to the fact that M type events are most intense along the auroral zone (see Figure 49a). It is not difficult to show that late M type events are associated with a typical polar substorm feature in the midnight sector (namely, the explosive motion of the aurora, a sharp onset of negative bays and the onset of the N type absorption). Figure 61 shows magnetic records from Siberian stations for the periods which correspond to the late morning M events in Figure 47b. It can be seen that each absorption event is clearly associated with a negative bay in the midnight sector.

The next step is to examine the development of late M type events in terms of the substorm time T. This problem has been studied by several workers (FEDYAKINA, 1963; JELLY and BRICE, 1967; BEWERSDORFF et al., 1967).

By using two Canadian riometer records and Kiruna records JELLY and BRICE (1967) showed that the N type absorption in Canada (the midnight sector) is associated with a late morning M type at Kiruna; Figure 62 shows their typical example. The first absorption which began at about 0130 UT in the two Canadian stations were clearly the E type which was associated with only a weak absorption at Kiruna. The second event which began a little after 03 UT in Canada was clearly associated with an M type absorption at Kiruna. There are at least two problems in determining an accurate time relationship between them. First of all, because the poleward expansion of the absorption region does not occur instantly, there will be an error of obtaining the onset time of the N event in the midnight sector; for example, if the Churchill record alone is used for this study, there would be an error of as large as 10 minutes in Figure 62. Secondly, by the nature of the M type events their onset is not often clearly defined; in fact, JELLY and BRICE (1967) showed that many cases have a double onset. In Figure 62, the absorption at Kiruna appeared to begin at the time indicated by D_1, but was suddenly intensified at the time indicated by D_2. They found that

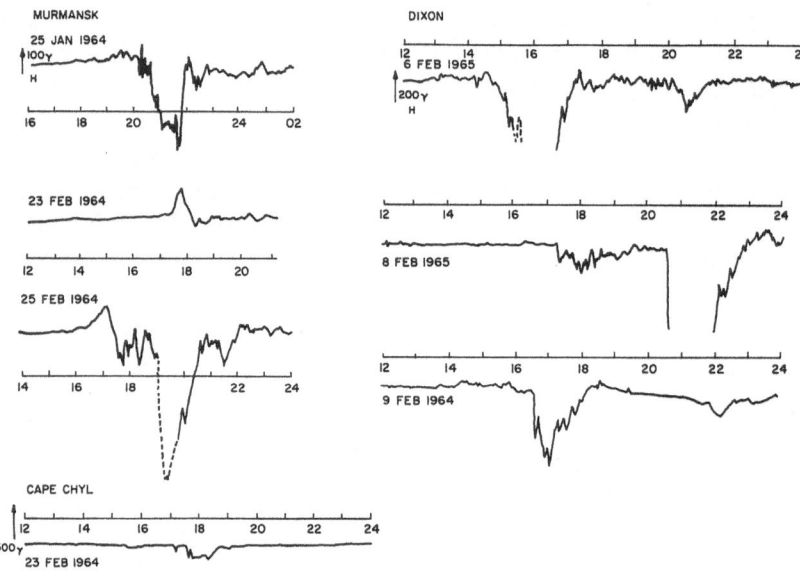

Fig. 61.   Magnetic records from Siberian stations for the periods which correspond to the late morning M event in Figure 47b.

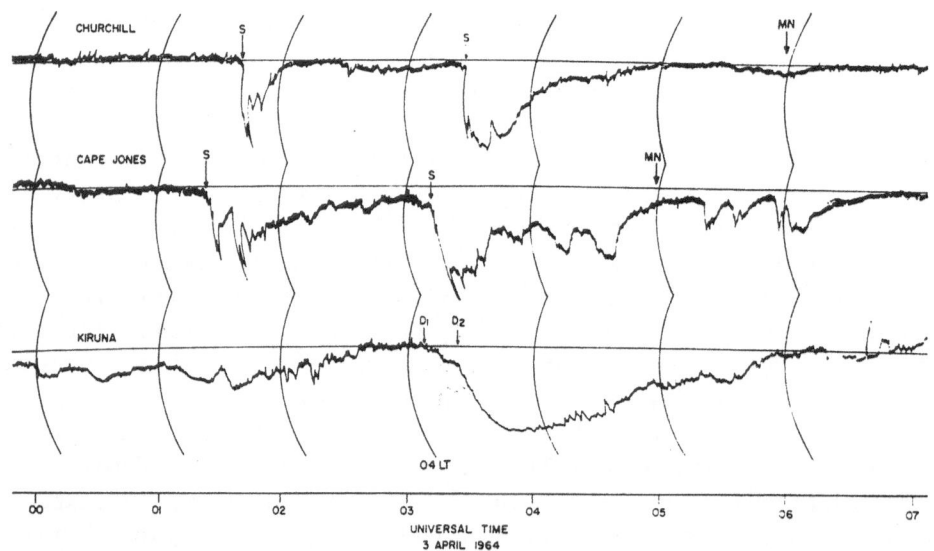

Fig. 62.   Comparison of the simultaneous cosmic radio noise absorption in the midnight and morning sectors (Jelly, D. and N. M. Brice: *J. Geophys. Res.* **72**, 5919, 1967).

for 14 of the 22 events, the D_1 onset at Kiruna was clearly either later than or at the same time as, the earlier of the onsets at the two Canadian stations. In 6 cases, D_1 preceded onsets at the Canadian stations, and in 2 cases both the D_1 and D_2 phases occurred before the substorm started at the Canadian stations.

BEWERSDORFF *et al.* (1967) also made an important study of the late morning M type events by using magnetometer records from several auroral zone stations, as well as riometer records. They concluded that late morning M type events are associated with a negative (magnetic) bay or N type absorption in the midnight sector. Recently, BEWERSDORFF *et al.* (1968) showed that the M type absorption starts North of $L=6.5$ and spreads southward to at least $L \simeq 4.5$ with speeds of about 500 m/sec.

4.7. Cosmic Radio Noise Absorption at Geomagnetically Conjugate Areas

The cosmic radio noise absorption at geomagnetically conjugate areas has been studied by HOOK (1962), LEINBACH and BASLER (1963), HARGREAVES and CHIVERS (1965); BROWN *et al.* (1965a, b) and others. There exists an excellent conjugacy at conjugate pairs at the auroral latitudes. Figure 63 shows an example of the simultaneous riometer records from Kotzebue (Alaska) and Macquarie Island (BROWN *et al.*, 1965a). The degree of conjugacy decreases toward latitudes higher than dp lat 70°; the reason for this decrease is the same as that discussed in Section 3.5.

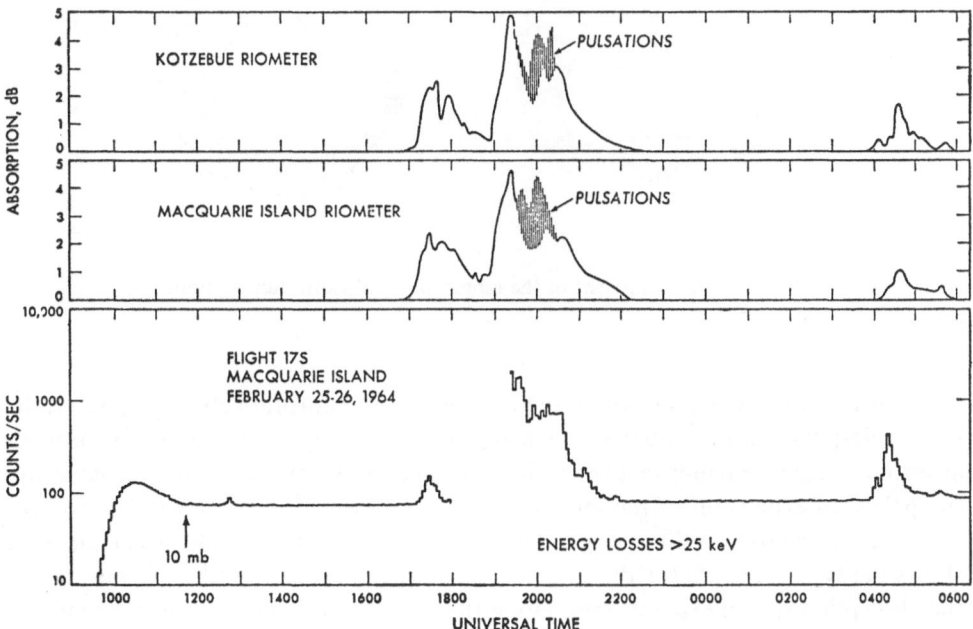

Fig. 63. Simultaneous cosmic radio noise absorption at the geomagnetically conjugate areas, Kotzebue and Macquarie Island (Brown, R. R., J. R. Barcus, and N. R. Parsons: *J. Geophys. Res.* **70**, 2579, 1965a).

4.8. Development of the Ionospheric Substorm (Absorption)

In this section, an attempt is made to construct the growth and decay of the ionospheric substorm (absorption) on the basis of what we learned in Chapter 2 and in the preceding sections.

Figure 64 illustrates the variations of the area of absorption during a single substorm. During a very early phase of the substorm, we infer that the absorption is

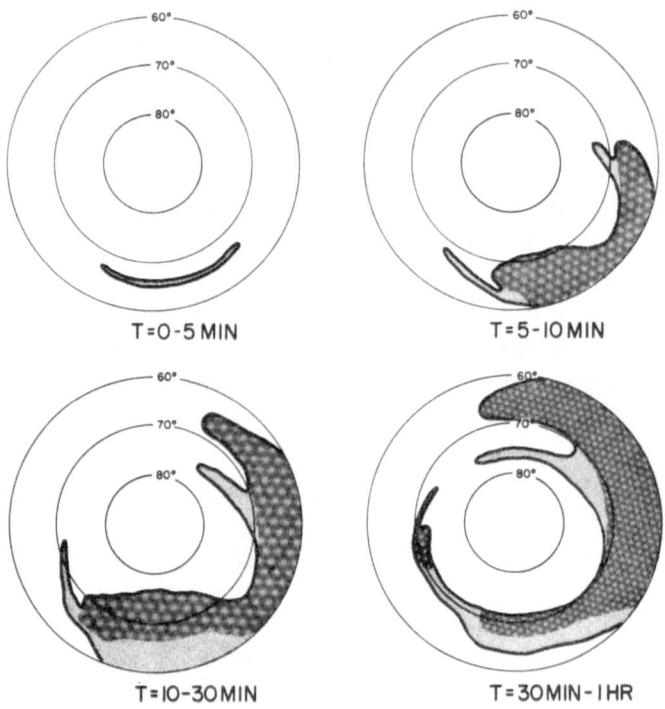

Fig. 64.   Development of the ionospheric substorm (absorption).

confined in a narrow region where the first sign of the auroral substorm (the sudden brightening of an arc) is observed. During the explosive phase of the substorm, the absorption region expands in all directions. The poleward expansion is associated with the poleward expansion of the auroral bulge. The westward expansion is associated with the westward traveling surge. The absorption is intense only when the surge shows the type B aurora (or the so-called 'red lower border'). In the morning sector, the absorption region expands both along the auroral oval and the auroral zone. The expansion along the oval is suggested by the polar blackout (PIGGOTT, 1964); see Section 1.4. This phenomenon indicates a complete absorption of radio waves transmitted by an ionospheric sounder. When this occurs, all the traces of the echoes in ionograms disappear. However, the absorption along the oval does not seem to be very strong, since none of the statistical studies by riometers detect such a region; note that an ionospheric sounder is more sensitive to the absorption than a riometer.

The other region expands along the auroral zone and is manifested by the M type absorption. HARGREAVES (1967) found that the absorption area expands from the midnight meridian; it moves eastward between 00 and 14 LT and westward between 00 and 14 LT; the typical speed is 4° of longitude/min or 3.7 km/sec.

There is still a great uncertainty in the morning hours about the expansion speed of the M type absorption in the morning hours, although this is indeed a key factor to determine whether or not energetic electrons (50 ~ 100 keV) are generated locally or whether or not they drift eastward from the midnight region. In order to determine accurately the onset time reckoned from the onset time of substorms in the midnight sector, we need at least two North-South chains of riometer stations, which are separated by about 90° or more in longitude. This is because the onset time of the absorption is significantly different in the same sector; in the midnight sector, the absorption region expands poleward, and in the morning sector equatorward. The origin of these electrons will be discussed further in Section 9.1.

What is clearly needed in future studies of riometer records is a synoptic study of the development of the absorption for individual substorm events.

4.9. Deformation of the Ionosphere

There is no doubt that in addition to the anomalous ionization in the lower ionosphere discussed above, the ionosphere is greatly affected by substorm activity. It is well known that the distribution of ionization in the ionosphere is greatly distorted during geomagnetic storms. This phenomenon has been studied by a number of workers (cf. MARTYN, 1953; KAMIYAMA, 1953; OBAYASHI, 1954; SATO, 1957; MATSUSHITA, 1959; SOMAYAJULU, 1963). There are at least two difficulties associated with studies of ionospheric disturbances. First of all, the reduction of ionospheric sounder data is not straightforward for a greatly disturbed ionosphere, and thus it is difficult to obtain a true profile of the electron density distribution. Secondly, an ionospheric sounder on the ground cannot see the upper part of the ionosphere. Further, the most commonly used quantities, such as f_0F2 (the critical frequency of the F2 region) or the virtual height is not necessarily the best parameter to express disturbed conditions. For these reasons, our understanding of ionospheric storms is still poor. Recently, a topside sounder carried by satellites has been used to examine ionospheric changes above the F2 peak (HIBBERD and ROSS, 1967; TITHERIDGE and ANDREWS, 1967; BAUER and KRISHNAMURTHY, 1968).

In Section 1.3, we have stressed the fact that the major phase of geomagnetic storms is essentially a period of frequent occurrence of intense polar substorms. Therefore, it is worthwhile to examine first the behavior of the ionosphere during the simplest, but fundamental, disturbance element, the polar substorms.

Unfortunately, there have been only a few studies on the variations of the ionosphere during polar substorms (KAMIYAMA, 1953, 1956; MAEDA and SATO, 1959; KOHL, 1960; RÜSTER, 1965). In this section, we shall review briefly Rüster's study. Figure 65 shows an example of ionospheric disturbances associated with a polar magnetic substorm. In the top diagram the computed true heights are plotted as contours of constant f_0/fc_0F2; fc_0F2 denotes the ordinary critical frequency. The middle part of the

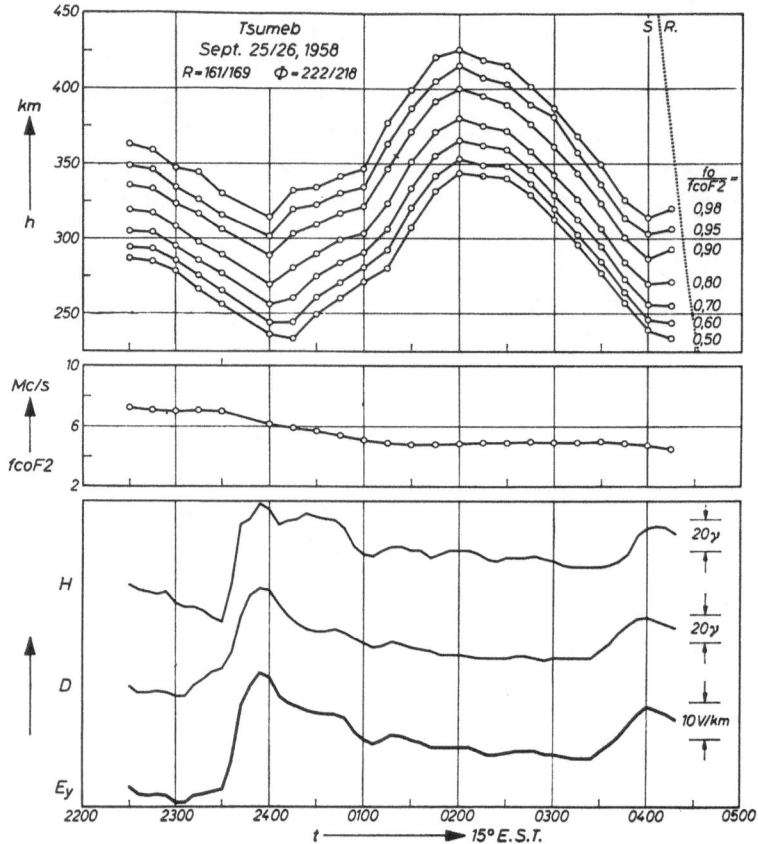

Fig. 65. Ionospheric disturbances during a polar magnetic substorm (Rüster, R.: *J. Atmospheric Terrest. Phys.* **27**, 1229, 1965).

figure shows the fc_0F2 and the bottom the three component magnetometer records; for details of the polar substorm field in middle and low latitudes, see Chapter 3. An intense polar substorm began at 2330 LT when the F2 layer had been moving downward. An upward motion began about 10 min after the peak period of the substorm. The upward motion continued till about the end of the polar substorm. Then, the F2 layer began to move downward. Figure 66 shows another example in which two polar substorms occurred successively. The first substorm began at about 2050 LT; the layer began to rise at 2100 LT and reached the maximum height near the end of the substorm. The second intense substorm began at 2335 LT; it was also associated with

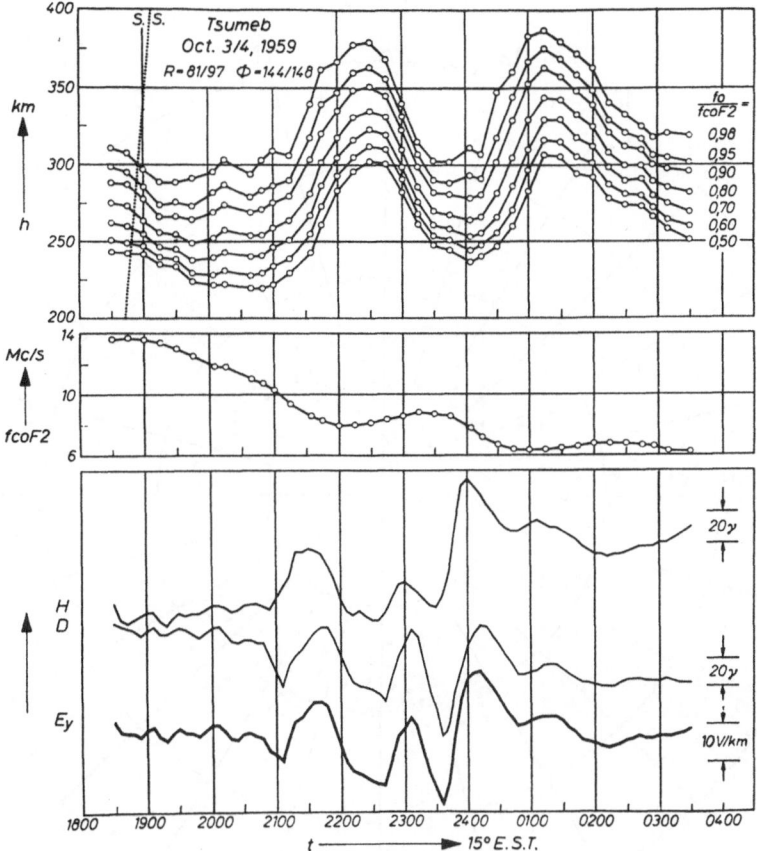

Fig. 66. Ionospheric disturbances during two successive polar magnetic substorms (Rüster, R.: *J. Atmospheric Terrest. Phys.* **27**, 1229, 1965).

an upward motion. Based on MARTYN's study (1953), RÜSTER (1965) assumed that the upward motion of the ionization was caused by an electric field (namely, the $(E \times B)$ drift) which also generated an ionospheric current and the observed polar substorm field. He showed that the computed height variation (based on the electric field deduced from the observed magnetic variations) and the observed height variation agree qualitatively. As mentioned in Chapter 3, the separation of the fields generated by ionospheric current systems and by extra-ionospheric current systems may be the first step toward the understanding of the processes associated with magnetospheric substorms. For this reason, it is essential to determine the world-wide behavior of the ionosphere during single polar substorms. The results will enable us to infer the electric field distribution associated with a magnetospheric substorm.

Another form of the deformation of the ionosphere is manifested by a considerable spread of the F region echo in ionograms and is called the spread F. There has been much discussion on the nature of the spread F phenomenon and its relation to mag-

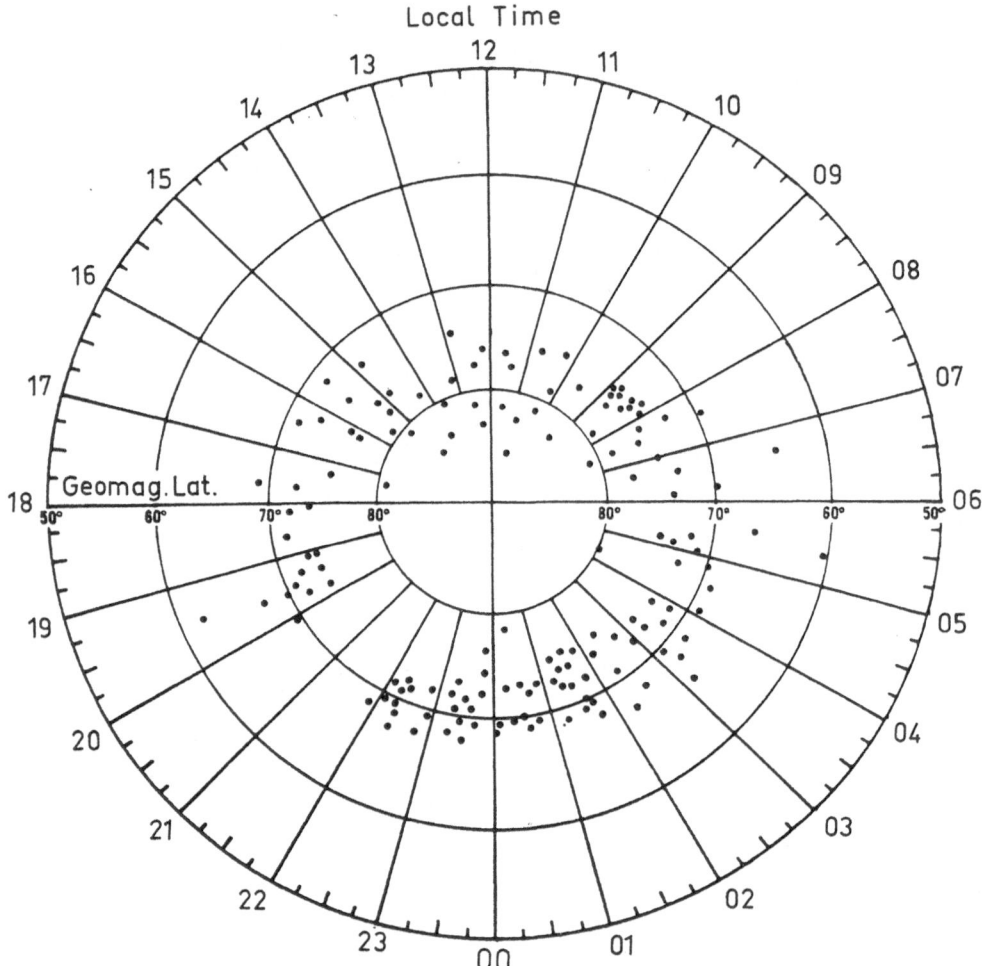

Fig. 67. Distribution of the spread F echoes observed by the Alouette 1 satellite (Petrie, L. E.:
Proc. AGARD, Agardograph 95, 1966).

netic and auroral activity. After eliminating the absorption effect by ionization in
the lower ionosphere, SHIMAZAKI (1962) established that the spread F phenomenon
has a positive correlation with auroral activity, although it appears to exist even
during periods of weak auroral activity. Figure 67 shows the distribution of the spread
F phenomenon on the polar map during quiet periods ($K_p < 2$); PETRIE (1964). It is
quite clear that the spread F appears along the quiet time auroral oval.

4.10. Radio Auroras

A considerable amount of work has, in the past, been devoted to the study of the
aurora by V.H.F. radio methods; for a review article, see AKASOFU *et al.* (1966).

On the basis of auroral radar echoes observed at 55 Mc/s at Bluff, New Zealand, UNWIN (1966a) showed that radio auroras can be associated with three separate spirals, E (evening), N (night) and M (morning) spirals; Figure 68. UNWIN (1966b) showed further that diffuse echoes are predominantely associated with the E spiral, diffuse echoes with structure and short discrete echoes with the M spiral. All echo types tend to persist for one to three hours. This lifetime is similar to that of polar substorms, and indeed radio auroras must also be one of the manifestations of the magnetospheric substorm. BATES (1966) and BATES *et al.* (1966) showed that radio

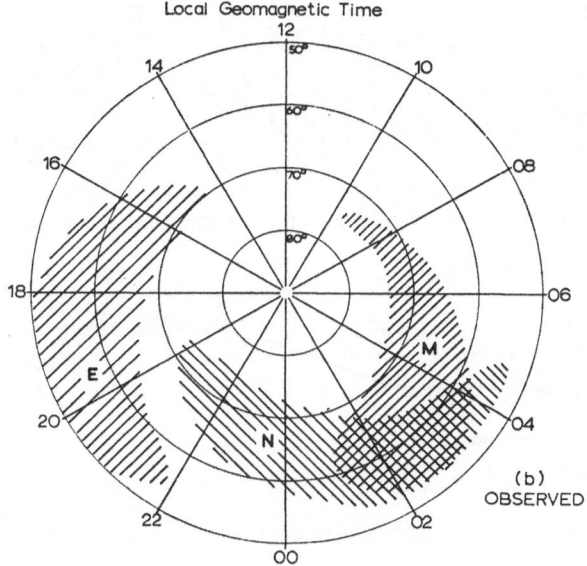

Fig. 68.   Distribution of three types of radio auroras (Unwin, R. S.: *J. Atmospheric Terrest. Phys.* **28**, 1167, 1966).

auroras appear along an oval band and that it coincides well with the auroral oval for individual events.

Radar studies of the aurora have at least two important applications in understanding the distribution of the electric field during polar substorms. It has been well established that a Doppler shift of the echoes can be used to determine the drift velocity of ionospheric electrons in the aurora. Figure 69 shows the drift speed of ionospheric electrons determined by using a 401 Mc/sec Doppler spectra taken at Fraserburgh, Scotland and the *SD* current system (LEADABRAND *et al.*, 1965). We have already discussed the validity of the *SD* current system at length in Chapters 2 and 3. On that basis, it is clear that the average curve in Figure 69 is dominated by the samples obtained along the auroral zone, not along the auroral oval.

Since the radar technique has proved useful in determining the drift speed of ionospheric electrons, it may become a powerful tool in mapping the distribution of

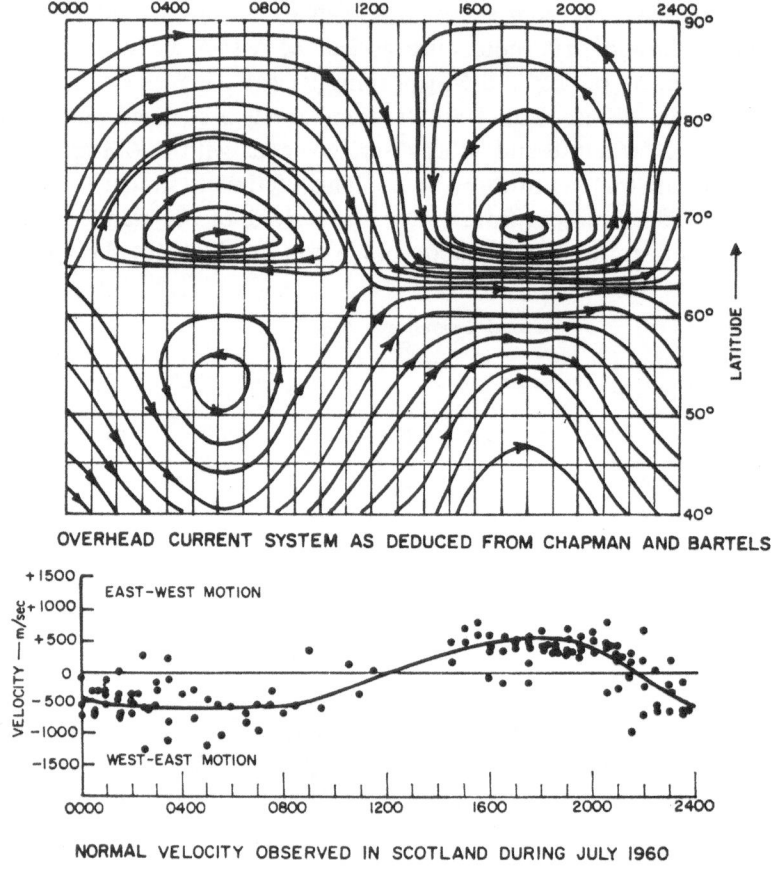

OVERHEAD CURRENT SYSTEM AS DEDUCED FROM CHAPMAN AND BARTELS

NORMAL VELOCITY OBSERVED IN SCOTLAND DURING JULY 1960

Fig. 69. Drift velocity of ionospheric electrons determined by using a 401 Mc/sec Doppler technique; the *SD* current system is shown for comparison (Leadabrand, R. L., J. C. Schlobohm, and M. J. Baron: *J. Geophys. Res.* **70**, 4235, 1965).

the motions of ionospheric electrons over the entire polar cap. A map thus constructed will enable us to determine the distribution of the electric field, since the drift motion is likely to be the $(E \times B)$ drift motion.

FARLEY (1963) proposed that if the electron drift speed exceeds a certain critical value, an instability develops in the ionosphere, which has a form of plane waves traveling perpendicular to the local magnetic field direction with a speed close to the thermal speed of ions. His theory was successfully applied to the ionosphere over the dip equator where the instability was detected by radar. UNWIN (1968) examined the possibility that auroral radar echoes are produced by a similar instability generated by the drift motion of electrons, which is associated with the auroral electrojet. He concluded that diffuse echoes in the evening sector (namely, the region of the eastward current) are reflected from the region in which the Farley instability is taking place, but the discrete types may be associated with energetic particle precipitation. We noted in Section 4.5 that positive bays are not associated with a significant absorption, so that Farley's instability may readily be observed. On the other hand, echoes may be produced by various other processes where energetic electrons precipitate, so that the Farley instability could be masked by those processes. This is in agreement with the conclusion given in Chapter 2 and Section 4.5 that the eastward current which causes a positive bay is quite different in nature from the polar electrojet which flows westward along the auroral oval.

4.11. Atmospheric Wave Substorm

A. TRAVELING DISTURBANCES

Horizontal traveling disturbances in the ionosphere (Section 1.2) were discovered by MUNRO (1950), and their possible cause was discussed by MARTYN (1953). Since then this phenomenon has been studied by a number of workers using various radio techniques. CHAN and VILLARD (1962) summarized and discussed earlier studies and compared their own study with them. They noted that four events out of nine occurred a few hours after storm sudden commencements (ssc) and remarked that if these two phenomena are related, the obvious questions which arise are how they are related and why some of them are not preceded by ssc's. BOWMAN (1965) found that after a storm sudden commencement, traveling disturbances are generated at two isolated locations in high latitude regions which are geomagnetically conjugate, and that their speed of travel is 361 m/sec.

WICKERSHAM (1964) examined Chan and Villard's results and demonstrated that they can be explained in terms of acoustic-gravity waves. Several observations are presented by DU CASTEL and FAYNOT (1964), LISZKA and TAYLOR (1965) and HUNSUCKER and TVETEN (1967) to show that traveling wave disturbances are generated in the auroral zone and propagated equatorward. HUNSUCKER and TVETEN (1967) noted that in twenty-two of thirty-seven cases (60%) a polar substorm began within ±15 min of the predicted time. They concluded that the horizontal speeds and periods of such disturbances are in agreement with a theoretical result of propagation of internal atmospheric gravity waves suggested by HINES (1964) and FRIEDMAN (1966).

B. INFRASONIC PRESSURE WAVES FROM ACTIVE AURORAS

Infrasonic waves with an amplitude of a few dynes/cm^2 and with periods of 40 to 80 sec have been studied by CAMPBELL and YOUNG (1963), WILSON and NICHPARENKO (1967), and WILSON (1967). WILSON (1967) suggested that the observed shock-like waves result from the amplification of a compression wave. The amplification is produced by the superposition of wave fronts in the direction parallel to the auroral motion, if the auroral form is moving at supersonic speed. Figures 70 and 71 show examples of infrasonic waves associated with rapid eastward motions of the aurora. However, the cause of the generation of the waves in an active aurora is not well understood.

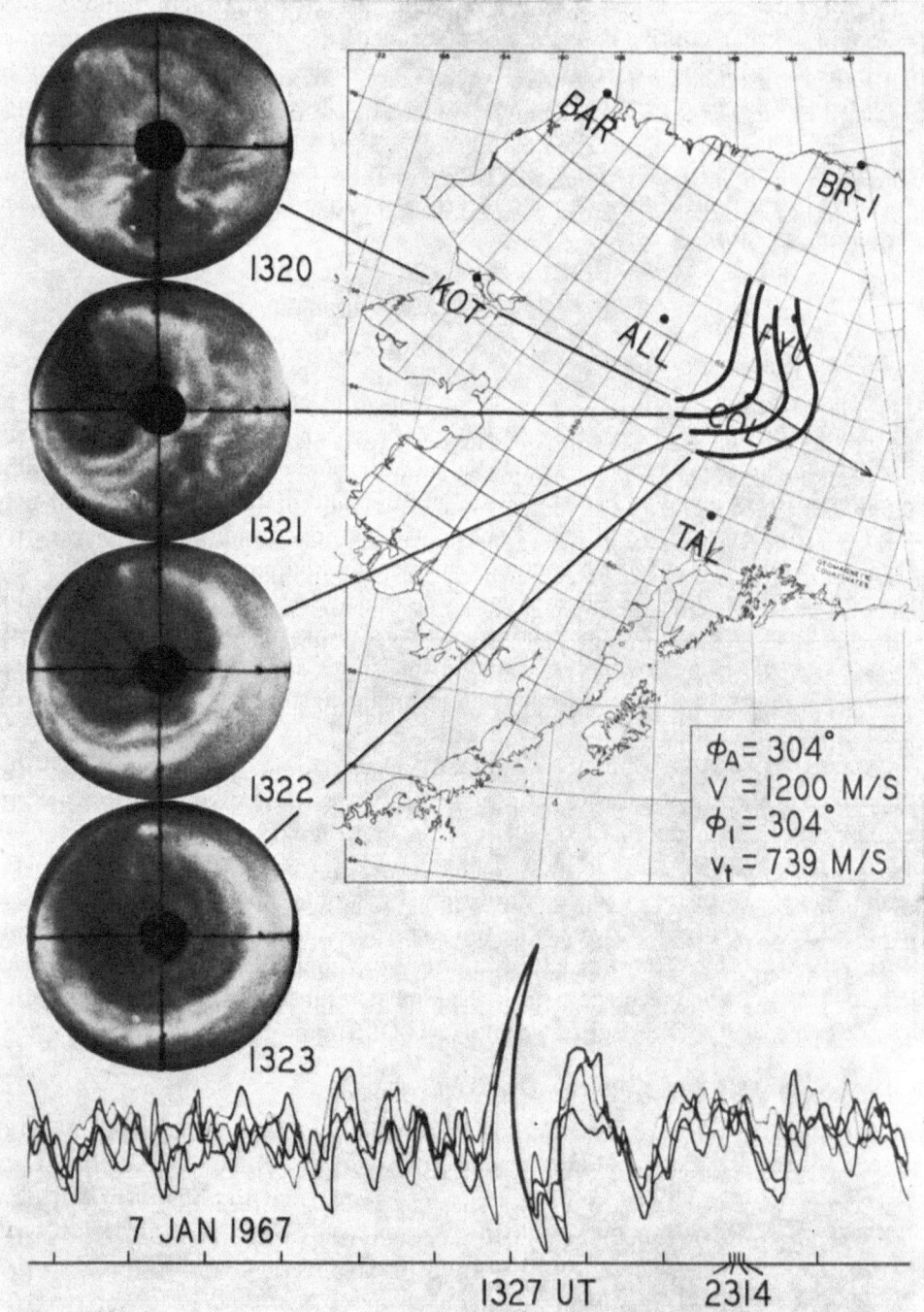

Figs. 70 and 71.   Infrasonic waves and the associated motions
(Wilson, C. R.: *Nature* **216,** 131, 1967).

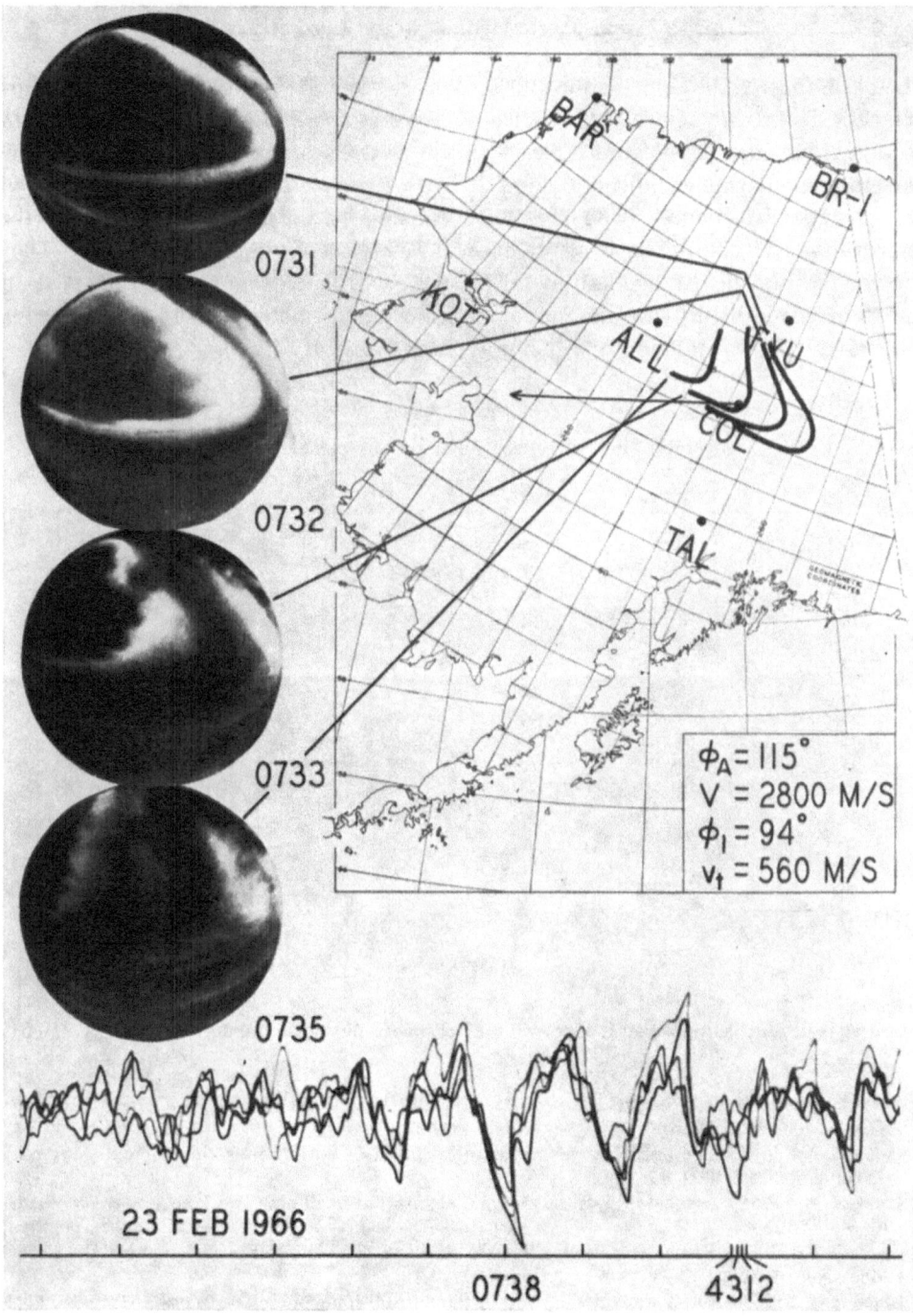

Fig. 71.

4.12. Atmospheric Heating in the Auroral Zone

The variations of the upper atmospheric density over the auroral zone during geomagnetic disturbances have been studied by JACCHIA and SLOWEY (1964) and JACCHIA et al. (1967). Their studies were based on the analysis of the drag of satellites; the planetary geomagnetic indices K_p and a_p are used as a measure of the degree of geomagnetic disturbance. They have also estimated the corresponding change of the temperature (ΔT) on the basis of the observed increase of atmospheric density. They showed that for the average latitude of 65° the mean time delay between the peak of the geomagnetic disturbance and that of the atmosphere is about 5 hours. The following table gives ΔT as a function of K_p and a_p (JACCHIA et al., 1967).

TABLE

Temperature Increment as a Function of Geomagnetic Indices

K_p	a_p	deg	K_p	a_p	ΔT
0_0	0	0	5—	39	134
0+	2	9	5_0	48	145
1—	3	19	5+	56	156
1_0	4	28	6—	67	167
1+	5	37	6_0	80	180
2—	6	47	6+	94	194
2_0	7	56	7—	111	210
2+	9	66	7_0	132	229
3—	12	75	7+	154	251
3_0	15	85	8—	179	279
3+	18	94	8_0	207	313
4—	22	104	8+	236	358
4_0	27	114	9—	300	417
4+	32	124	9_0	400	495

References

GENERAL

HULTQVIST, B.: 1966, 'Ionospheric absorption of cosmic radio noise', Space Sci. Rev. 5, 771–817.

REFERRED TO IN TEXT

AKASOFU, S.-I. CHAPMAN, S., and MEINEL, A. B.: 1966, 'The aurora', in Handbuch der Physik, Vol. XLIX–1 (ed. by S. Flügge), Springer-Verlag, Berlin, pp. 1–158.

ANSARI, Z. A.: 1964, 'The aurorally associated absorption of cosmic noise at College, Alaska', J. Geophys. Res. 69, 4493–4513.

ANSARI, Z. A.: 1965, 'A peculiar type of daytime absorption in the auroral zone', J. Geophys. Res. 70, 3117–3122.

BAILEY, D. K.: 1968, 'Some quantitative aspects of electron precipitation in and near the auroral zone', Rev. Geophys. 6, 289–346.

BASLER, R. P.: 1963, 'Radio wave absorption in the auroral ionosphere', J. Geophys. Res. 68, 4665–4681.

BATES, H. F.: 1966, 'Latitude of the dayside aurora', J. Geophys. Res. 71, 3629–3633.

BATES, H. F., BELON, A. E., ROMICK, G. J., and STRINGER, W. J.: 1966, 'On the correlation of optical and radio auroras', J. Atmospheric Terrest. Phys. 28, 439–446.

BAUER, S. J. and KRISHNAMURTHY, B. V.: 1968, 'Behavior of the topside ionosphere during a great magnetic storm', *Planetary Space Sci.* **16**, 653–663.

BERKEY, F. T.: 1968, 'Coordinated measurements of auroral absorption and luminosity using the narrow beam technique', *J. Geophys. Res.* **73**, 319–337.

BEWERSDORFF, A., KREMSER, G., RIEDLER, W., and LEGRAND, J. P.: 1967, 'Some properties of the slowly varying ionospheric absorption events in the auroral zone', *Arkiv Geofysik* **5**, 115–127.

BEWERSDORFF, A., KREMSER, G., STADSNES, J., TREFALL, H., and ULLALAND, S.: 1968, 'Simultaneous balloon measurements of auroral X-rays during slowly varying ionospheric absorption events', *J. Atmospheric Terrest. Phys.* **30**, 591–607.

BOWMAN, G. G.: 1965, 'Traveling disturbances associated with ionospheric storms', *J. Atmospheric Terrest. Phys.* **27**, 1247–1261.

BROWN, R. R.: 1964, 'A study of slowly varying and pulsating ionospheric absorption events in the auroral zone', *J. Geophys. Res.* **69**, 2315–2321.

BROWN, R. R., BARCUS, J. R., and PARSONS, N. R.: 1965a, 'Balloon observations of auroral zone X-rays in conjugate regions. 1. Slow time variations', *J. Geophys. Res.* **70**, 2579–2598.

BROWN, R. R., BARCUS, J. R., and PARSONS, N. R.: 1965b, 'Balloon observations of auroral zone X-rays in conjugate regions. 2. Microbursts and pulsations', *J. Geophys. Res.* **70**, 2599–2612.

CAMPBELL, W. H. and YOUNG, J. M.: 1963, 'Auroral-zone observations of infrasonic pressure waves related to ionospheric disturbances and geomagnetic activity', *J. Geophys. Res.* **68**, 5909–5916.

CHAN, K. L. and VILLARD, Jr., O. G.: 1962, 'Observation of large-scale traveling ionospheric disturbances by spaced-path high-frequency instantaneous-frequency measurements', *J. Geophys. Res.* **67**, 973–988.

CHIVERS, H. J. A.: 1967, Private communication.

DRIATSKIY, V. M.: 1966, 'Auroral radio wave absorption in the vicinity of the geographic North Pole', *Geomagnetizm i Aeronomiya* **6**, 285–287.

DRIATSKIY, V. M.: 1966, 'Study of the space and time distribution of auroral absorption according to observations of the riometer network in the Arctic', *Geomagnetizm i Aeronomiya* **6**, 828–834.

DU CASTEL, F. and FAYNOT, J. M.: 1964, 'Some irregularities observed simultaneously in the upper and lower ionosphere at middle latitudes', *Nature, London* **204**, 984–985.

EATHER, R. H. and JACKA, F.: 1966, 'Auroral absorption of cosmic radio noise', *Australian J. Phys.* **19**, 215–239.

FARLEY, Jr., D. T.: 1963, 'A plasma instability resulting in field-aligned irregularities in the ionosphere', *J. Geophys. Res.* **68**, 6083–6097.

FEDYAKINA, N. I.: 1963, 'Type-II absorption and its relationship to magnetic field disturbance', *Geomagnetizm i Aeronomiya* **3**, 393–395.

FRIEDMAN, J. P.: 1966, 'Propagation of internal gravity waves in a thermally stratified atmosphere', *J. Geophys. Res.* **71**, 1033–1054.

GUSTAFSSON, G.: 1964, 'Ionization in the D-region during auroral break-up events', *Planetary Space Sci.* **12**, 195–208.

HARGREAVES, J. K.: 1967, 'Auroral motions observed with riometers: movements between stations widely separated in longitude', *J. Atmospheric Terrest. Phys.* **29**, 1159–1164.

HARGREAVES, J. K. and CHIVERS, H. J. A.: 1965, 'A study of auroral absorption events at the south pole. 2. Conjugate properties', *J. Geophys. Res.* **70**, 1093–1102.

HARGREAVES, J. K. and COWLEY, F. C.: 1967, 'Studies of auroral radio absorption events at three magnetic latitudes. I. Occurrence and Statistical Properties of the Events', *Planetary Space Sci.* **15**, 1571–1583.

HARTZ, T. R., MONTBRIAND, L. E., and VOGAN, E. L.: 1963, 'A study of auroral absorption at 30 Mc/s', *Canadian J. Phys.* **41**, 581–595.

HEPPNER, J. P., BYRNE, E. C., and BELON, A. E.: 1952, 'The association of absorption and Es ionization with aurora at high latitudes', *J. Geophys. Res.* **57**, 121–134.

HIBBERD, F. H. and ROSS, W. J.: 1967, 'Variations in total electron content and other ionospheric parameters associated with magnetic storms', *J. Geophys. Res.* **72**, 5331–5337.

HINES, C. O.: 1964, 'Comments on paper by A. F. Wickersham Jr.: Identification of ionospheric motions detected by the high-frequency backscattering technique', *J. Geophys. Res.* **69**, 2395–2396.

HOLT, O. and OMHOLT, A.: 1962, 'Auroral luminosity and absorption of cosmic radio noise', *J. Atmospheric Terrest. Phys.* **24**, 467–474.

HOLT, O., LANDMARK, B., and LIED, F.: 1962, 'Analysis of riometer observations obtained during polar radio blackouts', *J. Atmospheric Terrest. Phys.* **23**, 229–243.

HOOK, J. L.: 1968, 'Morphology of auroral zone radiowave absorption', *J. Atmospheric Terrest. Phys.* **30**, 1341–1351.

HUNSUCKER, R. D. and TVETEN, L. H.: 1967, 'Large traveling-ionospheric-disturbances observed at midlatitudes utilizing the high resolution h.f. backscatter technique', *J. Atmospheric Terrest. Phys.* **29**, 909–916.

JACCHIA, L. G. and SLOWEY, J.: 1964, 'Atmospheric heating in the auroral zones: A preliminary analysis of the atmospheric drag of the Injun 3 satellite', *J. Geophys. Res.* **69**, 905–910.

JACCHIA, L. G., SLOWEY, J., and VERNIANI, F.: 1967, 'Geomagnetic perturbations and upper-atmosphere heating', *J. Geophys. Res.* **72**, 1423–1434.

JELLY, Doris H.: 1968, 'Apparent poleward motion of onsets of auroral absorption events', *Canadian J. Phys.* **46**, 33–37.

JELLY, D. and BRICE, N.: 1967, 'Changes in Van Allen radiation associated with polar substorms', *J. Geophys. Res.* **72**, 5919–5931.

JOHANSEN, O. E.: 1965, 'Variations in energy spectrum of auroral electrons detected by simultaneous observation with photometer and riometer', *Planetary Space Sci.* **13**, 225–235.

KAMIYAMA, H.: 1953, 'The disturbance in the ionosphere accompanying the geomagnetic storm on April 18, 1951', *Sci. Rep. Tohoku Univ. 5th Ser. Geophysics* **5**, 1–9.

KAMIYAMA, H.: 1953, 'Disturbances in the ionosphere during the geomagnetic bay', *Sci. Rep. Tohoku Univ. 5th Ser. Geophysics* **5**, 101–107.

KAMIYAMA, H.: 1956, *Sci. Rept. Tohoku Univ.* **7**, 125–135.

KOHL, H.: 1960, 'Movement of the F-layer of the ionosphere during bay disturbances of the earth', *Arch. Elekt. Übertr.* **14**, 169–176.

LEADABRAND, R. L., SCHLOBOHM, J. C., and BARON, M. J.: 1965, 'Simultaneous very high frequency and ultra high frequency observations of the aurora at Fraserburgh, Scotland', *J. Geophys. Res.* **70**, 4235–4284.

LEINBACH, H. and BASLER, R. P.: 1963, 'Ionospheric absorption of cosmic radio noise at magnetically conjugate auroral zone stations', *J. Geophys. Res.* **68**, 3375–3382.

LIN, W. C., McDIARMID, I. B., and BURROWS, J. R.: 1968, 'Electron fluxes at 1000 km altitude associated with auroral substorms', *Canadian J. Phys.* **46**, 80–83.

LISZKA, L. and TAYLOR, G. N.: 1965, 'A synoptic study of large scale ionospheric irregularities using observations of the Faraday rotation of satellite signals', *J. Atmospheric Terrest. Phys.* **27**, 843–854.

LITTLE, C. G. and LEINBACH, H.: 1958, 'Some measurements of high-latitude ionospheric absorption using extraterrestrial radio waves', *Proc. Inst. Radio Engrs.* **46**, 334–348.

MAEDA, K. I. and SATO, T.: 1959, 'The F region during magnetic storms', *Proc. Inst. Radio Engrs.* **47**, 232–239.

MARTYN, D. F.: 1953, 'The morphology of the ionospheric variations associated with magnetic disturbance. I. Variations at moderately low latitudes', *Proc. Roy. Soc. London* **A218**, 1–18.

MATSUSHITA, S.: 1959, 'A study of the morphology of ionospheric storms', *J. Geophys. Res.* **64**, 305–321.

MUNRO, G. H.: 1950, 'Traveling disturbances in the ionosphere', *Proc. Roy. Soc. London* **A202**, 208–223.

OBAYASHI, T.: 1954, 'On the world-wide disturbance of the ionosphere', *Rep. Ionos. Space Res. Japan* **8**, 135–142.

PARTHASARATHY, R. and BERKEY, F. T.: 1965, 'Auroral zone studies of sudden-onset radiowave absorption events using multiple-station and multiple-frequency data', *J. Geophys. Res.* **70**, 89–98.

PETRIE, L. E.: 1966, 'Preliminary results on mid and high latitude topside spread F, spread F and its effects upon radiowave propagation and communication' (ed. by P. Newman), W. and J. Mackay and Co. Ltd., London.

PIGGOTT, W. R.: 1964, 'Studies of ionospheric absorption', *Res. Geophysics*, Vol. I (ed. by H. Odishaw), the MIT Press, pp. 277–297.

RATCLIFFE, J. A.: 1959, *The magneto-ionic theory and its applications to the ionosphere*, Cambridge University Press, England.

RÜSTER, R.: 1965, 'Height variations of the F2-layer above Tsumeb during geomagnetic bay-disturbances', *J. Atmospheric Terrest. Phys.* **27**, 1229–1245.

SATO, T.: 1957, 'Disturbances in the ionospheric F2 region associated with geomagnetic storms. I. Equatorial zone', *J. Geomag. Geoelec.* **8**, 129–135. II. Middle Latitudes, **9**, 1–22.

SHIMAZAKI, T.: 1962, 'A statistical study of occurrence probability of spread F at high latitudes', *J. Geophys. Res.* **67**, 4617–4634.

SOMAYAJULU, Y. V.: 1963. 'Changes in the F region during magnetic storms', *J. Geophys. Res.* **68**, 1899–1922.

TITHERIDGE, J. E. and ANDREWS, M. K.: 1967, 'Changes in the topside ionosphere during a large magnetic storm', *Planetary Space Sci.* **15**, 1157–1167.

UNWIN, R. S.: 1966a, 'The morphology of the VHF radio aurora at sunspot maximum. I. Diurnal and seasonal variations', *J. Atmospheric Terrest. Phys.* **28**, 1167–1181; 1966b; II. 'The behaviour of different echo types', *ibid.* 1183–1194.

UNWIN, R. S. and KNOX, F. B.: 1968, 'The morphology of the VHF radio aurora at sunspot maximum. IV. Theory', *J. Atmospheric Terrest. Phys.* **30**, 25–46.

WICKERSHAM, Jr., A. F.: 1964, 'Analysis of large-scale traveling ionospheric disturbances', *J. Geophys. Res.* **69**, 3235–3243.

WILSON, C. R.: 1967, 'Infrasonic pressure waves from the aurora: A shock wave model', *Nature* **216**, 131–133.

WILSON, C. R. and NICHPARENKO, S.: 1967, 'Infrasonic waves and auroral activity', *Nature* **214**, 1299–1302.

CHAPTER 5

X-RAY SUBSTORM

5.1. Introduction

Energetic electrons which impinge into the polar upper atmosphere generate brems-strahlung X-rays when they collide with upper atmospheric particles. Balloon-borne detectors have been used to observe X-rays at about the 30 km level. However, because of the layer of atmosphere between the height at which the X-rays are generated and the height of the balloon, the X-rays suffer both photoelectric absorption and Compton scattering; further, these effects are energy-dependent. For these reasons, the procedure of obtaining the original energy spectrum of incoming electrons from the observed X-ray spectrum is not necessarily straightforward. Since our interest in this monograph is a synoptic aspect of the X-ray bursts, we shall not be concerned with details of the data reduction procedure (for details, see cf. BROWN, 1966; BARCUS

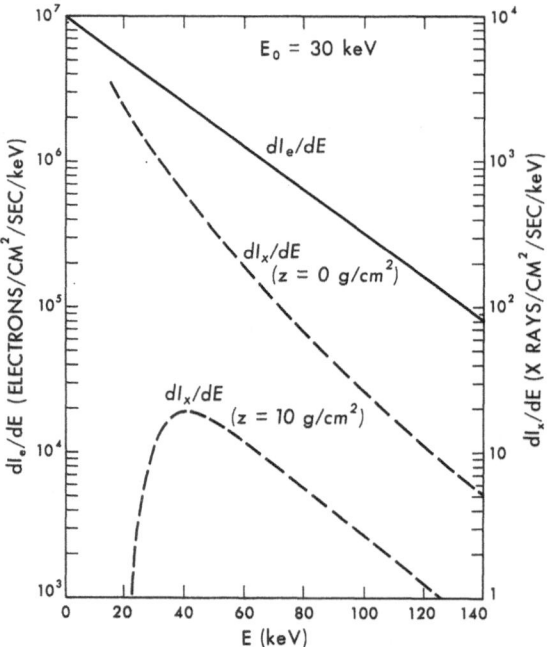

Fig. 72. Model calculations of the differential energy spectrum for bremsstrahlung X-rays produced in the upper atmosphere by precipitating electrons with an e-folding energy of 30 keV (Barcus, J. R. and T. J. Rosenberg: *J. Geophys. Res.* **71**, 803, 1966).

and ROSENBERG, 1966). Nevertheless, it is worthwhile to have some idea as to the effects of the atmosphere. For this purpose, we show in Figure 72 a model calculation by BARCUS and ROSENBERG (1966); they assumed the differential intensity of an incident exponential electron spectrum $dI_e/dE = (\kappa/E_0) \exp(-E/E_0)$ with $E_0 = 30$ keV and computed the thick-target bremsstrahlung spectrum dI_x/dE at the production height. Then, by taking into account photoelectric absorption and Compton scattering, they computed the X-ray spectrum dI_x/dE at a typical balloon depth 10 g cm^{-2}; a uniform isotropic precipitation over an area 10^4 km^2 above the balloon is assumed. It can be seen that there occurs a sharp cut-off of the X-rays for energies below 40 keV and thus that balloon data must be carefully analyzed for energies less than 50 keV.

In spite of this drawback, the balloon observations provide a more direct way of observing energetic electrons than a riometer, and have provided important information on incoming energetic electrons, which cannot be easily detected by either satellite, rocket, or ground equipment; in particular, various types of time fluctuations of the flux.

A notable feature of the X-ray precipitation in the polar upper atmosphere is its burst-like appearance of lifetime of order one to three hours. In this chapter, we show that such a burst-like appearance is indeed an important manifestation of the magnetospheric substorm.

5.2. Typical Daily Variation

Figure 73 shows an example of a simultaneous balloon observation of X-rays at Fort Yukon and Fort Wainwright (near Fairbanks) and the corresponding riometer records from Fort Yukon and College (BARCUS and BROWN, 1966). The burst-like appearance of X-rays is obvious in the records. Further, there is a great similarity in the time variation of both the X-ray records and the riometer records. Therefore, from what we have learned in the previous chapters, there is no doubt that the X-ray bursts are associated with the magnetospheric substorms.

In Figure 73, the first large burst, which began at about 1210 UT, coincided with the N type absorption recorded at Fort Yukon. There was then a quiet period between 14 and 15 UT, during which no significant X-ray burst and absorption occurred. New activity began at about 15 UT; it is not difficult to infer from the corresponding riometer records that the major peaks of the X-ray activity were associated with a successive occurrence of the M type absorption events.

Figure 74 shows another X-ray observation and the corresponding magnetic record from College; the observation period covers a period of about 36 hours. A definite correlation between the College magnetic record and X-ray burst exists in the midnight hours; for example, X-ray bursts which began at about 1030 UT on both September 16 and 17 were associated with intense negative bays. However, for the same reason given for the poor correlation between negative bays and the late M type absorption, the correlation between the bursts and negative bays becomes much less clear as the morning progresses; the polar auroral electrojet flows along the oval which is located well poleward of an auroral zone station in the late morning hours,

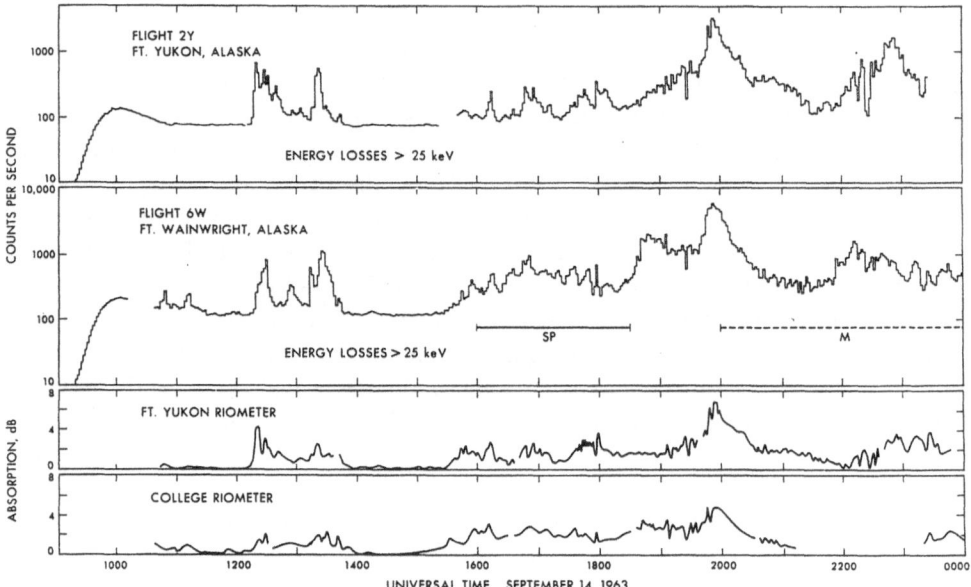

Fig. 73.   Typical X-ray activity recorded by a simultaneous balloon observation at Fort Yukon and Fairbanks; the corresponding riometer records from the two locations are shown (Barcus, J. R. and R. R. Brown: *J. Geophys. Res.* **71**, 825, 1966).

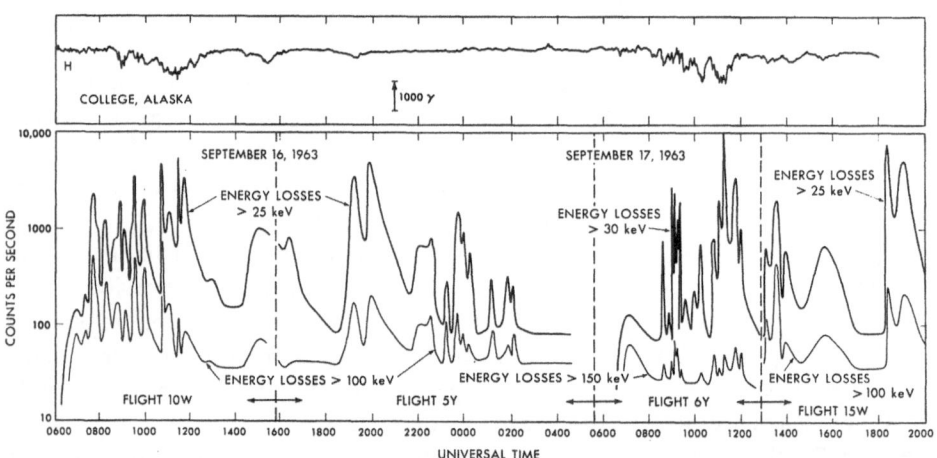

Fig. 74.   Typical X-ray activity and the corresponding magnetic record at College, Alaska (Barcus, J. R. and T. J. Rosenberg: *J. Geophys. Res.* **71**, 803, 1966).

whereas energetic electrons appear along the auroral zone. Indeed, little correlation exists for bursts which occur in the afternoon and early evening hours.

In addition to the burst-like appearance, the X-rays show considerable short period fluctuations which cannot properly be illustrated in the time scale used in Figures 73 and 74.

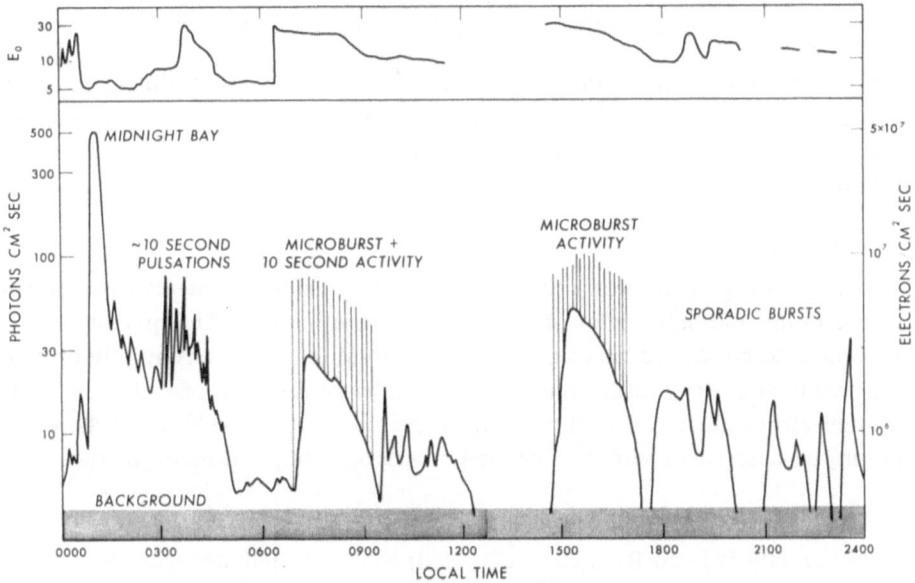

Fig. 75. Schematic daily variation of X-ray activity; the variation of the e-folding energy is also shown (Anderson, K. A.: University of California, Berkeley, Report 1966).

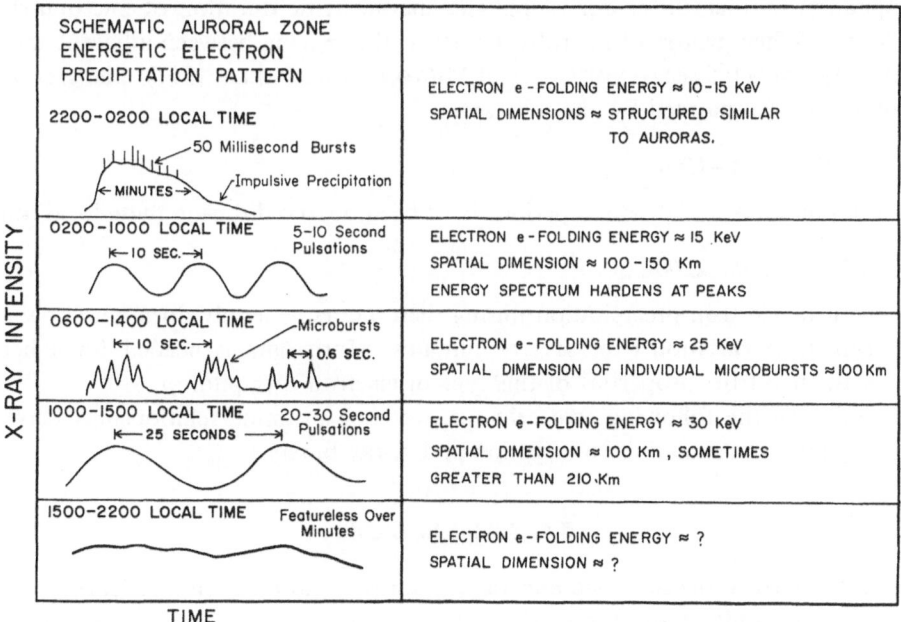

Fig. 76. Tables of the characteristics of X-ray bursts at different local times (Parks, G. K., F. V. Coroniti, R. L. McPherron, and K. A. Anderson: *J. Geophys. Res.* **73**, 1685, 1968).

5.3. Statistical Daily Variation Pattern

Figure 75 illustrates schematically characteristics of the X-ray bursts during the course of a day (ANDERSON, 1966) and Figure 76, details of the fine structures associated with the bursts (PARKS *et al.*, 1968). Here, we summarize the major features, following the latter authors.

1. *Local Time 2200–0200*

Energetic electron precipitation during this period is usually associated with auroral substorms in the midnight sector and negative (magnetic) bays. The precipitation has been reported to be featureless over times of a few minutes, but recent balloon-borne measurements at 2 g/cm^2 atmospheric depths indicate the presence of 5–50 millisec bursts of electrons during auroral breakup (PARKS *et al.*, 1967). These rapidly varying bursts may be a persistent feature of energetic electron precipitation during these times.

2. *Local Time 0200–1000*

BROWN *et al.* (1965b) and BARCUS *et al.* (1966) have reported numerous observations of X-ray pulsations of 5–10 sec periods during these local hours. These X-ray pulsations may be closely related to pulsating auroras (ROSENBERG *et al.*, 1967).

3. *Local Time 0600–1400*

In this region of local time, microbursts are the predominant type of energetic electron precipitation (ANDERSON *et al.*, 1966). The characteristic duration of microbursts is 0.1–0.5 sec. When groups of microbursts occur, the spacing between adjacent microbursts is typically 0.5 sec (ANDERSON and MILTON, 1964; PARKS, 1967). The groups of microbursts are separated by about 10 sec.

4. *Local Time 1000–1500*

Electron pulsations of 20–30 sec periods have been observed during these local times.

5. *Local Time 1600–2200*

The energetic electron precipitation during these hours generally does not show any significant time variation over several minutes. More data is needed for a better understanding of the properties of this type of electron precipitation.

In this and the following two sections, we shall examine relationships between auroral activity, cosmic noise absorption and X-ray bursts.

5.4. Midnight Bursts

Since the explosive phase of the auroral substorm is the most violent feature of the substorms, it is quite natural to expect intense bursts of electrons in association with the poleward explosive motion of the aurora, a sharp onset of negative bays, the N type absorption and other polar upper atmospheric phenomena. Figure 77 shows a classical example to illustrate the simultaneous burst-like appearance of cosmic noise absorption, X-ray burst and negative (magnetic) bay (WINCKLER *et al.*, 1959). This particular example occurred during the great storm of February 11, 1958, and coincided with the period during which the storm was rapidly growing. An important point to be noted here is that a surprisingly quiet period can exist during the growing phase of such a great storm; in our particular example, it is between the two substorms (namely, between 0730 UT and 0830 UT).

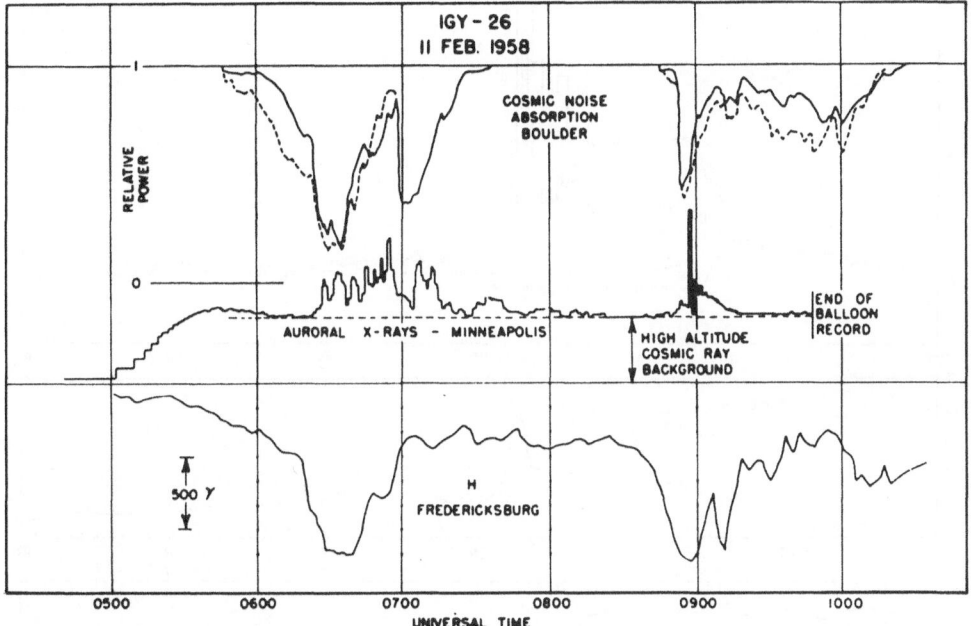

Fig. 77. Simultaneous records of the cosmic radio noise absorption, X-ray bursts and the *H* component magnetic record during the growing phase of the great magnetic storm of February 11, 1958 (Winckler, J. R., L. Peterson, R. Hoffman, and R. Arnoldy: *J. Geophys. Res.* **64**, 597, 1959).

Figure 78 shows another example to illustrate the simultaneous occurrence of the three phenomena, a X-ray burst, cosmic radio noise absorption and negative bay (BROWN and CAMPBELL, 1962).

One of the most interesting problems in the midnight sector is the distribution of X-ray precipitation within the auroral bulge or with respect to the auroral electrojet. BROWN and CAMPBELL (1962) showed that the X-rays are most intense in the vicinity of the auroral electrojet; since the electrojet appears to be concentrated near the poleward front of the expanding bulge, it may be inferred that X-rays are most intense there. Such an inference is also in agreement with the riometer observations (Section 4.4). GOSLING (1966) made a detailed study of the location of the X-ray precipitation by using a set of two directional detectors, vertical and horizontal, and confirmed that the motion of the region of the X-ray precipitation follows closely that

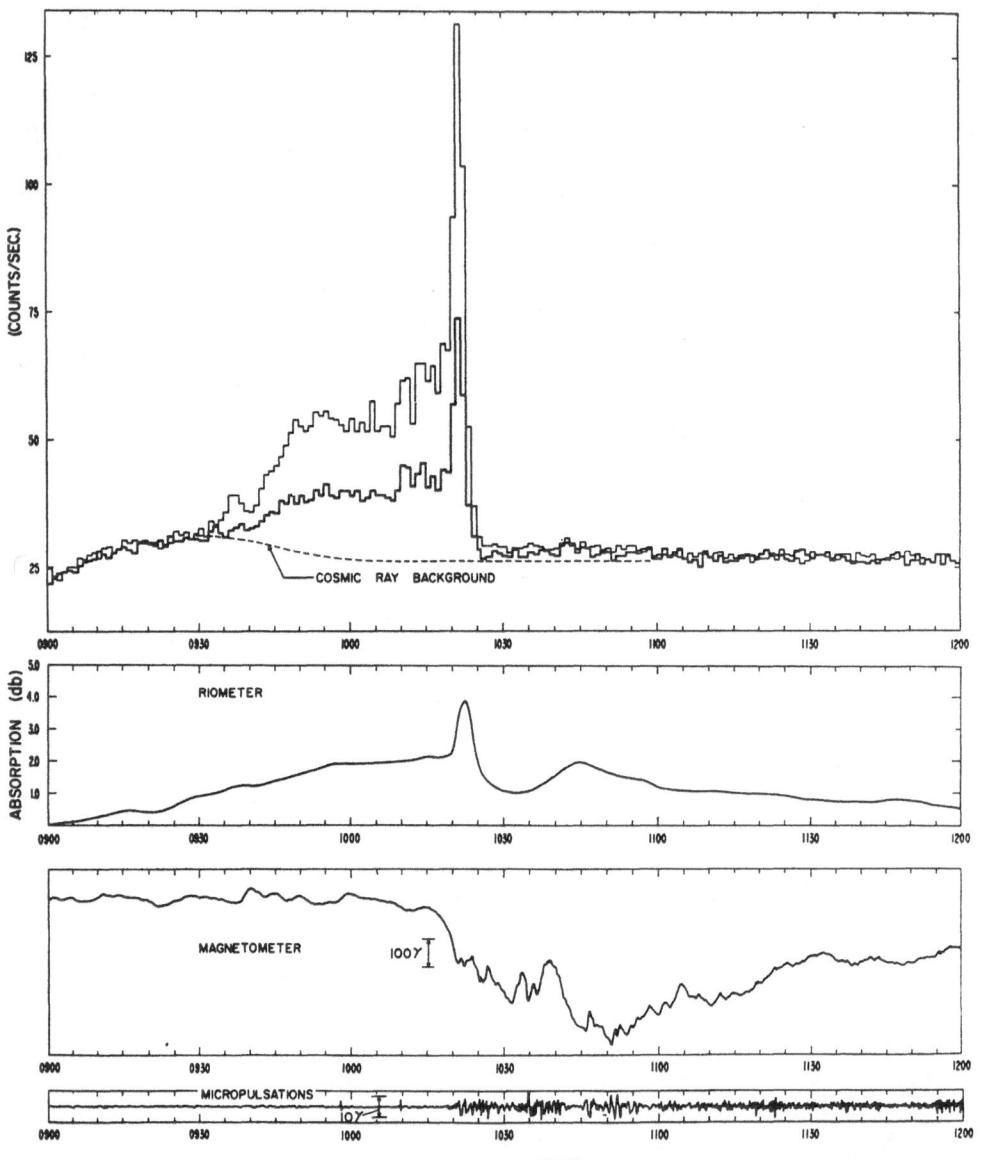

Fig. 78. Simultaneous records of the X-ray bursts, cosmic radio noise absorption, *H* component magnetic record and micropulsations (Brown, R. R. and W. H. Campbell: *J. Geophys. Res.* **67**, 1357, 1962).

of the auroral (westward) electrojet. CLARK and ANGER (1967) showed that X-ray precipitation occurs directly under an active Type B band aurora (the so-called 'red lower border' aurora). Their study will be mentioned in detail in the next section.

Another important feature of the midnight bursts is the decrease of the e-folding energy (E_0) during the lifetime of the substorm. Figure 79 shows such an example.

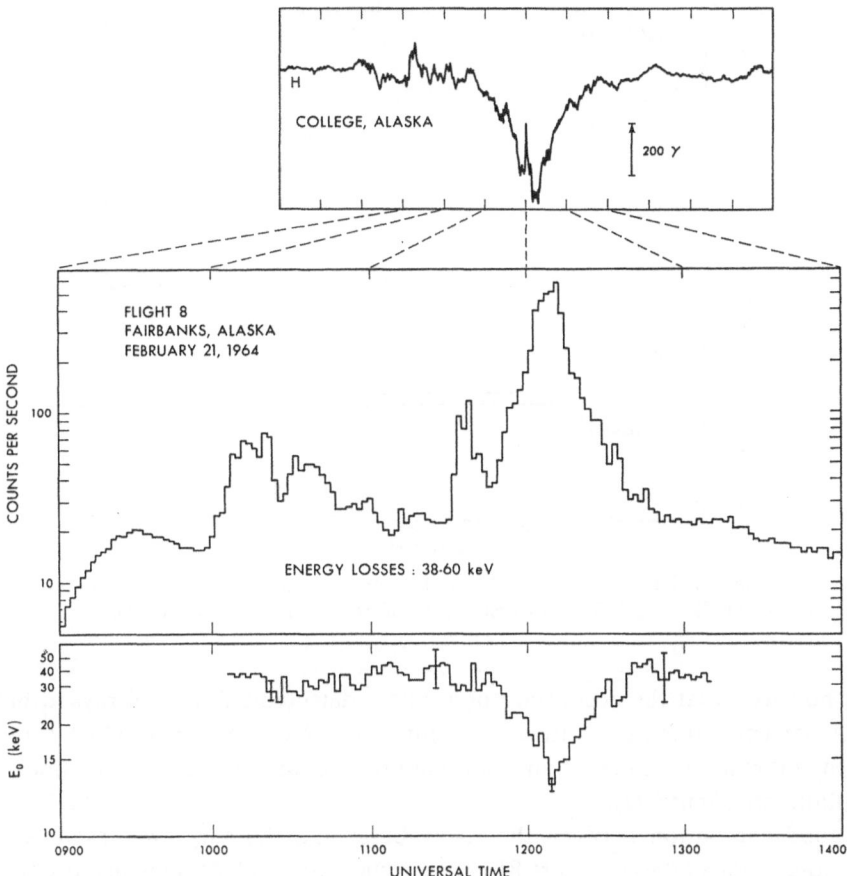

Fig. 79. Example showing that the midnight X-ray bursts are associated with a decrease of the
e-folding energy (Barcus, J. R. and T. J. Rosenberg: *J. Geophys. Res.* **71**, 803, 1966).

Before the sudden intensification of the negative bay (and thus of the auroral electrojet), E_0 was about 40 keV. However, E_0 began to decrease during the growth of the
electrojet and reached a minimum value $(E_0 \simeq 14 \text{ keV})$ when the magnetometer
registered the maximum deviation at about 1210 UT. During the recovery phase E_0
gradually increased and returned to the pre-substorm level at about the end of the
substorm.

5.5. X-Ray Bursts Associated with Westward Traveling Surges

Unfortunately, there have been few studies on the X-ray precipitation associated with
the westward traveling surges. Figure 80 shows one example in which the X-ray flux
associated with the surge was very small (Barcus, 1964). The surge passed over the
balloon between 1115 and 1125 UT, causing the riometer absorption and an increase
of the photometer output; the photometer was attached to the balloon. Barcus (1964)

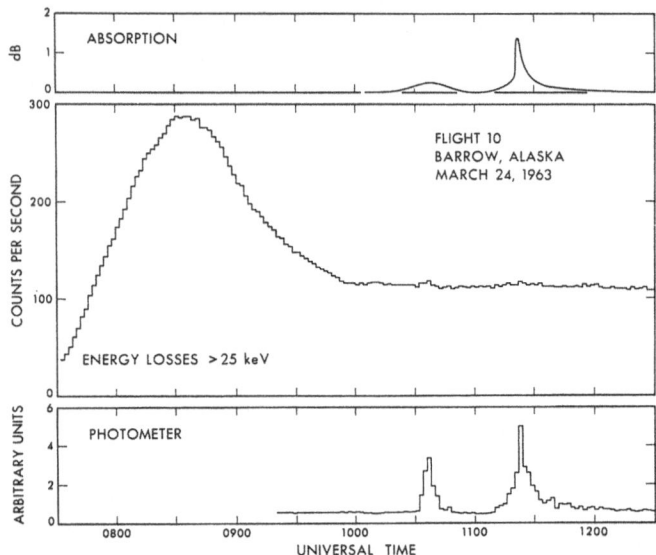

Fig. 80.   Example showing that a westward traveling surge was not associated with an intense X-ray
burst (Barcus, J. R.: University of California, Berkeley, Report 1964).

noted, however, that the occurrence of a barely detectable flux of X-rays at balloon
heights may correspond to a rather large influx of primary electrons which cannot be
detected at that level because of an inefficient bremsstrahlung process and atmospheric
absorption (see Figure 72).

A good correlation between the X-ray burst and the surge was established by
CLARK and ANGER (1967); Figure 81 shows a successive substorm which occurred on
October 12/13, 1963, over Churchill (dp lat 68.7°), Canada. From the top, the diagram
shows the riometer record, X-ray record, the relative location (deduced from the
Churchill magnetometer record) of the auroral electrojet with respect to the balloon,
the intensity of the electrojet and the flow direction of the electrojet. The first event,
denoted by A, was associated with the westward traveling surge which displayed Type
B aurora. The riometer showed a very sharp impulsive absorption during the event, so
that this is undoubtedly an E type event described in Sections 4.2 and 4.5. A high
time resolution examination of the burst indicated that it consisted of three very
intense X-ray bursts, each lasting less than one minute. The balloon was located
near the electrojet during the event. There was then a short quiet period during which
the electrojet moved equatorward. The electrojet and active auroras began to move
poleward again at 2212 LT; a brilliant red lower border was associated with the aurora.
When the approaching electrojet came within a distance of about 100 km from the
balloon, X-ray bursts were recorded (denoted by B). The current reached its poleward-
most location at 2218 LT and immediately began to return equatorward. When the
current passed directly above the balloon, an X-ray burst was observed (denoted by C).

At 2223 LT, a sudden intensification and poleward motion of the current began

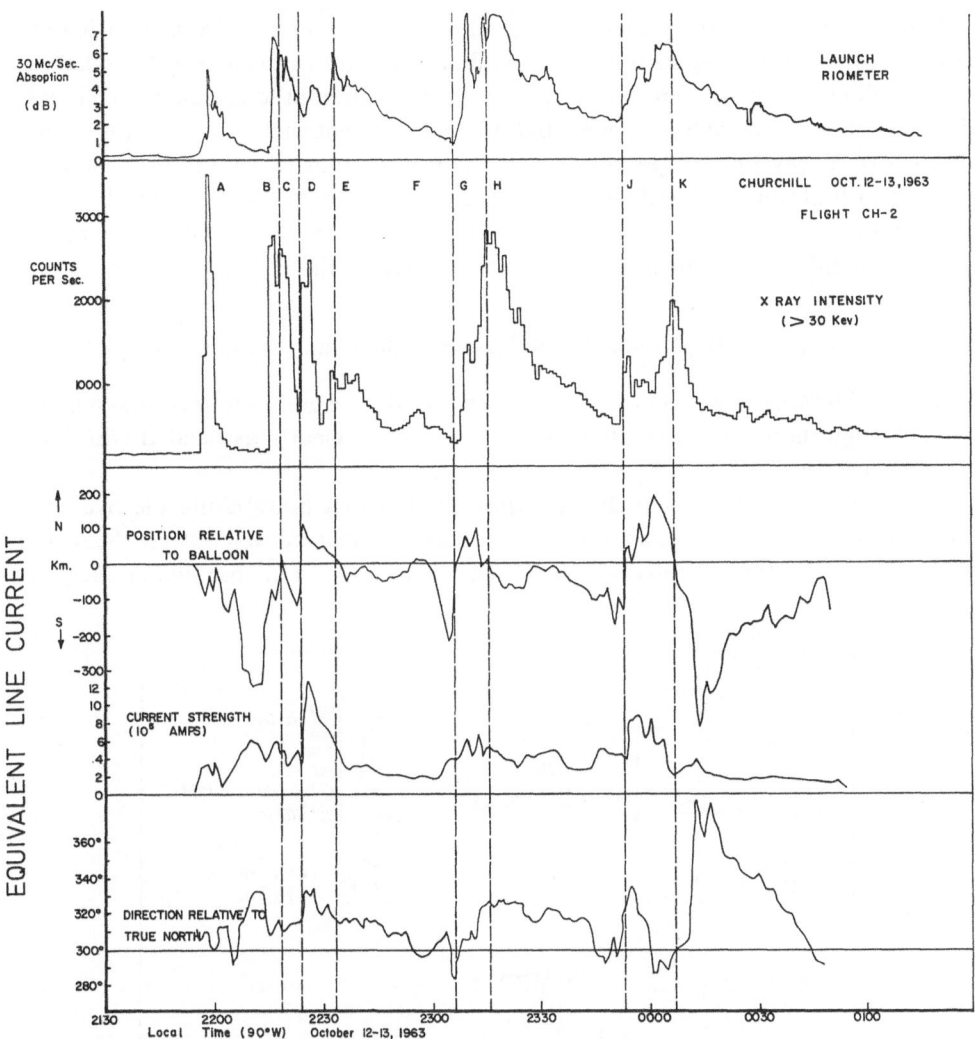

Fig. 81. Example showing that westward traveling surges (with the type B aurora) were associated with intense X-ray bursts (Clark, T. A. and C. D. Anger: *Planetary Space Sci.* **15**, 1287, 1967).

(denoted by D). An intense X-ray burst was observed when the current passed directly above the balloon. The next major activity began at 2304 LT when the current increased in intensity and moved poleward. The X-ray bursts were observed when active bands passed directly above the balloon (denoted by G). The X-ray flux continued to increase after the zenith passing and when the peak intensity was reached (denoted by H), a bright diffuse glow covered the entire sky. From an inspection of the riometer records, these two events may be identified as the N type events. The last two events (denoted by J and K) were also associated with another substorm; two peaks occurred when the electrojet was crossing the zenith.

In their study of the daily characteristics for the X-ray bursts, BARCUS and ROSEN-BERG (1966) noted that bursts in the evening hours are not associated with a decrease of the e-folding energy E_0. Thus far, there has been no attempt to correlate such bursts with auroral or magnetic phenomena, but it is unlikely that such bursts are associated with westward traveling surges. Since there is an important indication that energetic electrons drift around the earth (Section 9.2. C), such evening events may well be due to rather energetic electrons which drift over a considerable longitude range across the noon meridian, after being generated or injected in the midnight sector.

5.6. X-Ray Bursts in the Morning and Midday Hours

It was noted by several workers that X-ray bursts in the auroral zone are quite common in the daylight hours and that they are not necessarily clearly associated with local magnetic activity.

In Section 3.1., we learned that the auroral electrojet flows along the oval, not along the auroral zone. Thus, an auroral zone station is located well outside the region of the electrojet in the morning hours. On the other hand, an intense flux of energetic

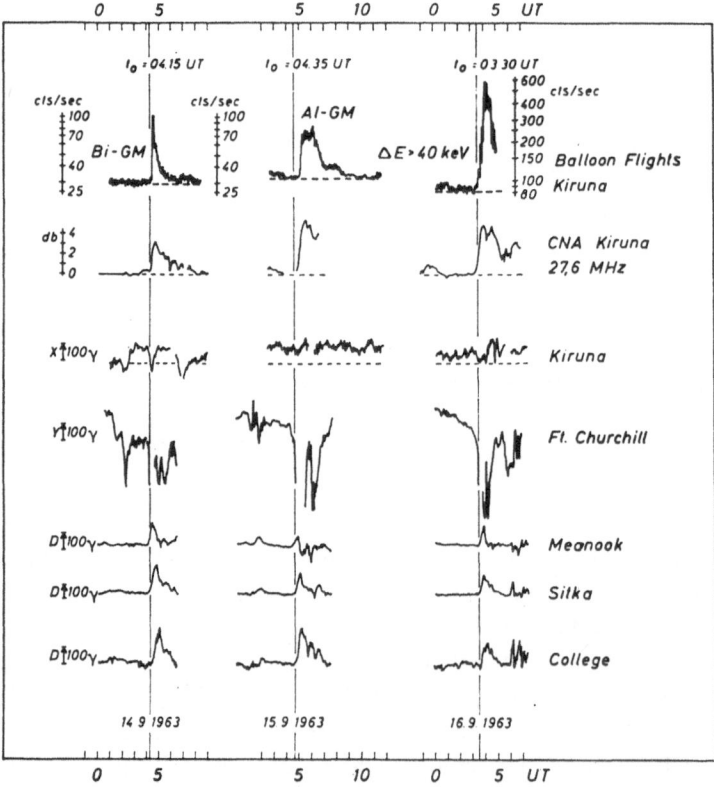

Fig. 82.   Three examples showing that the morning X-ray bursts are associated with polar magnetic substorms (Bewersdorff, A., G. Kremser, W. Riedler, and J. P. Legrand: *Arkiv Geofysik* **5**, 115, 1967).

electrons appears along the auroral zone in the morning sector. Therefore, it is not surprising to see a rather poor correlation between X-ray bursts and magnetic activity in the late morning hours.

In fact, BEWERSDORFF *et al.* (1967) showed conclusively that late morning X-ray bursts are associated with the polar substorm; they showed that late morning X-ray bursts occur when a sharp negative (magnetic) bay or the N type absorption begins in the midnight sector. Figure 82 shows three examples of the morning bursts, the associated M type absorption, magnetic activity at Kiruna (dp lat 65.3°), together with the corresponding magnetic records from Churchill (dp lat 68.7°), Meanook (dp lat 61.8°), Sitka (dp lat 60.0°) and College (dp lat 64.7°). Figure 83 shows the same late morning bursts, the associated M type absorption at Kiruna and the corresponding riometer records from Cape Jones, Val d'Or (Canada) and College (Alaska); it is not surprising to see no obvious absorption at College, since it is located well outside the oval in the early evening hours (05 UT = 19 LT at College).

BEWERSDORFF *et al.* (1968) showed that the bursts start North of $L = 6.5$ first and then spread equatorward to at least $L = 4.5$ with speeds of about 500 m/s; after the

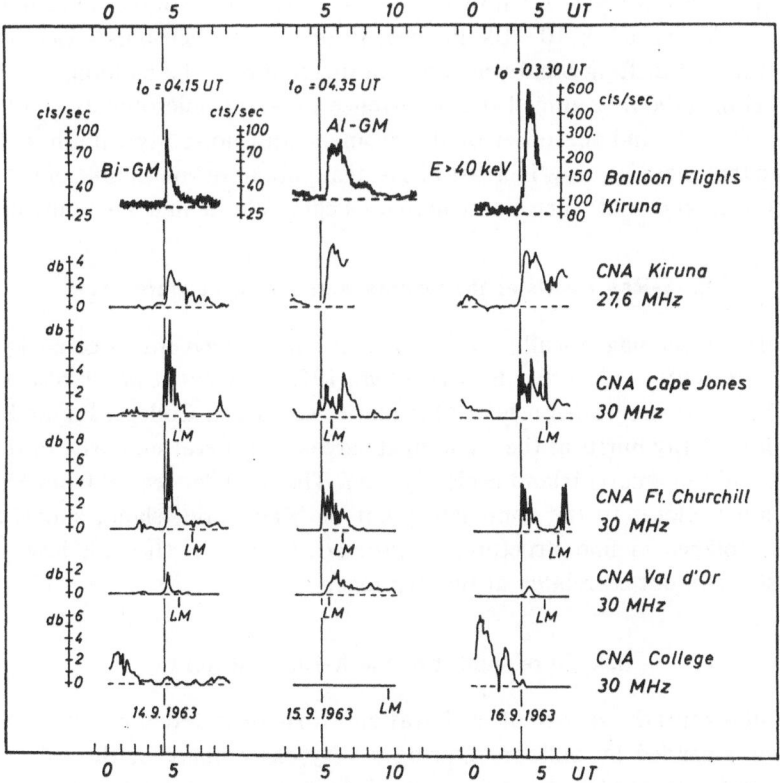

Fig. 83.   Three examples showing that the morning X-ray bursts are associated with the N type absorption in the midnight sector (Bewersdorff, A., G. Kremser, W. Riedler, and J. P. Legrand: *Arkiv Geofysik* **5**, 115, 1967).

expansion, the electron precipitation seems to decay uniformly over the area which was covered during the expanding phase.

The origin of the energetic electrons which cause both the daytime bursts and the late morning M type absorption is a controversial subject and will be discussed in Chapters 9 and 10.

5.7. Fine Structures of the X-Ray Bursts

As Figure 75 shows, semi-regular pulsations are often superposed on the morning bursts (BARCUS et al., 1966). In some cases, slow pulsating bursts are associated with similar pulsations of the riometer and magnetic records (BARCUS and ROSENBERG, 1966; ULLALAND and TREFALL, 1967).

ANDERSON and MILTON (1964) and VENKATESAN et al. (1968) reported a single, double, or multiple burst of lifetime of order less than 1 sec. Such short life bursts are called microbursts by ANDERSON and MILTON (1964). In many cases, the bursts have a very short rise time of order 30 millisec and a slow decay (~ 0.5 sec).

PARKS (1967), by using an array of four narrowly collimated detectors, showed that individual microburst precipitation regions are on the average 40 km radius at the X-ray production layer. He noted also that microburst sources appear to move sometimes with a speed of $10 \sim 100$ km/sec; in one case, microbursts were observed when an eastward drifting band was crossing the zenith of the balloon.

PARKS et al. (1966) examined the occurrence of X-ray microbursts at Flin Flon, Manitoba, Canada and the onset of the cosmic radio noise absorption at College. They found that there is a slow increase of X-rays at about the onset time of the absorption at College and that microburst activity begins about half an hour after that.

5.8. X-Ray Bursts at the Geomagnetically Conjugate Areas

X-ray bursts at geomagnetically conjugate areas have been investigated by Brown and others (BROWN et al., 1963; BROWN et al., 1965; BROWN et al., 1965a, b). They launched balloons simultaneously at Macquarie Island and College. Figure 84 shows an example of X-ray bursts at these conjugate areas. A general similarity of the bursts at College and Macquarie Island is clearly seen. The riometer record from Kotzebue, which is much closer to the conjugate point of Macquarie Island, shows an even better resemblance in fine structure. BROWN et al. (1965b) showed, however, that microbursts were not correlated at the two locations.

5.9. Development of the X-Ray Substorm

Figure 85 illustrates the growth of the X-ray substorm observed at about a 30 km level by a detector carried by balloon. During a very early phase of the substorm, the X-ray precipitation may be confined only along a narrow band where the first sign of the auroral substorm is seen. Then, the precipitation region expands in all directions. In particular, the expansion along the auroral zone in the morning hours continues

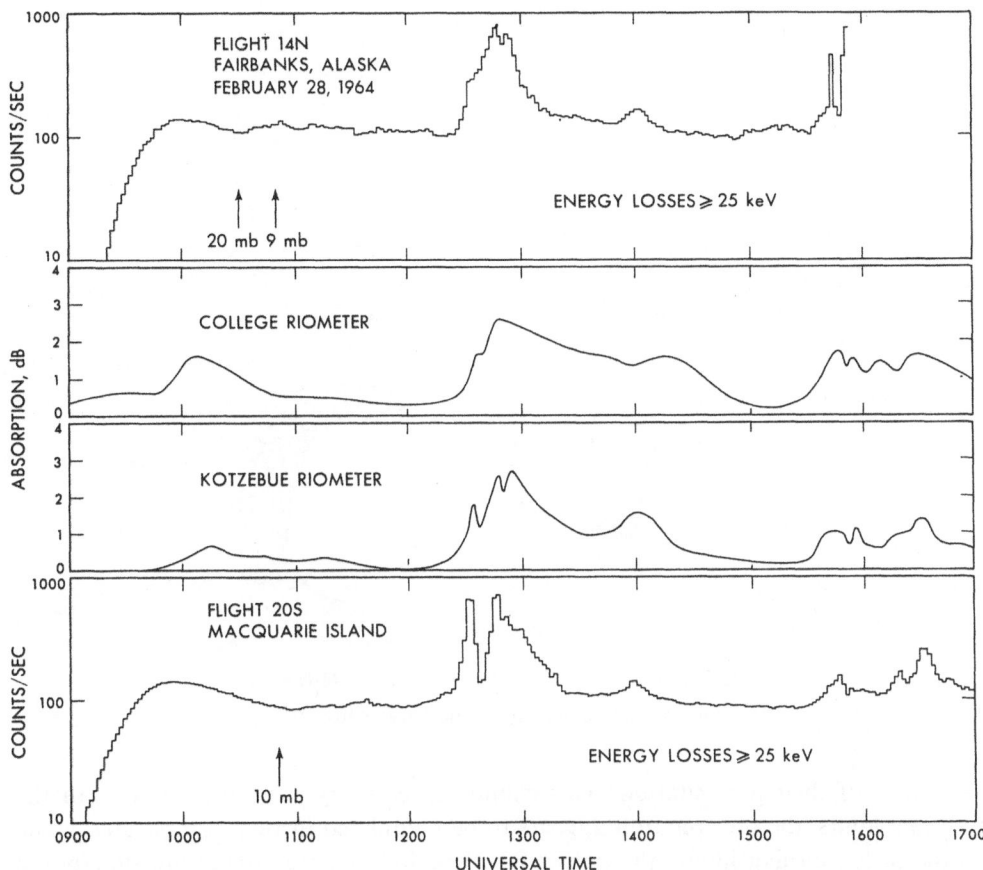

Fig. 84. Simultaneous observation of X-ray bursts at the geomagnetically conjugate areas, College, Alaska and Macquarie Island. Both College and Kotzebue riometer records are also shown (Brown, R. R., J. R. Barcus, and N. R. Parsons: *J. Geophys. Res.* **70**, 2579, 1965).

until the very end of the substorm. As mentioned in Section 4.8, it is essential to determine accurately the speed of the extension. BEWERSDORFF *et al.* (1968) reported that the X-rays start North of $L = 6.5$ first and then spread equatorward.

There may be one important difference between Figure 64 (the ionosphere substorm) and Figure 85. It is the lack of the precipitation of X-rays along the auroral oval. The X-ray precipitation occurs along the evening part of the oval only when a westward traveling surge is intense and shows the type B aurora. In Section 4.8, we noted that there is an indication that the absorption along the morning part of the oval is not intense. It is not known whether this is due to the lack of hard electrons or to a small flux of electrons. There are no X-ray observations to clarify this point. In Figure 85, the electrons which precipitate in the morning part of the oval are assumed to have a soft energy spectrum (Section 9.6).

The balloon observation of electrons has a great advantage of observing fine time

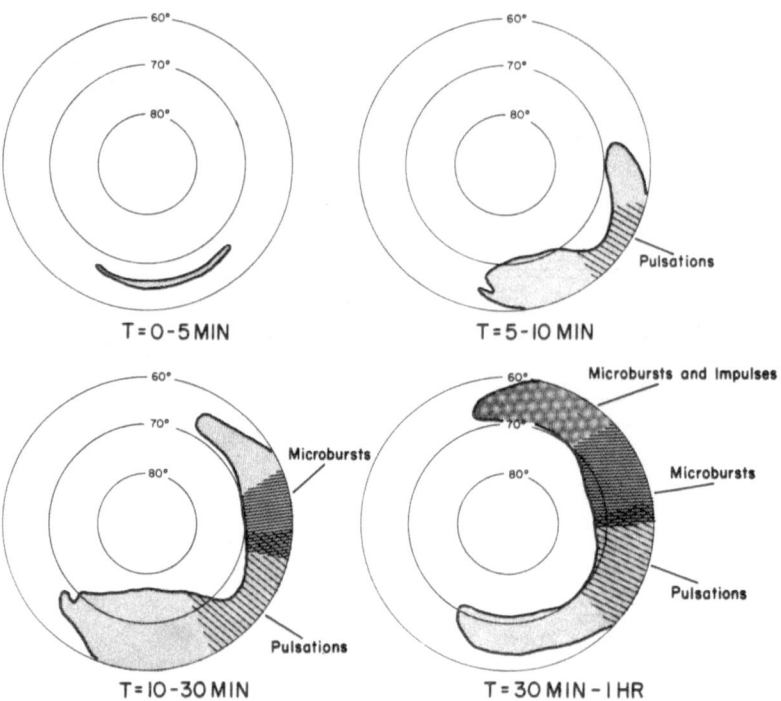

Fig. 85.   Development of the X-ray substorm.

variations of their precipitation. This cannot be done by any other means. In the morning hours, the precipitation appears to be modulated in very complicated ways. In the early morning hours, the pulsations of period of order 10 sec are superposed on the general increase; in this connection we note that pulsations of electron fluxes occur also in the morning sector of the tail region of the magnetosphere (LIN and ANDERSON, 1966; Section 9.4). Microbursts are a quite common feature during the morning bursts. They begin to develop about 30 min after the onset of the substorm (PARKS *et al.*, 1966). However, their latitudinal extent or variation is not known; it is assumed in Figure 85 that the microbursts occur over the entire cross-section of the morning extension. OLIVEN and GURNETT (1968) showed that microbursts of 40 keV electrons are always accompanied by a group of VLF chorus emissions (Section 7.6). In the midday sector, sharp impulsive bursts are superposed on microbursts.

References

GENERAL

BROWN, R. R.: 1966, 'Electron precipitation in the auroral zone', *Space Sci. Rev.* **5**, 311–387.

REFERRED TO IN TEXT

ANDERSON, K. A.: 1966, 'Energetic particles in the earth's magnetic field', Pub. Cosmic Ray Group, Dept. of Physics, Univ. of California, Berkeley, March.

ANDERSON, K. A. and MILTON, D. W.: 1964, 'Balloon observations of X-rays in the auroral zone. 3. High time resolution studies', *J. Geophys. Res.* **69**, 4457–4479.

ANDERSON, K. A., CHASE, L. M., HUDSON, H. S., LAMPTON, M., MILTON, D. W., and PARKS, G. K.: 1966, 'Balloon and rocket observations of auroral-zone microbursts', *J. Geophys. Res.* 71, 4617–4629.

BARCUS, J. R.: 1964, 'Observations on the relationship of energetic particle precipitation to auroral zone phenomena', Cosmic Ray Group, Dept. of Physics, Univ. of California, Berkeley, June 1964.

BARCUS, J. R. and BROWN, R. R.: 1966, 'Energy spectrum for auroral-zone X-rays. 2. Spectral variability and auroral absorption', *J. Geophys. Res.* 71, 825–834.

BARCUS, J. R. and ROSENBERG, T. J.: 1966, 'Energy spectrum for auroral-zone X-rays. 1. Diurnal and type effects', *J. Geophys. Res.* 71, 803–823.

BARCUS, J. R., BROWN, R. R., and ROSENBERG, T. J.: 1966, 'Spatial and temporal character of fast variations in auroral-zone X-rays', *J. Geophys. Res.* 71, 125–141.

BEWERSDORFF, A., KREMSER, G., RIEDLER, W., and LEGRAND, J. P.: 1967, 'Some properties of the slowly varying ionospheric absorption events in the auroral zone', *Arkiv Geofysik* 5, 115–127.

BEWERSDORFF, A., KREMSER, G., STADSNES, J., and TREFALL, H.: 1968, 'Simultaneous balloon measurements of auroral X-rays during slowly varying ionospheric absorption events', *J. Atmospheric Terrest. Phys.* 30, 591–607.

BROWN, R. R.: 1966, 'Electron precipitation in the auroral zone', *Space Sci. Rev.* 5, 311–387.

BROWN, R. R. and CAMPBELL, W. H.: 1962, 'An auroral-zone electron precipitation event and its relationship to a magnetic bay', *J. Geophys. Res.* 67, 1357–1366.

BROWN, R. R., BARCUS, J. R., and PARSONS, N. R.: 1965a, 'Balloon observations of auroral zone X-rays in conjugate regions. 1. Slow time variations', *J. Geophys. Res.* 70, 2579–2598.

BROWN, R. R., BARCUS, J. R., and PARSONS, N. R.: 1965b, 'Balloon observations of auroral zone X-rays in conjugate regions. 2. Microbursts and pulsations', *J. Geophys. Res.* 70, 2599–2612.

BROWN, R. R., ANDERSON, K. A., ANGER, C. D., and EVANS, D. S.: 1963, 'Simultaneous electron precipitation in the northern and southern auroral zones', *J. Geophys. Res.* 68, 2677–2684.

BROWN, R. R., BARCUS, J. R., REID, J., and PARSONS, N. R.: 1965, 'Observations of long-period pulsations of electron precipitation in conjugate regions of the auroral zones', *J. Geophys. Res.* 70, 1246–1249.

CLARK, T. A. and ANGER, C. D.: 1967, 'Morphology of electron precipitation during auroral substorms', *Planetary Space Sci.* 15, 1287–1301.

GOSLING, J. T.: 1966, 'Localization and motion of energetic electron precipitation during magnetic bays', *J. Geophys. Res.* 71, 835–848.

LIN, R. P. and ANDERSON, K. A.: 1966, 'Periodic modulations of the energetic electron fluxes in the distant radiation zone', *J. Geophys. Res.* 71, 1827–1835.

OLIVEN, M. N. and GURNETT, D. A.: 1968, 'Microburst phenomena. 3. An association between microbursts and VLF chorus', *J. Geophys. Res.* 73, 2355–2362.

PARKS, G. K.: 1967, 'Spatial characteristics of auroral-zone X-ray microbursts', *J. Geophys. Res.* 72, 215–226.

PARKS, G. K., McPHERRON, R. L., and ANDERSON, K. A.: 1966, 'Relation of 5- to 40-second-period geomagnetic micropulsations and electron precipitation to the auroral substorm', *J. Geophys. Res.* 71, 5743–5745.

PARKS, G. K., MILTON, D. W., and ANDERSON, K. A.: 1967, 'Auroral-zone X-ray bursts of 5- to 25-millisecond duration', *J. Geophys. Res.* 72, 4587–4589.

PARKS, G. K., CORONITI, F. V., McPHERRON, R. L., and ANDERSON, K. A.: 1968, 'Studies of the magnetospheric substorm. 1. Characteristics of modulated energetic electron precipitation occurring during auroral substorms', *J. Geophys. Res.* 73, 1685–1696.

ROSENBERG, T. J., BJORDAL, J., and KVIFTE, G.: 1967, 'Balloon observations of pulsating X-rays and aurora', *Trans. Am. Geophys. Union* 48, 72.

ULLALAND, S. L. and TREFALL, H.: 1967, 'Observations of large-scale coherent pulsating electron precipitation events in the auroral zone, accompanied by geomagnetic continuous pulsations', *J. Atmospheric Terrest. Phys.* 29, 395–410.

VENKATESAN, D., OLIVEN, M. N., EDWARDS, P. J., McCRACKEN, K. G., and STEINBOCK, M.: 1968, 'Microburst phenomena. 1. Auroral-zone X-rays', *J. Geophys. Res.* 73, 2333–2343.

WINCKLER, J. R., PETERSON, L., HOFFMAN, R., and ARNOLDY, R.: 1959, 'Auroral X-rays, cosmic rays, and related phenomena during the storm of February 10–11, 1958', *J. Geophys. Res.* 64, 597–610.

CHAPTER 6

PROTON AURORA SUBSTORM

6.1. Introduction

VEGARD (1939) and GARTLEIN (1950) were among the first to identify the Balmer lines, Hα, Hβ, and Hγ in auroral spectra. The discoveries of the Doppler broadening of the Hα line by GARTLEIN (1950) and of the Doppler shift of the Hα line by MEINEL (1951) marked important epochs in the development of auroral physics; their findings had been taken to be direct evidence to show that solar corpuscles produced auroras after entering the geomagnetic field. For this reason, a considerable effort was made to study the excitation of auroral luminosity by incoming protons. However, OMHOLT (1959) and others concluded that protons cannot be responsible for producing auroral arcs (CHAMBERLAIN, 1961, Chapter 7). Indeed, rocket observations have demonstrated that auroras with an arc or band structure are not produced by protons, but by electrons (cf. MCILWAIN, 1960; EVANS, 1968; PFISTER, 1967). Only very recently, however, ROMICK and BELON (1967a, b), MURCRAY (1967) and MOZER and BRUSTON (1967) have presented evidence which suggests that the λ5577 line may be produced by other processes than a simple excitation of oxygen atoms by secondary electrons.

GARTLEIN (1950) also noted that the hydrogen lines are strong early in an auroral display. MEINEL (1951), DAHLSTROM and HUNTEN (1951) concluded that strong hydrogen lines are observed during the quiet homogeneous arc phase, but are absent when the arc develops a rayed structure. Later, FAN and SCHULTE (1954), ROMICK and ELVEY (1958), GALPERIN (1959), YEVLASHIN (1961) and VEISSBERG (1962) confirmed the earlier observation and concluded that protons were responsible for the production of quiet arcs, but not for active forms.

More detailed observations, with good spatial and time resolutions, showed, however, that in the early evening hours, a diffuse band containing the hydrogen emissions appears equatorward of quiet arcs (REES et al., 1961; OMHOLT et al., 1962; STOFFREGEN and DERBLOM, 1962; MONTBRIAND and VALLANCE JONES, 1962; GALPERIN, 1963). This diffuse band of luminosity is called the proton aurora or the hydrogen aurora.

If incoming protons have energies greater than 200 keV, they ionize atmospheric particles. But if their energies are less than 200 keV or are slowed down by the ionization collision to about 200 keV, the charge exchange process becomes important

$$H^+ + X \rightarrow H^* + X^+ .$$

As a result, the protons become energetic neutral hydrogen atoms. Thus, protons spiral down along the geomagnetic field line until the above process takes place and then move in a straight line in the atmosphere until the next collision occurs. During this period, the hydrogen atoms can emit Lyman or Balmer radiation. The former can only be detected by rockets or satellites. If the next collision is the charge exchange type, the neutral hydrogen atoms become protons again, so that its motion is constrained again by geomagnetic field lines until the next charge exchange collision occurs.

$$H + X^+ \rightarrow H^+ + X.$$

In addition to the above process, the H atoms can attain an excited state by colliding with atmospheric particles. They return to the ground state by emission. ROMICK and SHARP (1967) concluded that this collisional excitation contributes significantly to the intensity of the hydrogen lines observed on the ground.

The above sequence of processes may be repeated hundreds of times before the hydrogen atom finally becomes a thermal particle; Figure 86 illustrates schematically the path of a proton injected into the polar upper atmosphere from outside.

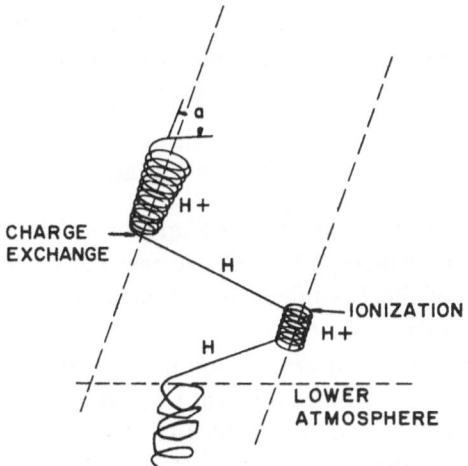

Fig. 86.   Schematic drawing of the path of a proton injected from the equatorial plane into the polar upper atmosphere (Davidson, G. T.: *J. Geophys. Res.* **70**, 1061, 1965).

Further, since the motion of energetic neutral hydrogen atoms is not constrained by the geomagnetic field, even an extremely thin sheet beam of protons injected into the polar atmosphere tends to spread, resulting in a very broad band in which the hydrogen emissions appear. In the band, other spectral lines excited by ionizations or excitations of atmospheric particles by protons or energetic neutral hydrogen atoms also appear. Figure 87 shows a model calculation by DAVIDSON (1965) showing the North-South distribution of the intensity of the Hα emission as a result of injection of 10 keV protons along the field line which originates at dp lat 67°.

The proton aurora shows definite changes during the auroral substorm. This phenomenon, the proton aurora substorm, is the subject of this chapter. As mentioned

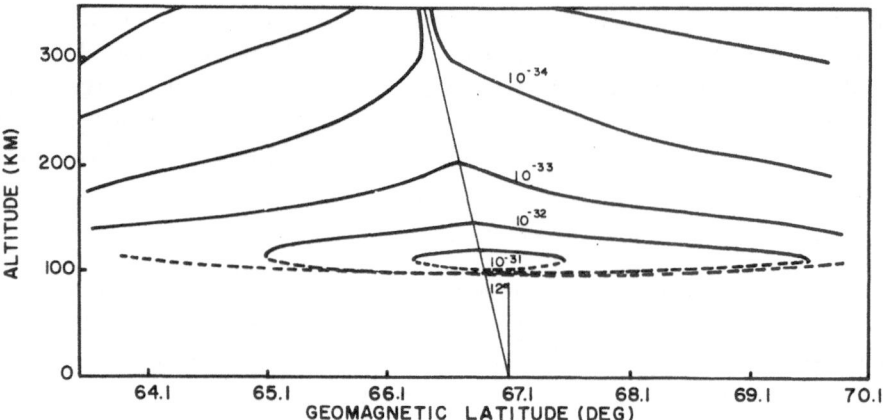

Fig. 87. Distribution of the intensity (ergs/proton-cm³) of the Hα intensity produced by 10 keV protons injected along the field line which originates at dp lat 67° (Davidson, G. T.: *J. Geophys. Res.* **70**, 1061, 1965).

in Section 1.3, protons of energies of order 1–50 keV appear to play a key role in the basic processes associated with the magnetospheric substorm. We are thus particularly interested in determining the behavior of the protons by studying the proton aurora substorm.

6.2. Typical Daily Variation

Figure 88 shows the daily variation of the meridian extent of the Hα line for 6 successive nights (December 2/3 – December 7/8, 1958), observed at College (REES *et al.*, 1961). The hourly mean value of the *H* component magnetic intensity is also added at the bottom for later discussions.

There is a definite equatorward progression in the evening hours and a poleward recession in the morning hours. Since this feature is very similar to electron auroras which lie in the auroral oval, it is quite possible that the proton aurora has an eccentric oval geometry surrounding the pole (REID and REES, 1961).

The extent of this systematic daily variation varies greatly day by day. For example, on December 3/4, 4/5, 5/6, the equatorward boundary of the proton aurora was located well South of College (dp lat 64.7°), but on December 2/3, 6/7, and 7/8, it was confined in the northern sky. Further, during certain periods on December 3/4 and 4/5, the proton emission appeared over a substantial part of the College sky.

Figure 89 shows another example of the daily variation of the proton aurora for 6 successive nights, observed at Churchill (dp lat 68.7°) (November 8/9 – November 13/14) (MONTALBETTI and McEWEN, 1961). The behavior of the proton aurora on the night of November 9/10 and 10/11 is quite similar to that shown in Figure 88. However, on the night of November 12/13, the hydrogen emission covered the entire sky at Churchill. It should be noted, however, that the intensity of the hydrogen emission varied considerably during the course of the night; in particular, there were four periods when the Hβ intensity was more than 100 *R*; at about 04, 08, 11 and 12 UT.

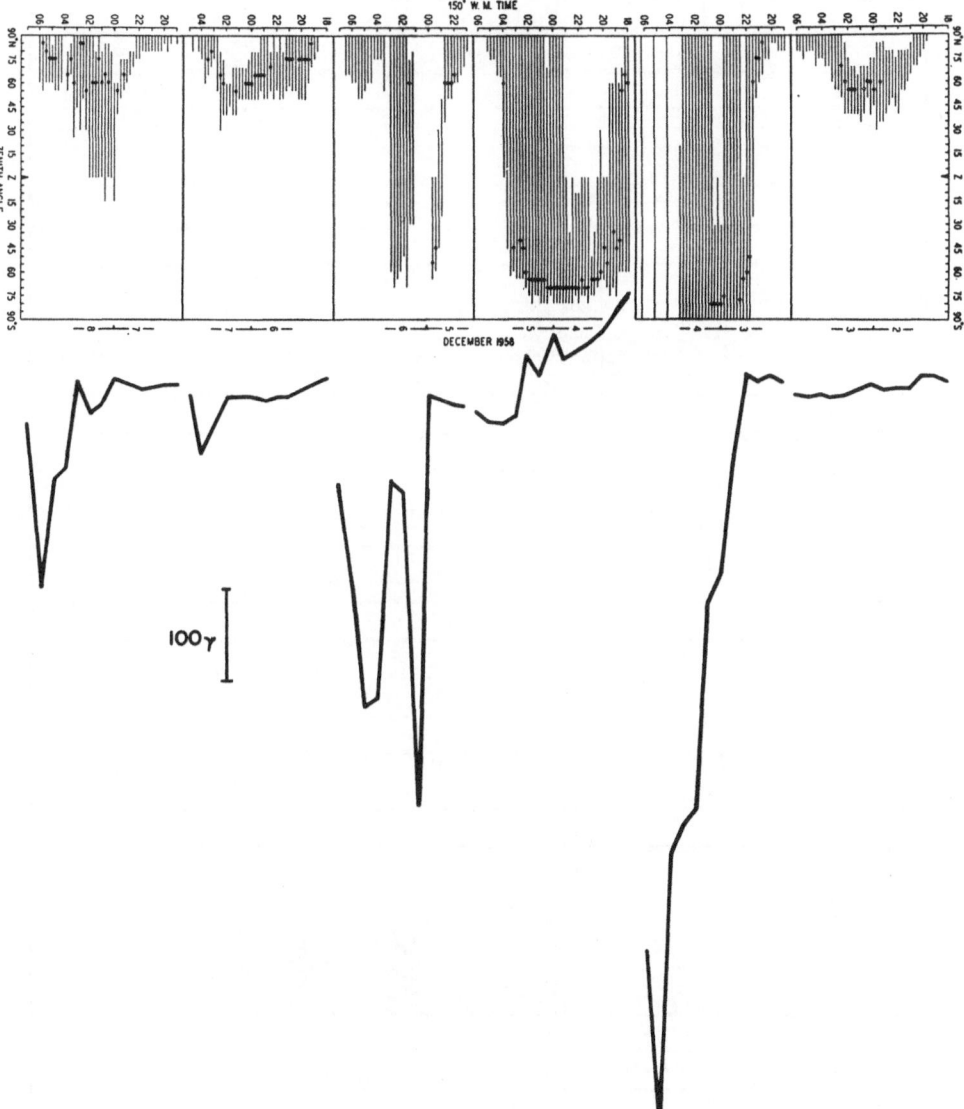

Fig. 88.   Daily variation of the meridian extent of the Hα emission for 6 successive nights at College; the corresponding hourly mean values of the *H* component magnetic variation are added (Rees, M. H., A. E. Belon, and G. J. Romick: *Planetary Space Sci.* **5**, 87, 1961).

6.3. Statistical Daily Variation Pattern

In Section 6.2 we noted that the proton aurora has an eccentric oval geometry surrounding the pole. The eccentric oval geometry of the proton aurora was first recognized by REID and REES (1961) on the basis of Figure 88. If this is actually the case, a station located at about dp lat 70° should 'cross' the proton aurora twice a day.

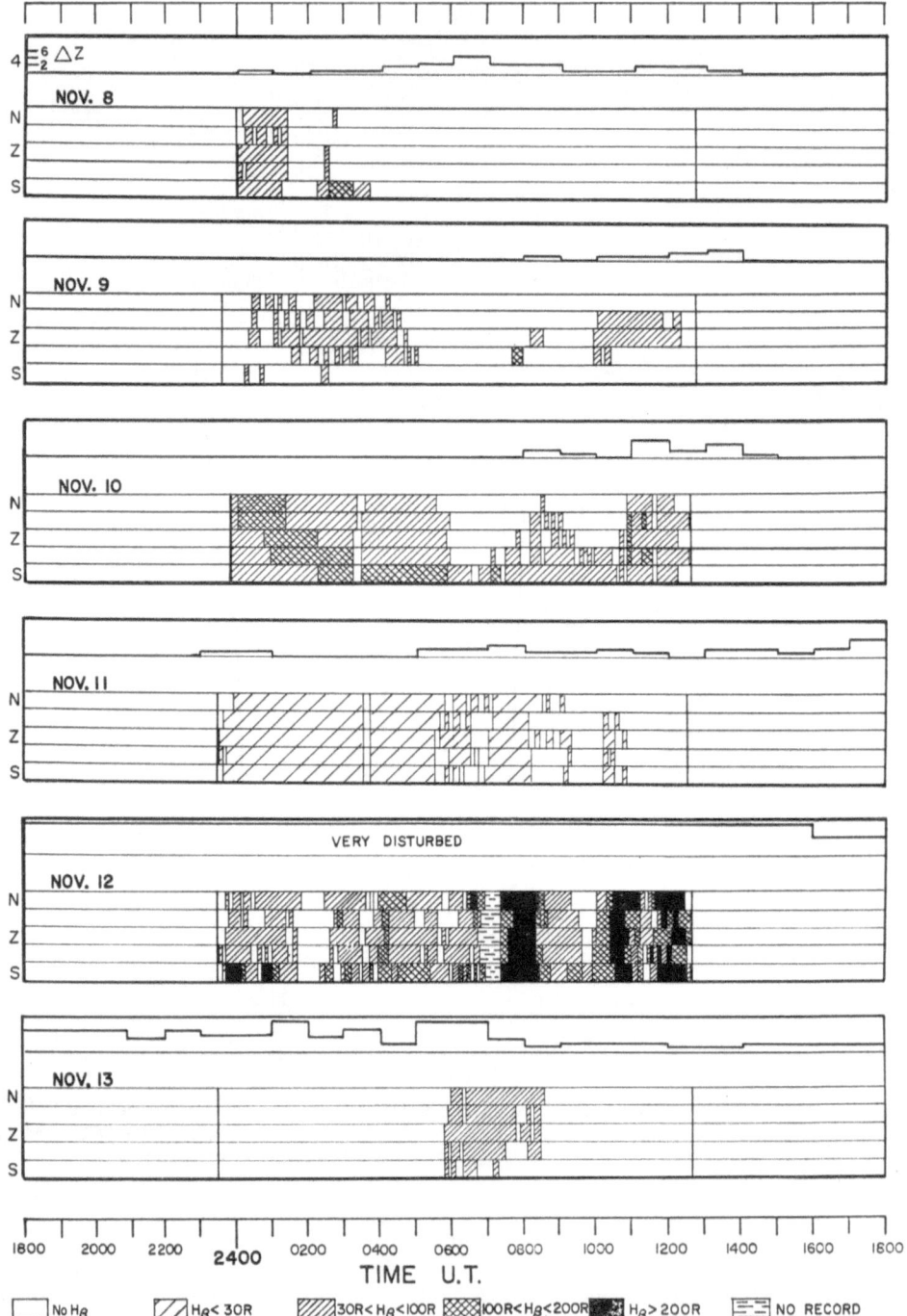

Fig. 89.   Daily variation of the meridian extent of the Hβ emission for 6 successive nights at Churchill
(Montalbetti, R. and D. J. McEwen: *Canadian J. Phys.* **39**, 617, 1961).

MONTALBETTI and McEWEN (1962) showed statistically that at Churchill (dp lat 68.7°) the Hβ intensity peaks twice a day, at about 20 LT and 04 LT; Figure 90. On the other hand, at College (dp lat 64.7°) there is only a single peak of the Hβ intensity in the midnight hours; Figure 90 (see also Section 1.4 and Figure 7). Further, MONTALBETTI and McEWEN (1962) showed that the occurrence of the retardation type sporadic

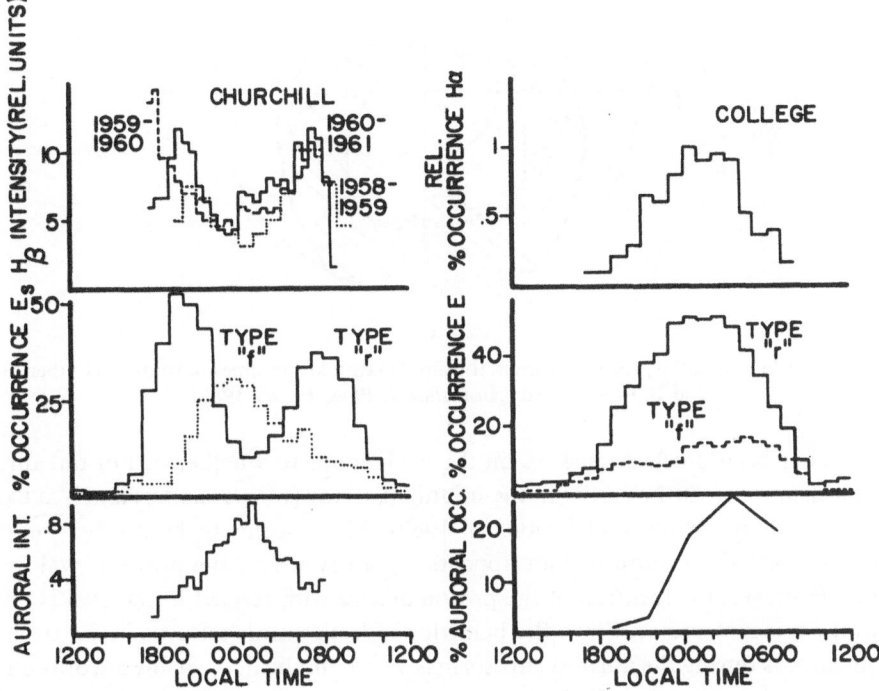

Fig. 90. Top: The daily variation of the Hβ intensity; Middle: the occurrence frequency of r-type and f-type sporadic E; Bottom: the intensity of the visual aurora at Churchill and College (Montalbetti, R. and D. J. McEwen: *J. Phys. Soc. Japan* 17, Suppl. A-1, 212, 1962).

E layer has a similar daily variation of occurrence, so that it is produced by the proton bombardment. An oval like structure of the proton aurora was also noted by YEVLASHIN (1963).

EATHER and SANDFORD (1966) combined all available information to date and constructed a polar plot of the hydrogen emission (Figure 91). There is no doubt that satellite observations of the protons or of the hydrogen emissions will refine such a pattern in the future. Indeed, SHARP *et al.* (1967) observed a narrow band of proton precipitation which is located at about dp lat 70° in the noon sector, as well as a narrow band at about dp lat 77° (corresponding to the location of the oval); Section 9.6. It is quite likely that the protons which precipitate into the former band (dp lat 70°) are leakage from the ring current belt. Unfortunately, the geometrical relationship between this band and the oval belt is not known.

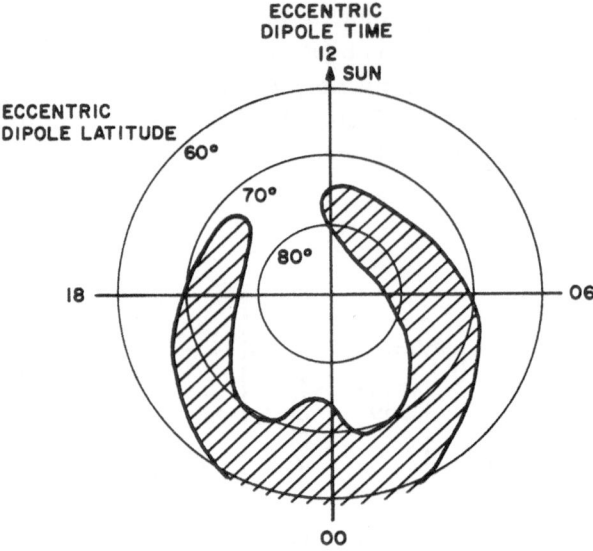

Fig. 91.   Distribution of the proton aurora in the dipole-latitude and time coordinates (Eather, R. H. and B. P. Sandford: *Australian J. Phys.* **19**, 25, 1966).

There has been much discussion on the problem as to whether or not the auroral oval and the proton aurora cross in the midnight sector (MONTALBETTI and VALLANCE JONES, 1957; STOFFREGEN and DERBLOM, 1962). There seems to be no disagreement on the fact that the proton aurora appears equatorward of the auroral oval in the evening. However, the location of the proton aurora with respect to the oval is uncertain in the morning sector. Since the behavior of both visible auroras in the oval and proton auroras during the auroral substorm is very complicated, visible auroras during the substorm do not necessarily provide a good reference to discuss the location of the proton aurora. The distribution of the intensity of the hydrogen emissions during a quiet period between two auroral substorms would be the best period to study in order to settle this important problem. It should also be kept in mind that there are two belts of proton precipitation (SHARP *et al.*, 1967; see Section 9.6. C.) and that their behavior during auroral substorms may be quite different.

6.4. Protons in the Auroral Bulge and Patches

As mentioned in Section 7.1, the proton aurora is most clearly recognized in the evening sector. However, there has been some disagreement as to how the proton aurora in the midnight sector behaves during an auroral substorm.

By using a meridian spectrograph operated at Saskatoon (dp lat 60°) MONTBRIAND and VALLANCE JONES (1962) showed that the hydrogen emissions slowly weakened during the period when the station was covered by the auroral bulge; they discussed in detail their records taken at Saskatoon on the night of April 28/29, 1960. In the early evening hours the zone of hydrogen emissions lay well to the South of the brightest auroral luminosity and was associated visually with a weaker homogeneous arc coincident with the maximum in the hydrogen emission. Then, an auroral substorm began at about 2335 LT, and active auroras covered the whole sky. MONTBRIAND and VALLANCE JONES (1962) noted that spectra taken during this period indicated that over a 20 min period the aurora remained bright from North to South, while the hydrogen emission slowly weakened.

On the other hand, by using a photometer at Mawson (dp lat 73.1°S) EATHER and JACKA (1966) reported that the $H\beta$ intensity often increased over the whole sky during and after an auroral breakup. Figure 92 shows an example of the daily variation (July 27, 1963) of the zenith intensity of λ 5577, $H\beta$ line and the cosmic radio noise absorption (27.6 Mc/sec); their statement refers to the increase of the three records at about 2218 UT. Further, the $H\beta$ intensity remained at a high level well after that of λ 5577 decreased considerably. EATHER (1967) noted that the $H\beta$ intensity also remains fairly constant in the region where pulsating patchy auroras exist.

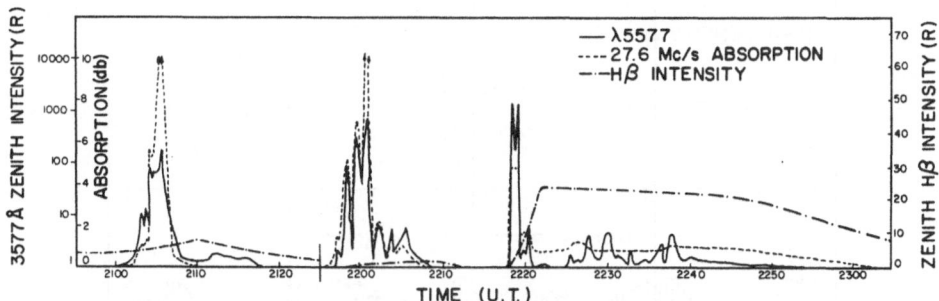

Fig. 92.   Example of the daily variation of the zenith intensity of λ 5577, $H\beta$ line and the cosmic radio noise absorption (27.6 Mc/sec) (Eather, R. H. and F. Jacka: *Australian J. Phys.* **19**, 215, 1966).

6.5. Protons and Westward Traveling Surges

EATHER and JACKA (1966) mentioned that it is the 'last breakup' in the course of a night which is associated with the hydrogen emission. In Figure 92, the $H\beta$ intensity shows little corresponding change for the first two events. It is easily inferred that their 'breakup', which is not associated with an increase of the $H\beta$ intensity, is auroral activity accompanied by a westward traveling surge or secondary auroral activity in the evening hours, which is generated by the expanding auroral bulge. In order to clarify this problem, we should note the location of Mawson (dp lat 73.1°) with respect to the proton aurora and the auroral oval in the evening hours. When the auroral oval is located near Mawson in the evening hours, the center-line of the proton aurora should be located equatorward of the station. Surges travel over Mawson along the oval in the evening hours, so that a λ 5577 photometer or riometer will be able to detect these surges. Therefore, unless the proton aurora spreads extensively in latitude or surges contain hydrogen emissions, the zenith photometer at Mawson will not be able to detect the $H\beta$ emission.

The reason for the enhancement of the Hβ intensity during the 'last breakup' at Mawson is then due to the fact that the auroral bulge expands poleward and covers Mawson and that an appreciable precipitation occurs in the bulge. The midnight bulge in Figure 91 illustrates this poleward expansion of the proton precipitation.

Figure 93 shows an example of the observation of the proton aurora in the evening hours at Kiruna (dp lat 65.3°) (OMHOLT *et al.*, 1962). There was a rapid equatorward motion of the proton aurora across the zenith of Kiruna, which began at about 2005 LT. At 2104 LT, there was a sudden increase of the *H* component magnetic intensity; a bright arc was seen a little North of Kiruna. It is not difficult to infer from our study in Section 2.3. (Figure 28) that this is the case when a westward traveling

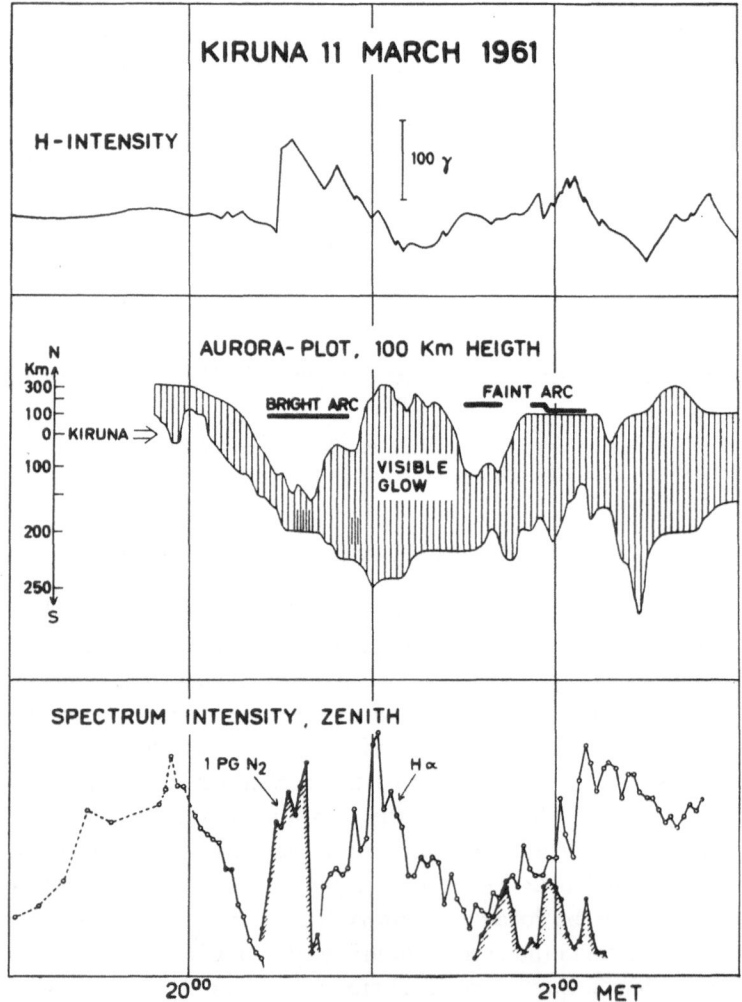

Fig. 93.   Proton aurora observed in the evening hours at Kiruna (Omholt, A., W. Stoffregen, and H. Derblom: *J. Atmospheric Terrest. Phys.* **24**, 203, 1962).

surge passed well North of Kiruna. A few minutes after the peak time of the *H* component variation, the proton aurora spread both northward and southward. This was detected by the spectral observation at Kiruna.

Figure 94 shows all-sky camera records from Bar I (dp lat 70°) and College (dp

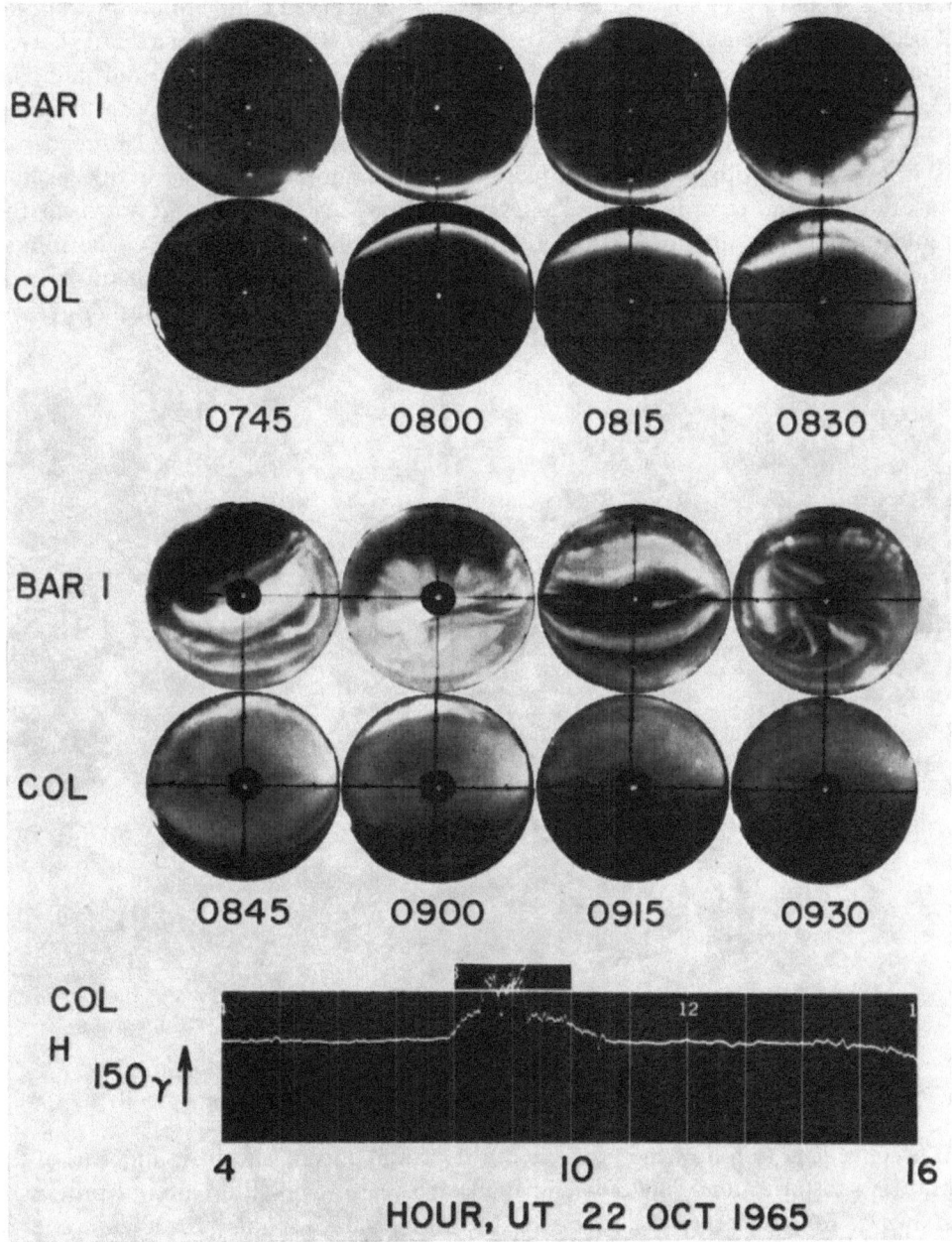

Fig. 94.   Relationship between the proton aurora and the westward traveling surge.

lat 64.7°) in the evening of October 21/22, 1965; Bar-I is located 600 km North (magnetic) of College. At 0800 UT, the same arc is seen South of Bar-I and North of College. The photographs taken at 0830, 0845, 0900, 0915 and 0930 indicate a westward traveling surge which traveled rapidly along the arc and covered the sky over Bar-I. At College, only the southern part of the surge was seen near the northern horizon. At 0815 UT, however, a faint luminosity appeared a little South of College; it became very intense at 0845 UT. The simultaneous Hβ photometer record showed that this luminosity contained hydrogen emission, but that the surge did not. The College magnetogram (H) shows a gradual positive change which began at about 0755 UT. The positive change was suddenly enhanced at about 0825 UT. This tendency is in agreement with Figure 93. Figure 95 shows another example of the proton aurora substorm, recorded at four Alaskan stations. An intense surge was near the zenith of Bar-I at about 0730 UT (2130 LT) and 1001 UT (2401 LT). The diffuse luminosity is clearly seen a little south of College at 0730 UT and 0930 UT. The corresponding College magnetic record (H) showed a positive change in both cases.

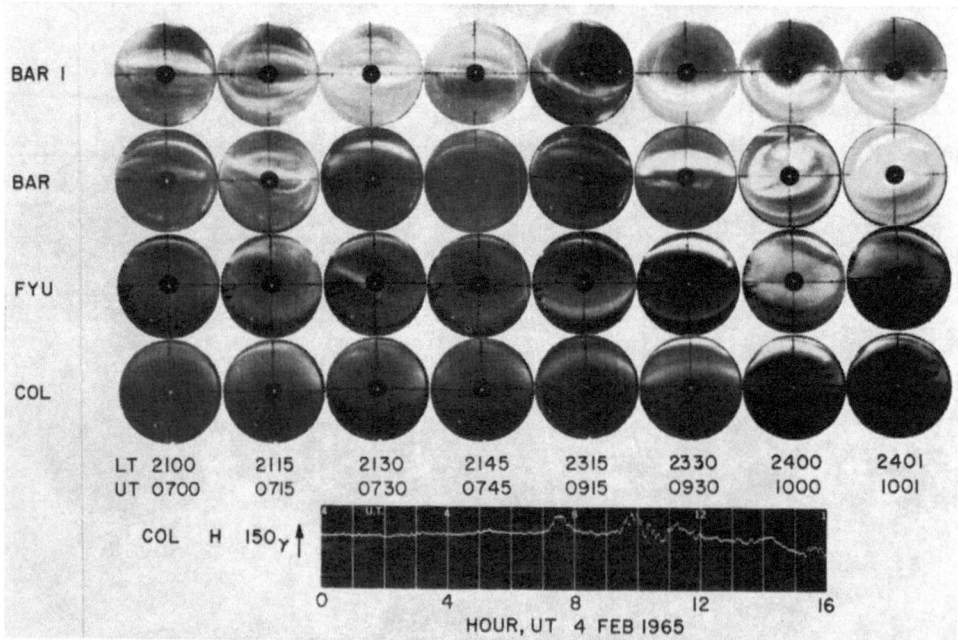

Fig. 95.   Relationship between the proton aurora and the westward traveling surge.

6.6. Development of the Proton Aurora Substorm

It is convenient to follow the progress of the proton aurora substorm in terms of the auroral substorm time. Therefore, in Figure 96, the precipitation area of protons is shown at four stages of the auroral substorm, $T=0$, 5–10 min, 10–30 min, and 30 min–1 hour.

During a quiet period between substorms, the proton precipitation occurs along an oval band which is located a little equatorward of the auroral oval. Here, we have also included the precipitation band found by SHARP *et al.* (1967), although the relationship between the proton aurora and the precipitation zone found by them is not known. In the midnight meridian, the precipitation occurs over the entire region

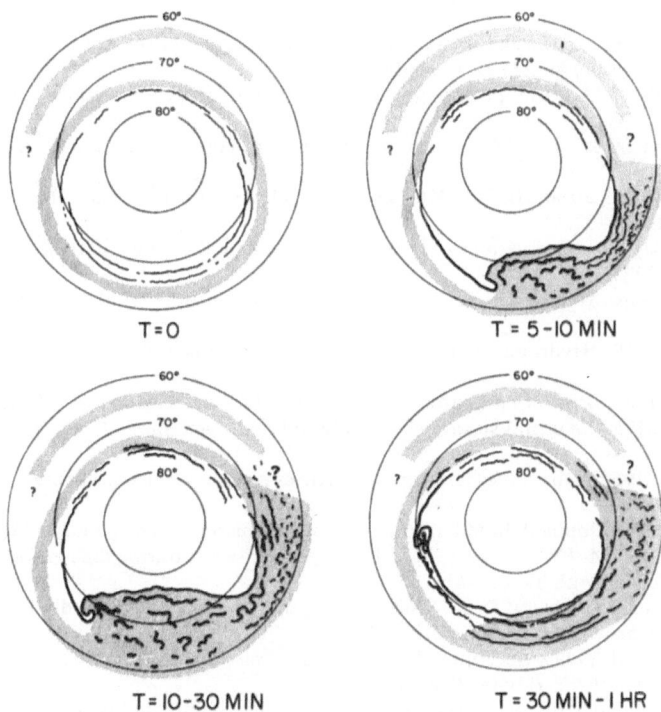

Fig. 96.   Development of the proton aurora substorm.

of the auroral bulge, so that the precipitation area expands rapidly poleward. In the evening sector, the proton aurora shifts rapidly equatorward during an early phase of the substorm. However, westward traveling surges which had advanced well into the evening sector do not seem to contain the hydrogen emissions. In the morning sector, the hydrogen emissions seem to appear over an extensive area. EATHER (1967) noted that the intensity remains fairly constant, even though embedded patches pulsate. However, the extent of the morning precipitation area is not known.

As we shall discuss in Section 9.2, protons in the midnight and early morning sector of the ring current belt suddenly disappear during a very early phase of the polar substorm. It may well be that the proton aurora substorm is caused partly by those ring current protons which disappear from the ring current.

References

GENERAL

CHAMBERLAIN, J. W.: 1961, *Physics of the aurora and airglow*, Academic Press, New York.

EATHER, R. H.: 1967, 'Auroral proton precipitation and hydrogen emissions', *Rev. Geophys.* **5**, 207–285.

REFERRED TO IN TEXT

CHAMBERLAIN, J. W.: 1961, *Physics of the aurora and airglow*, Academic Press, New York.

DAHLSTROM, C. E. and HUNTEN, D. M.: 1951, 'O_2^+ and H in the auroral spectrum', *Phys. Rev.* **84**, 378–879.

DAVIDSON, G. T.: 1965, 'Expected spatial distribution of low-energy protons precipitated in the auroral zones', *J. Geophys. Res.* **70**, 1061–1068.

EATHER, R. H.: 1967, 'Auroral proton precipitation and hydrogen emissions', *Rev. Geophys.* **5**, 207–285.

EATHER, R. H. and JACKA, F.: 1966, 'Auroral absorption of cosmic radio noise', *Australian J. Phys.* **19**, 215–239.

EATHER, R. H. and SANDFORD, B. P.: 1966, 'The zone of hydrogen emission in the night sky', *Australian J. Phys.* **19**, 25–33.

EVANS, D. S.: 1968, 'The observations of a near monoenergetic flux of auroral electrons', *J. Geophys. Res.* **73**, 2315–2323.

FAN, C. Y. and SCHULTE, D. H.: 1954, 'Variations in the auroral spectrum', *Astrophys. J.* **120**, 563–565.

GALPERIN, Y. I.: 1959, 'Hydrogen emission and two types of auroral spectra', *Planetary Space Sci.* **1**, 57–62.

GALPERIN, Y. I.: 1963, 'Proton bombardment in aurora', *Planetary Space Sci.* **10**, 187–193.

GARTLEIN, C. W.: 1950, 'Auroral spectra showing broad hydrogen lines', *Trans. Am. Geophys. Union* **31**, 18–20.

McILWAIN, C. E.: 1960, 'Direct measurement of particles producing visible auroras', *J. Geophys. Res.* **65**, 2727–2747.

MEINEL, A. B.: 1951, 'Doppler-shifted auroral hydrogen emission', *Astrophys. J.* **113**, 50–54.

MONTALBETTI, R. and McEWEN, D. J.: 1961, 'Hydrogen emissions during the period November 9–16, 1960', *Canadian J. Phys.* **39**, 617–619.

MONTALBETTI, R. and McEWEN, D. J.: 1962, 'Hydrogen emissions and sporadic E layer behaviour', *J. Phys. Soc. Japan* **17**, Supp. A-1, 212–215.

MONTALBETTI, R. and VALLANCE JONES, A.: 1957, 'Hα emissions during aurorae over west-central Canada', *J. Atmospheric Terrest. Phys.* **11**, 43–50.

MONTBRIAND, L. E. and VALLANCE JONES, A.: 1962, 'Studies of auroral hydrogen emissions in west-central Canada. 1. Time and geographical variations', *Canadian J. Phys.* **40**, 1401–1410.

MOZER, F. S. and BRUSTON, P.: 1967, 'Electric field measurements in the auroral ionosphere', *J. Geophys. Res.* **72**, 1109–1114.

MURCRAY, W. B.: 1967, 'Spatial relationship of auroral OI and N_2^+ emissions', *J. Geophys. Res.* **72**, 1047–1051.

OMHOLT, A.: 1959, 'Studies on the excitation of aurora borealis. 1. The hydrogen lines', *Geofys. Publikasjoner* **20**, 1–40.

OMHOLT, A., STOFFREGEN, W., and DERBLOM, H.: 1962, 'Hydrogen lines in auroral glow', *J. Atmospheric Terrest. Phys.* **24**, 203–209.

PFISTER, W.: 1967, 'Auroral investigations by means of rockets', *Space Sci. Rev.* **7**, 642–688.

REES, M. H., BELON, A. E., and ROMICK, G. J.: 1961, 'The systematic behaviour of hydrogen emission in the aurora. I.', *Planetary Space Sci.* **5**, 87–91.

REID, G. C. and REES, M. H.: 1961, 'The systematic behaviour of hydrogen emission in the aurora. II.', *Planetary Space Sci.* **5**, 99–104.

ROMICK, G. J. and BELON, A. E.: 1967a, 'The spatial variation of auroral luminosity. I. The behavior of synthetic model auroras', *Planetary Space Sci.* **15**, 475–493.

ROMICK, G. J. and BELON, A. E.: 1967b, 'The spatial variations of auroral luminosity. II. Determination of volume emission rate profiles', *Planetary Space Sci.* **15**, 1695–1710.

ROMICK, G. J. and ELVEY, C. T.: 1958, 'Variations in the intensity of the hydrogen emission line Hβ during auroral activity', *J. Atmospheric Terrest. Phys.* **12**, 283–287.

ROMICK, G. J. and SHARP, R. D.: 1967, 'Simultaneous measurements of an incident hydrogen flux and the resulting hydrogen Balmer alpha emission in an auroral hydrogen arc', *J. Geophys. Res.* **72**, 4791–4801.

SHARP, R. D., JOHNSON, R. G., SHEA, M. F., and SHOOK, G. B.: 1967, 'Satellite measurements of precipitating protons in the auroral zone', *J. Geophys. Res.* **72**, 227–237.

STOFFREGEN, W. and DERBLOM, H.: 1962, 'Auroral hydrogen emission related to charge separation in the magnetosphere', *Planetary Space Sci.* **9**, 711–716.

VEGARD, L.: 1939, 'Hydrogen showers in the auroral region', *Nature* **144**, 1089–1090.

VEISSBERG, O. L.: 1962, 'Spectro-electrophotometric studies of hydrogen emission in auroras', Aurora and Airglow, Results of Researches of the I.G.Y., *SSSR Acad. Sci.* **8**, 36–42.

YEVLASHIN, L. S.: 1961, 'Space-time variations of hydrogen in aurorae and their relationship to magnetic disturbances', *Geomagnetizm i Aeronomiya* **1**, 50–54.

YEVLASHIN, L. S.: 1963, 'Some patterns of behavior of auroral hydrogen emissions', *Geomagnetizm i Aeronomiya* **3**, 405–408.

CHAPTER 7

VLF EMISSION SUBSTORM

7.1. Introduction

WATTS (1957) and ELLIS (1957, 1959) were among the first to recognize atmospheric radio noise at kilocycle frequencies, which is closely associated with strong auroral and magnetic activity. ELLIS (1959) showed that noise appears as bursts and that some of the bursts were well correlated with an enhancement of the red oxygen line (λ 6300) in airglow. He found that the duration of most of such bursts is of order 3 hours, although some lasted for 15 hours. These facts strongly suggest that VLF emissions are associated with polar substorms.

VLF and ELF emissions have been studied extensively by a number of workers. Here, after GURNETT (1967), emissions with frequencies less than 2 kc/sec are called the ELF emissions and those above 2 kc/sec the VLF emissions. Much of the work

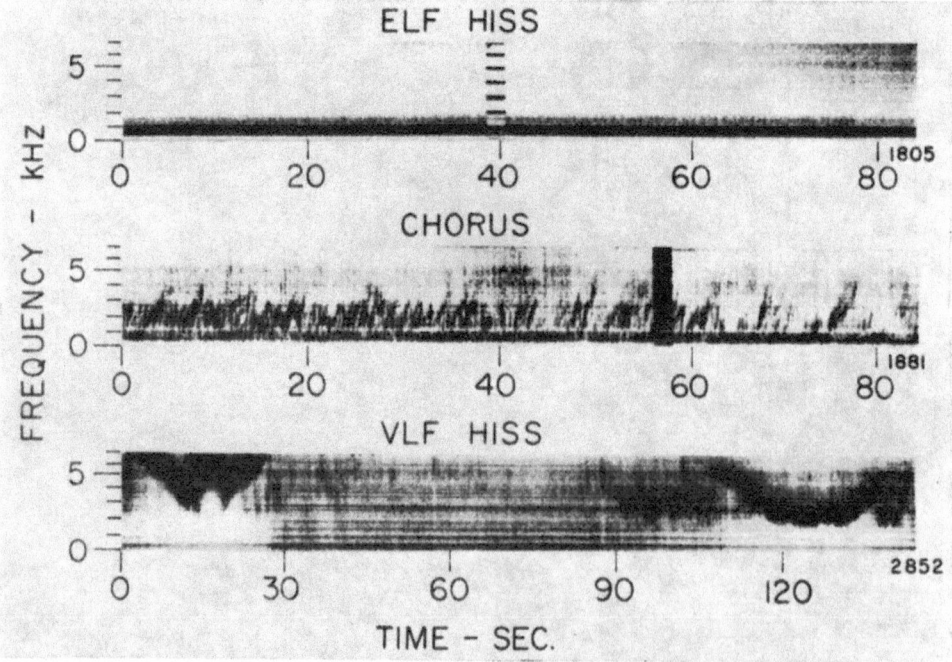

Fig. 97.   Examples of ELF hiss, chorus and VLF hiss recorded by the Injun 3 satellite (Taylor, W. W. L. and D. A. Gurnett: Dept. of Phys. and Astronom., University of Iowa, Rep. 68-6, January 1968).

prior to 1965 is summarized by HELLIWELL (1965); various types of VLF and ELF emissions are classified and illustrated in his Table 7-1.

In this chapter, we shall be concerned mainly with two types of VLF and ELF emissions, *hiss* and *chorus*. Their definitions are (HELLIWELL, 1965):

Hiss: An emission whose spectrum resembles that of band-limited thermal or fluctuation noise. It can be identified aurally by a hissing sound.

Chorus: A sequence of closely spaced, discrete events, often overlapping in time; some forms of chorus have been compared to the sound of a flock of birds. The most common form consists of a multitude of rising tones in the range 1 to 5 kc/sec, with rates of change of frequency with time averaging about 3 kc/sec.

Figure 97 shows examples of ELF hiss, chorus and VLF hiss observed by the Injun 3 satellite (TAYLOR and GURNETT, 1968).

The propagation of electromagnetic waves of frequency of an order of a few kc/sec in the magnetosphere has been studied by a number of workers. Figure 98 shows an example of ray paths of a 5 kc/sec wave generated in the equatorial plane at $r_e = 3a$ (SHAWHAN, 1967) constructed by assuming a diffusive equilibrium distribution of electrons and H^+, He^+ and O^+ ions. Their distribution specifies the distribution of the refractive index along the path of the waves; in this particular model the refractive

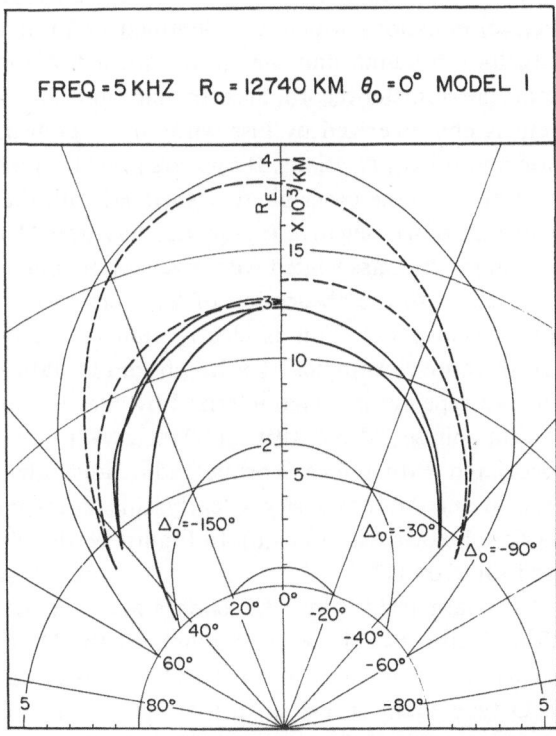

Fig. 98.   Ray paths of a 5 kc/sec wave generated at $r_e = 3a$ for different initial values of the wave normal directions with respect to the radius vector ($\varDelta_0$) (Shawhan, 1967).

index is a slowly varying function of position. The calculation was made for three waves whose initial wave normal directions (Δ_0) with respect to the radius vector are $-30°$, $-90°$ and $-150°$; these can penetrate to an altitude of 300 km. The other waves are refracted well above the ionosphere.

If the distribution of electrons and ions are not nearly as homogeneous as is assumed in the above calculation, the ray path of the waves can be drastically different from those illustrated in Figure 98. SMITH *et al.* (1960) showed how the waves can be trapped in field-aligned columns of enhanced ionization. Further, many other processes can occur along the path; for example, non-thermal particles along the path interact with some of the VLF waves by the cyclotron-resonance mechanism (cf. LIEMOHN, 1967). Unfortunately, however, the mechanisms responsible for such VLF emissions are not well understood.

7.2. Typical Daily Variation

MOROZUMI (1965) classified the VLF emissions into N and D types; the former occurs in the night hours and the latter in the daylight hours. The N type is further sub-classified into three, N1, N2, and N3. Figure 99 shows a typical example of the N type event, together with other simultaneous phenomena. The top record shows the VLF emission at 1 kc/sec with a band width of 100 c/sec. MOROZUMI (1965) identified this type of emission as chorus emission. The second and third records show VLF broad band (1–25 kc/sec) emissions which are identified as hiss; the second record with the North-South loop antenna and the third with a horizontal antenna. The fourth record represents the riometer record, the fifth, micropulsations X (0.02–5 c/sec).

The N1 type event is characterized by hiss without a significant chorus, cosmic radio noise absorption and micropulsations. It appears most frequently in the evening hours. The N2 type event is, on the other hand, associated with the auroral break-up (the poleward explosive motion), negative (magnetic) bay and N type cosmic radio noise absorption. It is, in general, associated with a very short burst of hiss; this short life of hiss is likely to be due to the absorption of VLF emissions by the increase in ionization of the lower ionosphere. This is clearly seen in the cosmic radio noise absorption record. In the above example, VLF hiss began at 0519 UT and lasted for only 1 min or so. The N3 type event is characterized by chorus emissions and occurs during the post-break-up phase. MOROZUMI and HELLIWELL (1966) noted that the N3 type event is associated with intense cosmic radio noise absorption without a corresponding increase in auroral luminosity. Clearly, this corresponds to the M type event of the ionospheric substorm (Section 4.6). In Figure 99, the N3 type event began at 0530 UT and lasted till 0700 UT.

The D type event is characterized by VLF chorus and is associated with cosmic radio noise absorption. Figure 100 shows an example of the D type event in which chorus (bottom record) and cosmic noise absorption (R) are well correlated. It is then quite likely that the D type event is related to the late M type absorption; indeed, MOROZUMI (1965) noted that the N3 type event and the D type event have many characteristics in common.

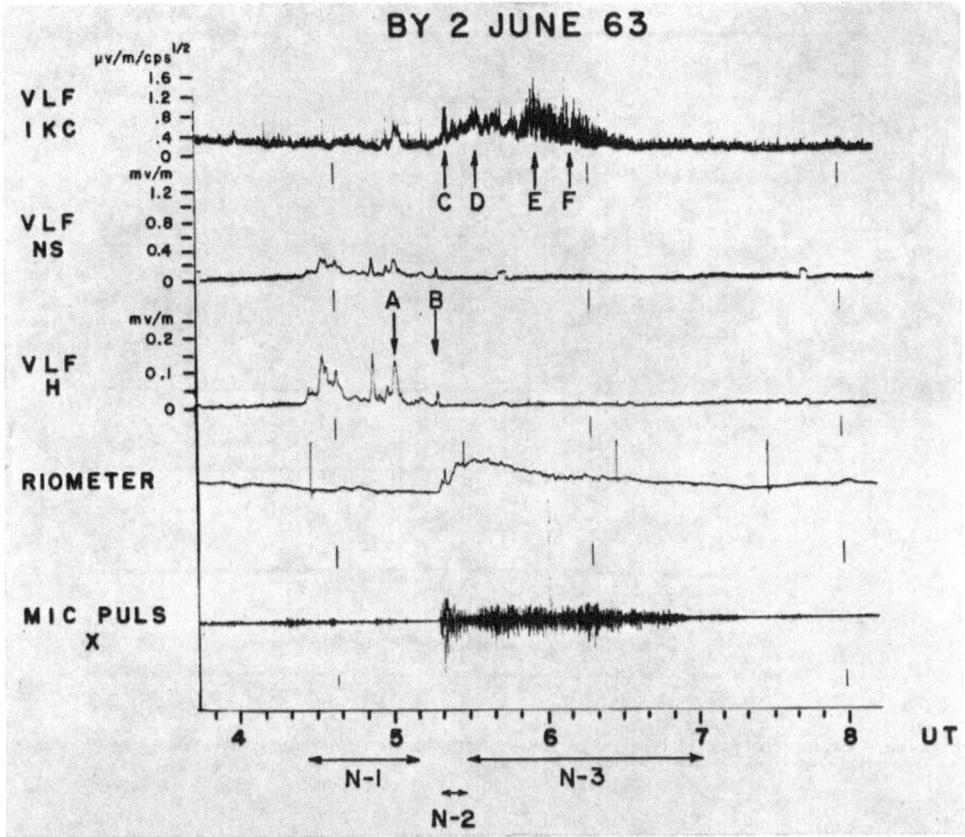

Fig. 99.   Typical example of the N type event (Morozumi, H. M.: *Rep. Ionosphere Space Res. Japan*, **19**, 286, 1965).

7.3. Statistical Daily Variation Pattern

Based on data taken from the South Pole, Byrd and Eights stations in the Antarctic, Morozumi (1967) constructed the pattern of the daily variation of the characteristics of VLF emission; Figure 101a. The N1 event occurs mostly in the evening hours and occurs appreciably higher in dp latitudes than the N3 or D type events. The N2 type event occurs during a brief interval when the N1 and N3 type events overlap. The N3 event occurs in the early morning hours of the auroral zone, and the D type event in the late morning and midday hours. Jorgensen (1966) also made an extensive study of wide-band hiss in the frequency range 4–9 kc/s at 13 stations and showed that hiss is most often observed at about dp lat 70° shortly before magnetic midnight; Figure 101b. He noted that hiss appears to be a rare phenomenon at very high latitudes.

As already mentioned in Section 7.2, low frequency electromagnetic waves such as hiss suffer considerable attenuation when they propagate through the ionosphere. For this reason, it is interesting to compare Morozumi's or Jorgensen's daily pattern

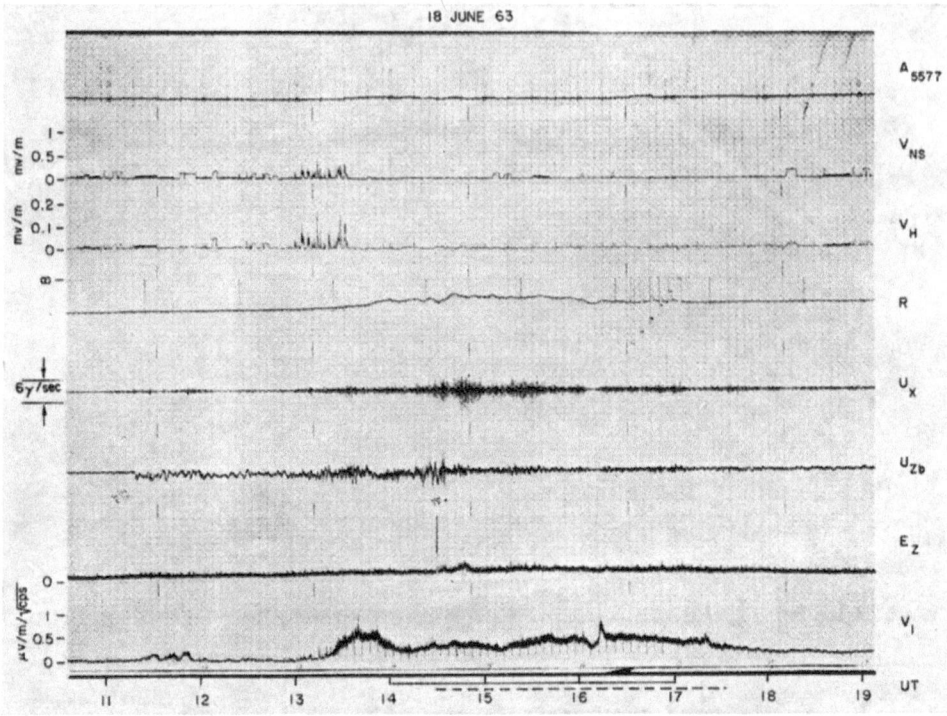

Fig. 100.   Example of the D type event (Morozumi, H. M.: *Rep. Ionosphere Res. Japan*, **19**, 286, 1965).

with those constructed by GURNETT (1966, 1967) and TAYLOR and GURNETT (1968) who made a detailed study of the VLF and ELF records received by the Injun 3 satellite.

Figure 102 shows the distribution of a flat noise spectrum extending from a lower frequency limit of about 2 kc/sec to several tens of kilocycles (GURNETT, 1966). It is interesting to note, first of all, that this particular type of hiss appears predominantly in the afternoon and evening hours, but infrequently in the morning hours. Secondly, the distribution lies approximately along the poleward boundary of the auroral oval. There is no doubt that Figure 102 represents a refined version of the distribution of the N1 type in Figure 100.

Another type of broadband hiss, ELF hiss, and chorus occurs predominantly in the morning hours in the latitude range 50° and 70° (Figure 103). The distribution is similar to that of the D type in Figure 100. JORGENSEN (1968) also made a similar study on the basis of the data obtained by the OGO-2 satellite. His pattern agrees with Figures 101a, b, 102, and 103. McEWEN and BARRINGTON (1967) noted that auroral hiss has a very sharp lower-frequency cutoff in the records obtained by the Alouette-1 satellite. They suggested that the sharp cutoff frequency is the lower hybrid resonance (LHR) frequency of the ambient plasma.

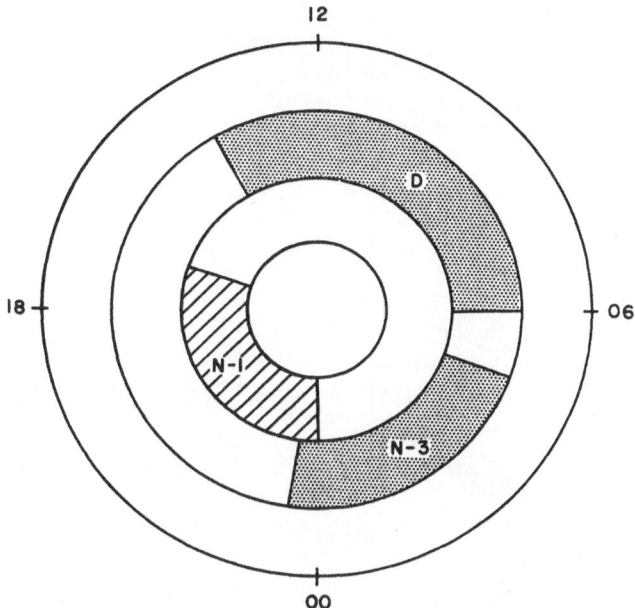

Fig. 101a. Average daily variation of characteristics of VLF emissions (Morozumi, H. M.: Proc. 'Symposium on Pacific Antarctic Sciences', JARE Scientific Reports, Special Issue No. 1, 53-64, 1967).

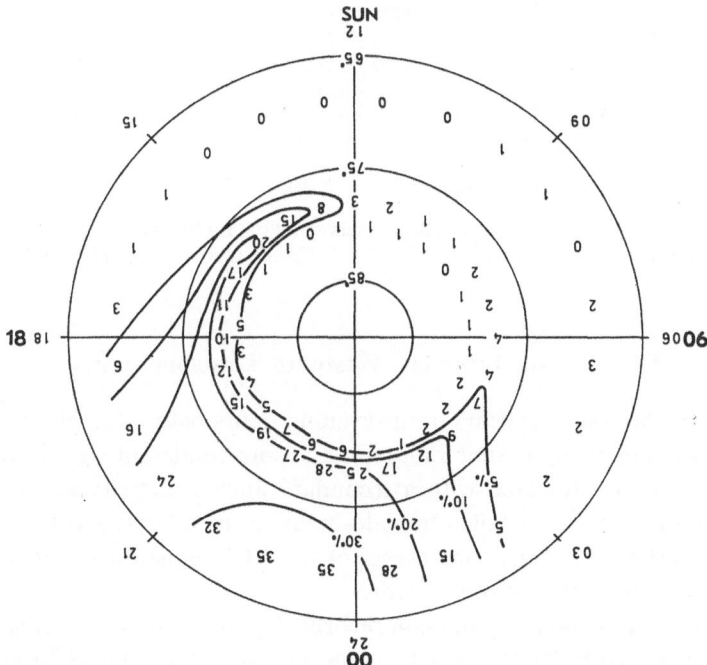

Fig. 101b. Contour map of the 8 kc/sec hiss on moderately disturbed days in 1964 in the Mayaud latitude ($\sim$ dipole latitude) and time coordinates (Jorgensen, T. S.: J. Geophys. Res. 71, 1367, 1966).

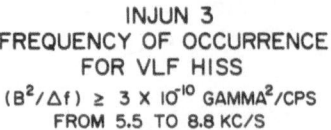

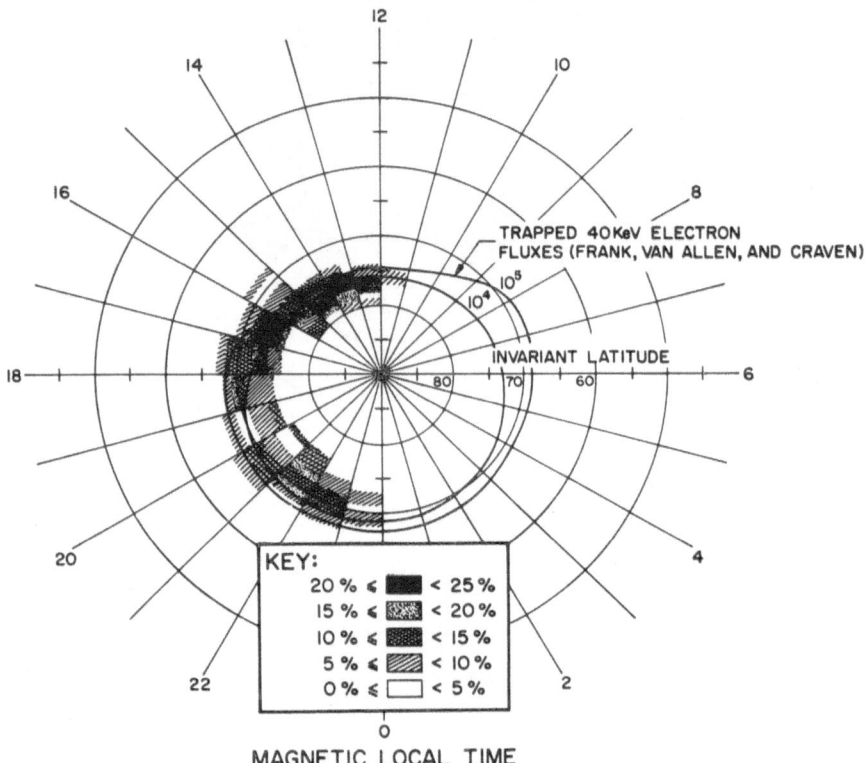

Fig. 102. Occurrence frequency of auroral hiss observed by the Injun 3 satellite in the dipole latitude and time coordinates (Gurnett D. A.: *J. Geophys. Res.* **71**, 5599, 1966).

7.4. N1 Type Event and Westward Traveling Surges

The fact that the N1 event appears in the evening hours before the N2 type emission and that the accompanying cosmic radio noise absorption is not significant, suggests that the N1 event is associated with a secondary auroral activity generated by the auroral bulge, such as the so-called 'pseudo-breakup' or a westward traveling surge. Therefore, the N1 type event is one aspect of the VLF emission substorm, which is most frequently seen in the evening sector.

An inspection of many VLF emission records, together with corresponding all-sky camera films taken from Byrd, suggests, however, that the relative location of the station with respect to *active* auroras plays an essential role in determining the types of events and subsequent development of the VLF emissions.

If the station is far from active auroras, hiss seems to be the only VLF emission received at the station. As we learned in Section 2.3, westward traveling surges are most often seen near the poleward horizon of a typical auroral zone station in the evening hours. Therefore, at an auroral zone station, hiss is the most common VLF emission; this is the N1 type event. Figure 104 shows an example obtained by HARANG and LARSEN (1965) in which hiss was received when positive bays were recorded at

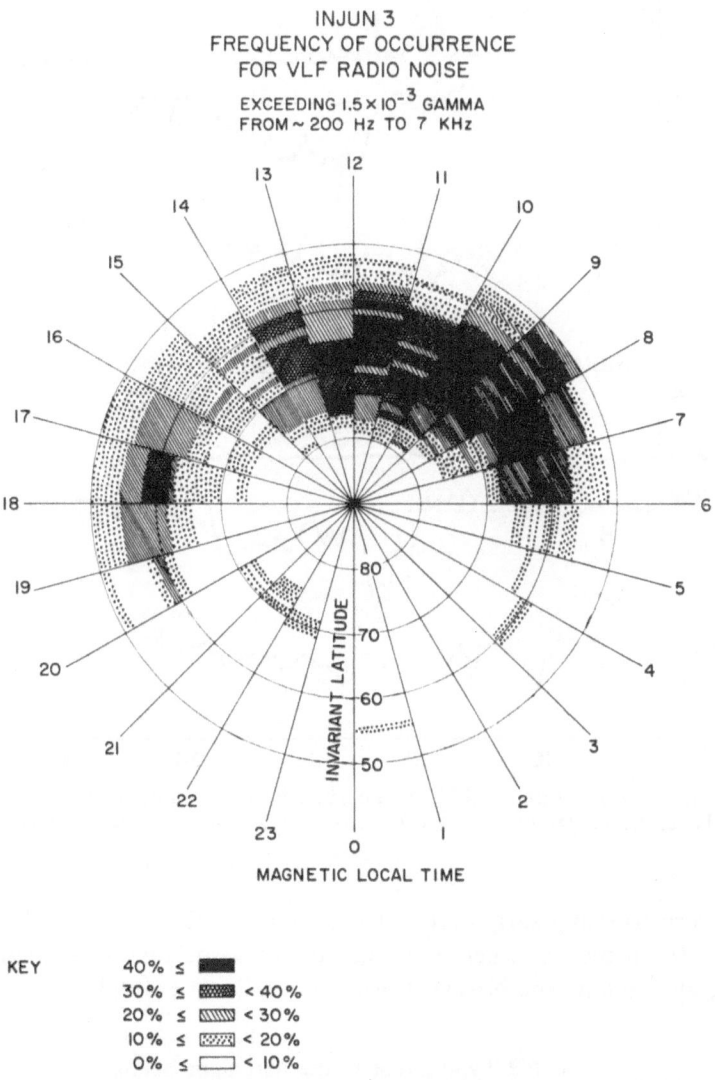

Fig. 103. Occurrence frequency of ELF hiss and chorus observed by the Injun 3 satellite in the dipole latitude and time coordinates (Taylor, W. W. L. and D. A. Gurnett: Dept. of Phys. and Astronom., University of Iowa, Rep. 68-6, January 1968).

Tromsø. Figure 104 is supplemented by the corresponding magnetic record from
Dixon Island which was located in the midnight sector when hiss was recorded at
Tromsø. It is quite clear that hiss (thus the N1 type event) was associated with the
polar magnetic substorm in the midnight sector and thus that hiss is one of the aspects
of the VLF emission substorm in the evening sector.

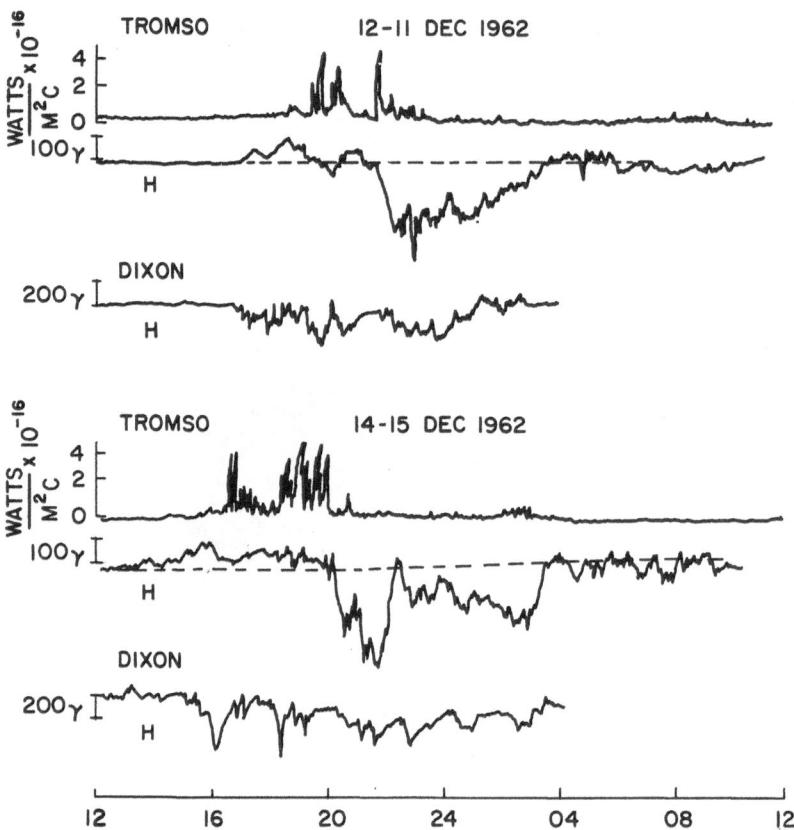

Fig. 104. Simultaneous occurrence of hiss and positive bays at Tromsø, together with the magnetic
record from Dixon Island. (Harang, L. and R. Larsen: *J. Atmospheric Terrest. Phys.* **27**, 481, 1965).

If westward traveling surges are not very intense, hiss may be observed directly
under them; this is the case when the surges are not associated with an intense cosmic
radio noise absorption. The N1 event in Figure 99 is an example of this type.

7.5. N2 Type Event in the Midnight Sector

The simultaneity of the onset of VLF hiss, an increase of the λ 5577 intensity, cosmic
radio noise absorption and ULF emissions (micropulsations) was studied by MORO-

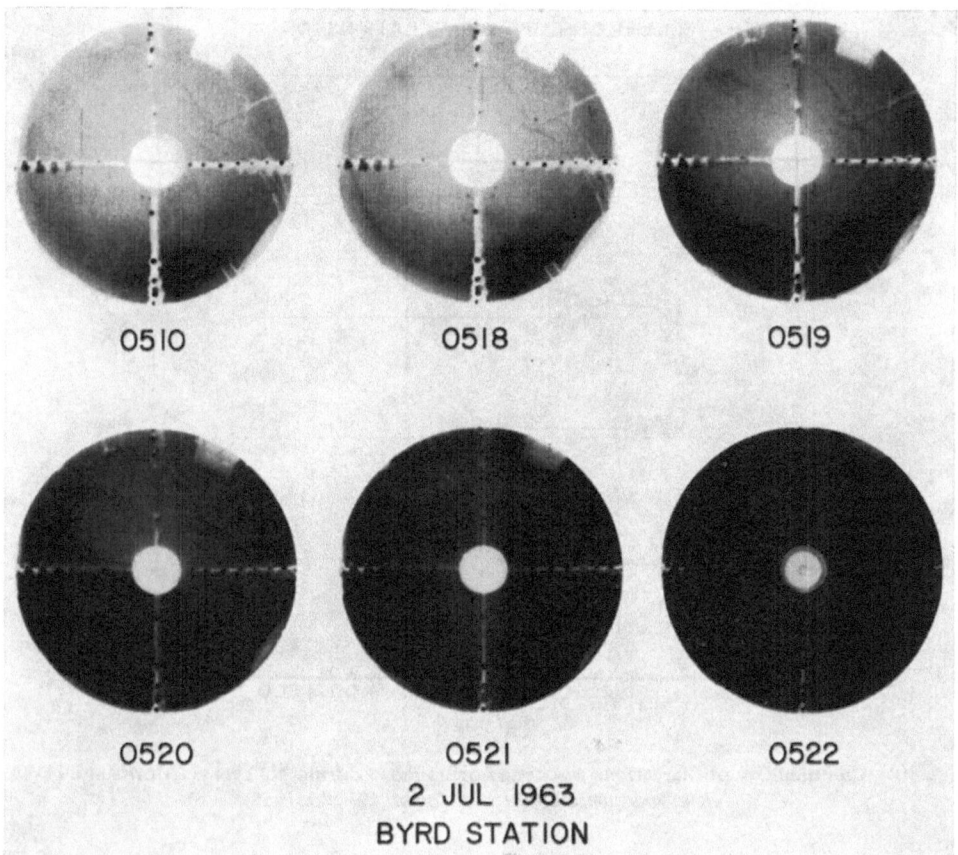

Fig. 105. Poleward explosive motion of the aurora during the N2 event in Figure 99.

ZUMI (1965) by using twenty-four examples obtained at Byrd Station. The increase of the λ 5577 intensity may be due either to a sudden increase of the brightness of auroras or to the explosive motion of auroras. Figure 105 (in negative) shows all-sky photographs taken from Byrd Station during the N2 event in Figure 99. An arc, which is seen near the equatorial horizon at 0510 UT, suddenly became active at 0518 UT and moved rapidly toward the station (0518–0521 UT). It can be seen that hiss was recorded during a very brief period between 0519 and 0520 UT.

Morozumi's statistical results on the simultaneity of the various associated phenomena during N2 events are summarized in Figure 106. Both VLF and ULF emissions appear to precede the onset of cosmic radio noise absorption and the λ 5577 intensity by about 1 to 2 min. But by the time both the λ 5577 intensity and cosmic radio noise absorption peak, hiss almost fades. In most cases, this is so when active auroras are rapidly approaching the station. Hiss can be received until active auroras reach the station. However, the arrival of active auroras is associated with heavy ionization in the lower ionosphere which absorb hiss.

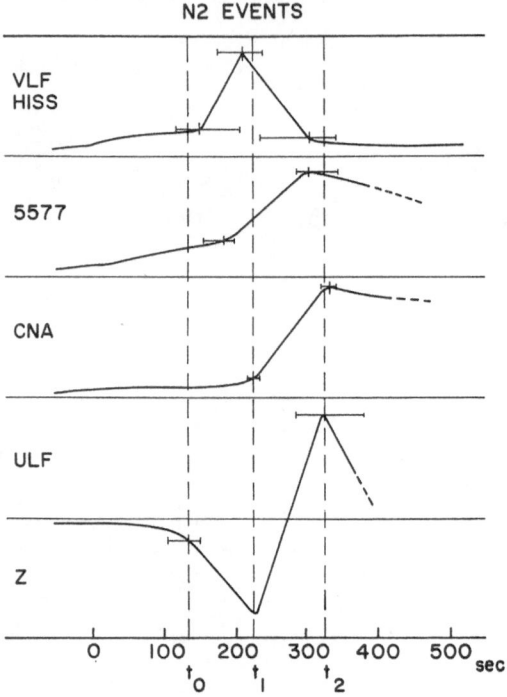

Fig. 106. Simultaneity of the various associated phenomena during N2 events (Morozumi, H. M.: *Rep. Ionosphere Space Res. Japan*, **19**, 286, 1965).

This point has been studied further by HARANG and LARSEN (1965) who showed that there is a critical value of cosmic noise absorption, above which hiss emission cannot appear (thus there is a negative correlation between the absorption and hiss) and below which both exist (thus a positive correlation between them).

7.6. N3 and D Type Events

The N3 type event tends to occur immediately after the N2 event, that is, when a station is within the auroral bulge or under active auroras (Figure 99 and 105). MOROZUMI (1965) noted that the N3 and D events have many characteristics in common, so that apparently the auroral bulge and late morning auroras provide essentially the same conditions for the generation and subsequent propagation of VLF emissions.

The D type events occur most frequently in the late morning hours and are well correlated with the M type absorption (Section 4.6), so that they are undoubtedly one of the aspects of the VLF emission substorm, which is most commonly seen in the morning hours. OLIVEN and GURNETT (1968) showed that microbursts of electrons are always accompanied by a group of VLF emissions.

Recently, HARANG *et al.* (1967) reported that during periods of great magnetic activity VLF emissions appear simultaneously at Lycksele (dp lat 62.50), Oslo (60°.0), Rude Skov (dp lat 55.8) and Chambon-la-Forêt (dp lat 50.50). Such low latitude emissions occur, however, only during daytime. Figure 107 shows the occurrence time for both such low latitude emissions and polar VLF (hiss). The time of the occurrence for the low latitude emissions is quite similar to that of the D event (Figure 101a) or Figure 103.

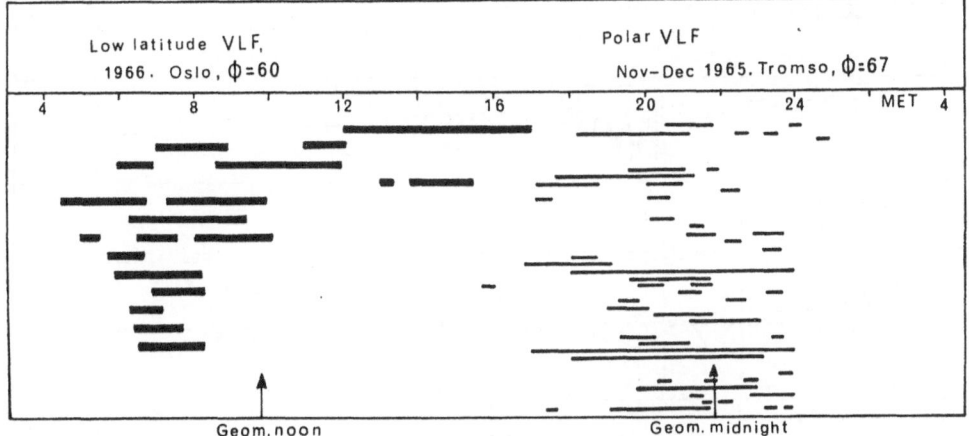

Fig. 107.   Occurrence time of the low latitude emissions and hiss (Harang, L., R. Larsen, and J. Skogtvedt: *Physica Norvegica* **2**, 271, 1967).

HARANG *et al.* (1967) noted by examining simultaneous Tromsø magnetic records that such emissions tend to appear during the last phase of storms. However, we should note that the distance between the auroral oval and Tromsø is increasing in the morning hours, so that their 'last phase' appears to be only an apparent one. Indeed, an examination of simultaneous records from Great Whale River, Canada (which was located in the midnight sector), shows that intense substorms were in progress when the low latitude emissions were observed. In Figure 108 the *H* component records from Great Whale River are added for comparison. Their statement that the emissions are associated with strong cosmic radio noise absorption at Tromsø also supports such a view. It appears that this interesting type of emission is closely related to the D type event. The fact that this particular emission appears only during intense storms may be due to the expansion of the auroral oval toward the equator, so that mid-latitude stations become temporarily, subauroral zone stations; Section 1.6.

7.7. Development of the VLF Emission Substorm

During an early phase of the VLF emission substorm (corresponding to the explosive phase of the auroral substorm), hiss is observed both a little poleward and equator-

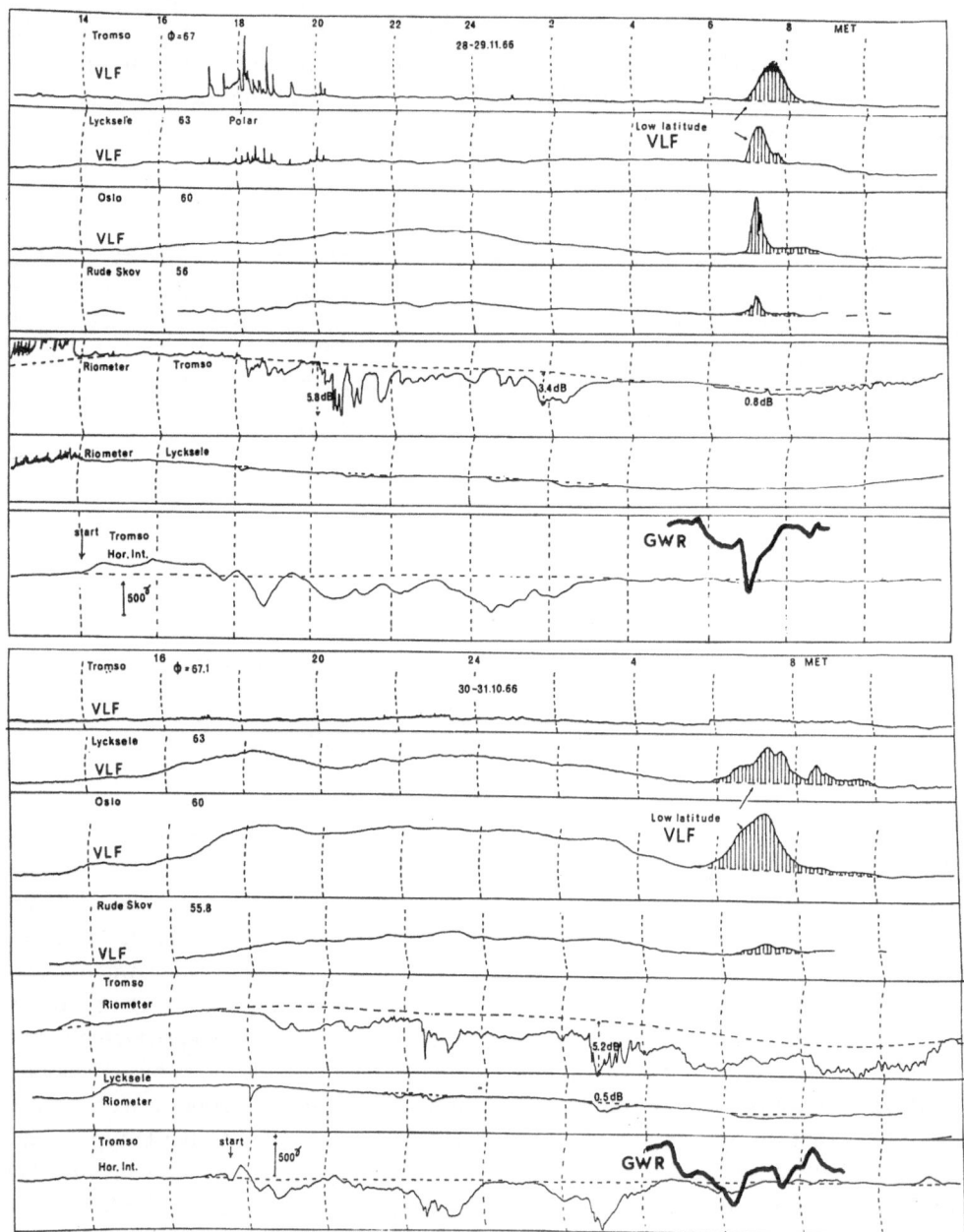

Fig. 108. Examples of low-latitude emissions (8 kc/sec) along the North-South chain of stations in Europe; the magnetic records from Great Whale River are added to the Tromsø magnetic records. (Harang, L., R. Larsen, and J. Skogtvedt: *Phys. Norvegica* **2,** 271, 1967).

ward of the oval which is located along the path of westward traveling surges; the two regions are indicated by NI in Figure 109. It is assumed that significant ionization occurred in the lower ionosphere along the oval, so that hiss cannot be recorded at stations directly under the oval. However, if the ionization is weak or westward traveling surges are weak, hiss may be observed in the area which includes the two patchy regions.

Hiss is also observed in a narrow region a little ahead of the expanding auroral bulge; this narrow region is indicated by N2 in Figure 109. The equatorward boundary of N2 may be considered to be the front of the expanding bulge.

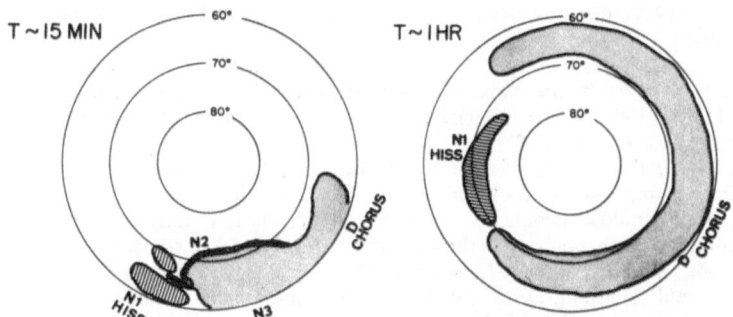

Fig. 109.   Development of the VLF emission substorm.

Chorus is observed within the bulge and along the auroral zone in the morning sector (N3 and D); we note that Morozumi's N3 and D type events are essentially the same.

During a later phase of the VLF emission substorm (corresponding to the recovery phase of the auroral substorm), hiss may be observed along the path of the surge; by this time, the surge may be greatly weakened, so that the lower ionosphere does not absorb hiss. The area in which chorus is observable is considerably extended to as far as the noon sector. This expansion is closely associated with that of the M type absorption and X-ray bursts (Sections 4.8 and 5.9).

References

GENERAL

GENDRIN, R.: 1967, 'Progrès récents dans l'étude des ondes T.B.F. et E.B.F.', *Space Sci. Rev.* **7**, 314–395.
HELLIWELL, R. A.: 1965, *Whistlers and related ionospheric phenomena*, Stanford Univ. Press, Stanford, Calif.

REFERRED TO IN TEXT

ELLIS, G. R.: 1957, 'Low-frequency radio emission from aurorae', *J. Atmospheric Terrest. Phys.* **10**, 302–306.

ELLIS, G. R.: 1959, 'Low-frequency electromagnetic radiation associated with magnetic disturbances', *Planetary Space Sci.* **1**, 253–258.

GURNETT, D. A.: 1966, 'A satellite study of VLF hiss', *J. Geophys. Res.* **71**, 5599–5615.

GURNETT, D. A.: 1967, Satellite observations of VLF emissions and their association with energetic charged particles, Dept. of Phys. and Astronomy, Univ. of Iowa Rep. 67–53, Sept.

HARANG, L. and LARSEN, R.: 1965, 'Radio wave emissions in the VLF-band observed near the auroral zone. I. Occurrence of emissions during disturbances', *J. Atmospheric Terrest. Phys.* **27**, 481–497.

HARANG, L., LARSEN, R., and SKOGTVEDT, J.: 1967, 'VLF-emissions in the 8-kc/s band observed at stations close to the auroral zone and at stations on lower latitudes', *Phys. Norvegica* **2**, 271–292.

HELLIWELL, R. A.: 1965, *Whistlers and related ionospheric phenomena*, Stanford Univ. Press, Stanford, Calif.

JORGENSEN, T. S.: 1966, 'Morphology of VLF hiss zones and their correlation with particle precipitation events', *J. Geophys. Res.* **71**, 1367–1375.

JORGENSEN, T. S.: 1968, 'Interpretation of auroral hiss measured on OGO 2 and at Byrd Station in terms of incoherent Cerenkov radiation', *J. Geophys. Res.* **73**, 1055–1069.

LIEMOHN, H. B.: 1967, 'Cyclotron-resonance amplification of VLF and ULF whistlers', *J. Geophys. Res.* **72**, 39–55.

McEWEN, D. J. and BARRINGTON, R. E.: 1967, 'Some characteristics of the lower hybrid resonance noise bands observed by the Alouette I satellite', *Canadian J. Phys.* **45**, 13–19.

MOROZUMI, H. M.: 1965, 'Diurnal variation of aurora zone geophysical disturbances', *Rep. Ionos. Space Res. Japan* **19**, 286–298.

MOROZUMI, H. M.: 1967, 'Auroral-zone geophysical events and their relationship to the magnetosphere', Proc. Symp. on Pacific-Antarctic Sci., JARE Sci. Rep. Sp. Issue No. 1, 53–64.

MOROZUMI, H. M. and HELLIWELL, R. A.: 1966, 'A correlation study of the diurnal variation of upper atmospheric phenomena in the southern auroral zone', Radioscience Lab., Stanford Electronics Labs., Stanford Univ., SU-SEL-66-124.

OLIVEN, M. N. and GURNETT, D. A.: 1968, *Microburst phenomena. 3. An association between microbursts and VLF chorus*, Vol. 73, 2355–2362.

SHAWHAN, S. D.: 1967, Behavior of VLF ray paths in the ionosphere, University of Iowa Rept. Dept. of Physics.

SMITH, R. L., HELLIWELL, R. A., and YABROFF, I. W.: 1960, 'A theory of trapping of whistlers in field-aligned columns of enhanced ionization', *J. Geophys. Res.* **65**, 815–823.

TAYLOR, W. W. L. and GURNETT, D. A.: 1968, 'The morphology of VLF emissions observed with the Injun 3 satellite', Dept. of Phys. and Astron., Univ. of Iowa, Rep. 68–6.

WATTS, J. M.: 1957, 'An observation of audio-frequency electromagnetic noise during a period of solar disturbance', *J. Geophys. Res.* **62**, 199–206.

CHAPTER 8

MICROPULSATION SUBSTORM

8.1. Introduction

It has long been known that in middle latitudes one particular type of micropulsation occurs during an early phase of the 'bay' disturbances. This type of pulsation was once called the Pt type pulsation, but is now classified as Pi-2 pulsations. As we learned in Chapter 3, the low latitude bay disturbance is a part of the polar magnetic substorm, so that Pi-2 pulsations are undoubtedly another manifestation of the magnetospheric substorm.

In higher latitudes, particularly along the auroral oval and zone, characteristics of micropulsations are much more complicated than those in lower latitudes.

The present classification scheme for micropulsations is as follows:

Type	Range of periods (sec)
Continuous Pulsations	
Pc-1	0.2– 5
Pc-2	5 – 10
Pc-3	10 – 45
Pc-4	45 –150
Pc-5	150 –600
Irregular Pulsations	
Pi-1	1 – 40
Pi-2	40 –150

Some of the pulsations classified in the above have common names; for examples, the so-called 'pearl' of 'PP' type is now classified as Pc-1 type in the above scheme. The so-called 'IPDP' (=intervals of pulsations diminishing periods) or the 'sweeper' type is a mixture of Pi-1 and Pc-1 types. The 'giant' pulsations belong to Pc-4 or Pc-5 and the 'sip' (=short irregular pulsations) to Pi-1.

Micropulsations are ultra-low-frequency (ULF) electromagnetic waves. When such waves propagate in an ionized gas (the magnetosphere in our case), they are considered to be hydromagnetic waves, since the frequency of micropulsations is less than the gyro-frequency of positive ions.

If such waves are generated in the magnetosphere and transmitted toward the earth, their propagation characteristics can be obtained by assuming the propagation of such waves occurs in a hot plasma (of temperature of order 10^4 K). However, they must propagate through the magnetospheric medium in which propagation charac-

teristics change considerably (cf. JACOBS and WATANABE, 1964). Furthermore, inter-
actions between such waves and energetic particles in the Van Allen belts may result
in the cyclotron·resonance amplification of the waves (LIEMOHN, 1967). Figure 110
shows the normalized refractive index versus the angle θ (= the wave normal direction
with respect to B) for various values of the normalized wave frequency λ (= wave
frequency/proton gyro-frequency); $n(0) = c/V_A$, where V_A denotes the Alfvén velocity

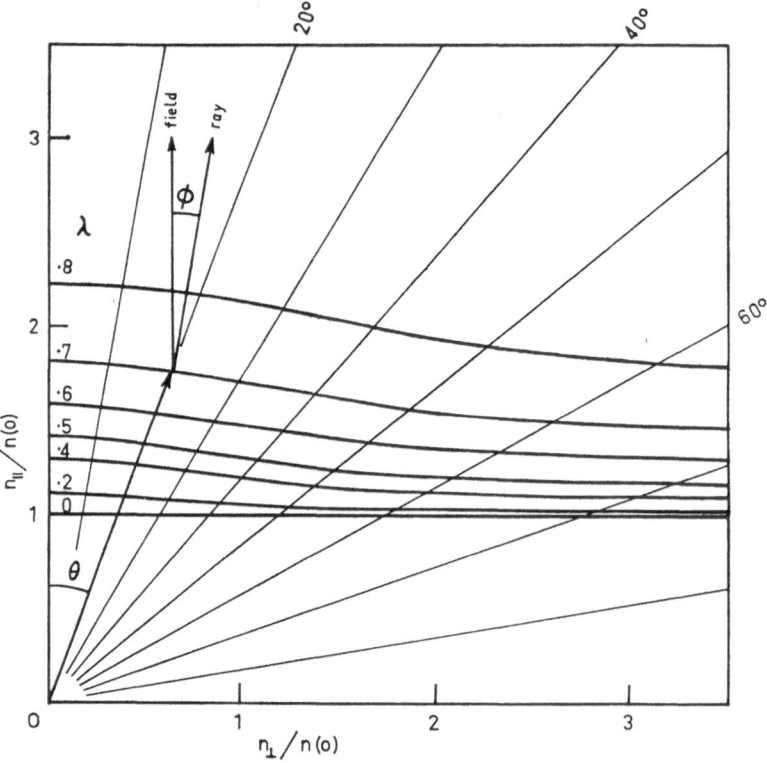

Fig. 110.   Polar diagram of normalized refractive index versus wave normal angle θ parametric in
normalized frequency λ. The ray direction (ϕ) is normal to these curves as shown (Dowden, R. L.:
Planetary Space Sci. **13**, 761, 1965).

(DOWDEN, 1965). The ray direction (ϕ) is given by the normal to the curves. It is seen
that for all values of θ and λ the ray direction is close to the magnetic field direction,
indicating a strong guiding of the waves by the magnetic field.

The propagation becomes more complicated in the lower magnetosphere, where
plasma is only partially ionized. In the ionosphere, which may be regarded as the base
of the magnetosphere, plasma is only very weakly ionized. Figure 111 shows how the
basic characteristics of transverse waves must be altered when the waves propagate
through the ionosphere (WATANABE, 1962); the abscissa gives the wave frequency,

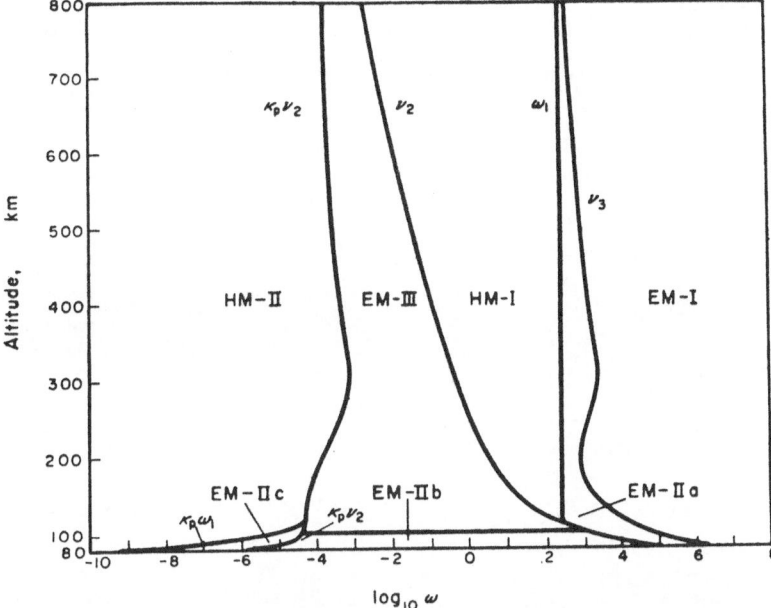

Fig. 111. Characteristics of electromagnetic waves of different frequencies in the ionosphere
(Watanabe, T.: *J. Atmospheric Terrest. Phys.* **24**, 117, 1962).

and the ordinate the altitude above the earth's surface. In the region (EM-I) bounded
by the curve $v_3 (= v_{ie} + v_{en} + (m_e/m_i) (v_{in}/2))$, the frequency of the waves ω is high
enough to regard the propagation to be magneto-ionic. In the region (EM-IIa) bounded
by v_3 and ω_i (= ion gyro-frequency), the waves behave as if they were electromagnetic
waves in an anisotropic and non-dispersive metallic media. In the region (HM-I)
bounded by ω_i and $v_2 (= (v_{in}/2) + (m_e/m_i) v_{en})$, the wave characteristics are hydromagnetic, without participation of neutral particles. In the region (EM-III), the wave

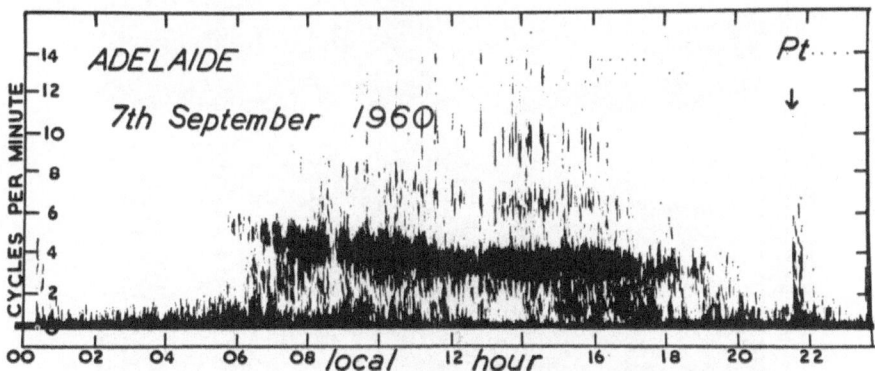

Fig. 112a. Daily sonagraph record of geomagnetic micropulsations from Adelaide, Australia
(Duncan, R. A.: *J. Geophys. Res.* **66**, 2087–2094, 1961).

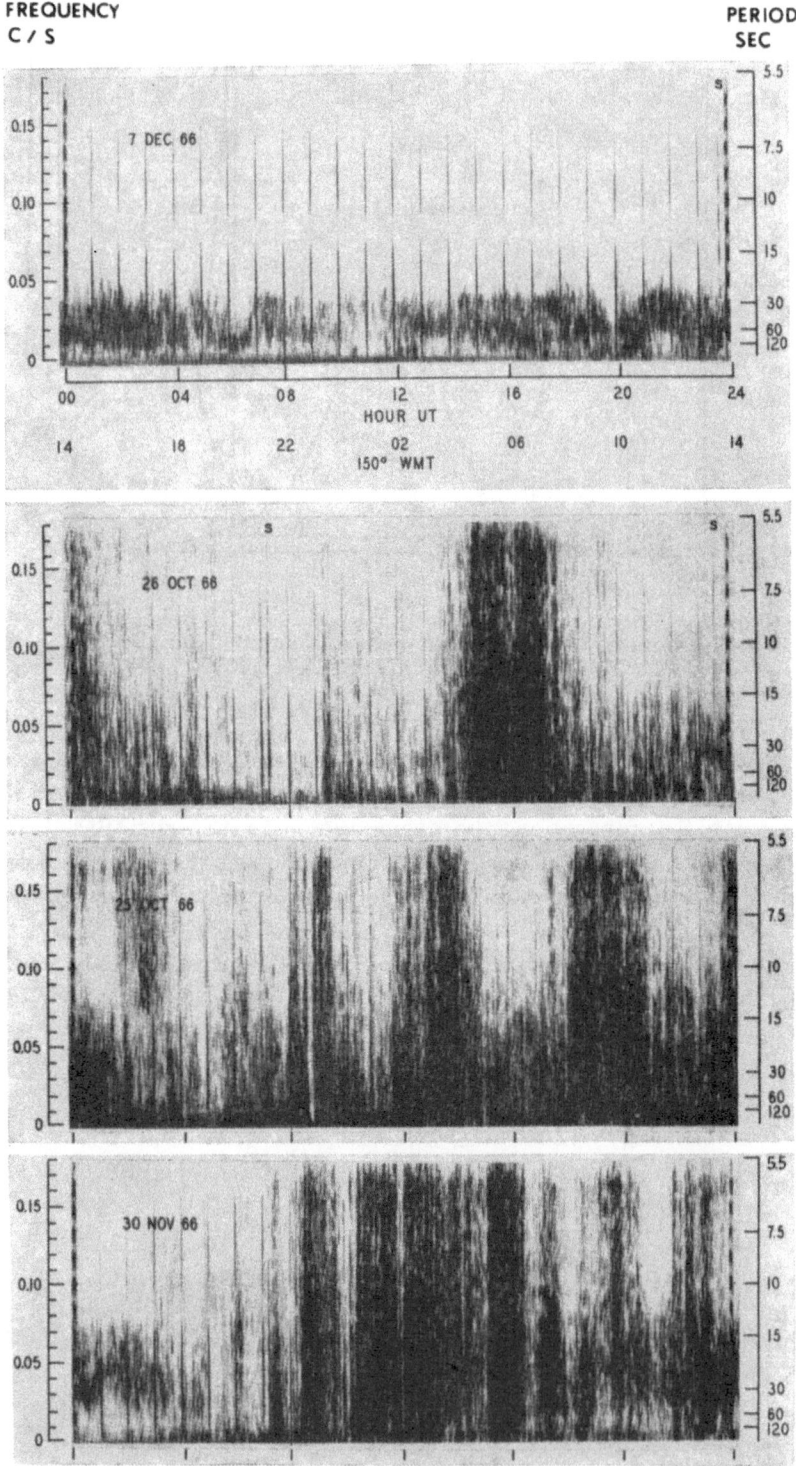

Fig. 112b. Daily sonagraph records of geomagnetic micropulsations from College, Alaska.

characteristics are similar to those for the region EM-II. In the region (HM-II), the wave characteristics are hydromagnetic; in this region, neutral particles participate in the oscillatory motions (DUNGEY, 1958, Chapter 9).

8.2. Typical Daily Variations

Figure 112a shows a sonagraph record of geomagnetic micropulsations from a typical middle latitude station, Adelaide ($L = 2.0$), Australia (DUNCAN, 1961). Two distinct features are obvious in the record. There are pulsations of frequency of about 3 cycles/ min (20 sec period) throughout the daylight hours; they are the Pc-3 type pulsations. Then, at about 2140 LT (1210 UT), there was a burst of micropulsations which lasted for about 30 min. This burst is the Pi-2 type; an intense bay was widely observed in the Pacific sector at that time. Figure 113 shows in detail how Pi-2 pulsations are superposed on the field of a polar magnetic substorm (ROSTOKER, 1967a).

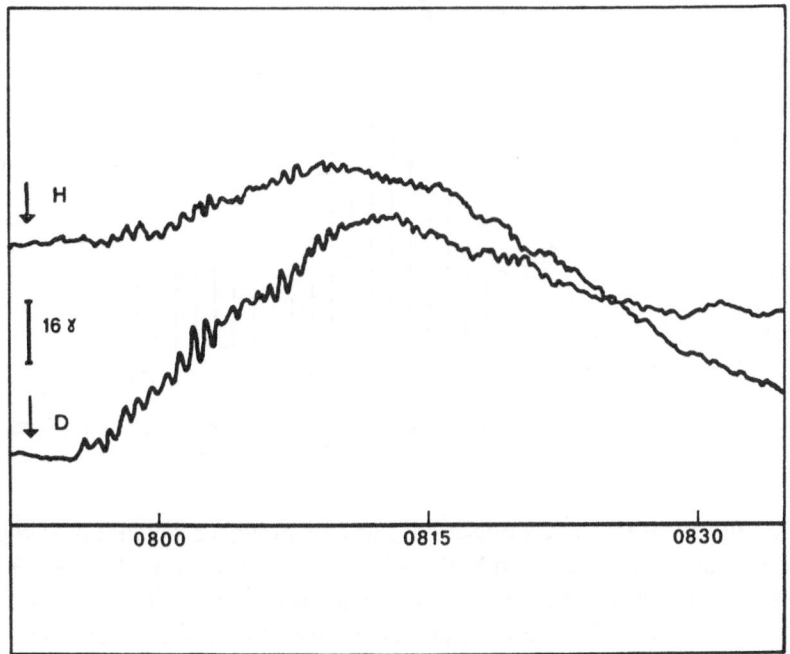

Fig. 113.   Example of Pi-2 micropulsations superposed on a bay disturbance (Rostoker, G.: *Canadian J. Phys.* **45**, 1319, 1967).

Figure 112b shows a sonagraph record of geomagnetic micropulsations from College, Alaska. The top record (December 7, 1966) is a typical quiet day record. There was almost a continuous pulsation activity throughout the day; the lower limit of the periods was about 20 sec. During more disturbed days, micropulsations appear as a burst (the second record) or a succession of bursts (the third and fourth records).

Each burst contains micropulsations with a wide frequency (or period) range. The typical life time of these bursts is of order 1 ∼ 3 hours.

8.3. Statistical Daily Variation Pattern

A. PI-2 PULSATIONS

The daily occurrence of Pi-2 pulsations has been studied by a number of workers. Figure 114 shows the results obtained by YANAGIHARA (1960); the occurrence frequency has a pronounced peak at about 22 LT.

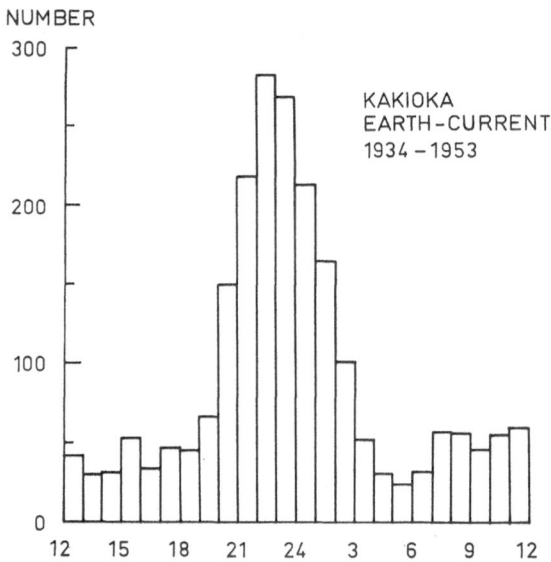

Fig. 114.   Occurrence frequency of Pi-2 micropulsations as a function of local time (Yanagihara, K.: *Mem. Kakioka Mag. Obs.* 9, No. 2, 15, 1960).

The amplitude of Pi-2 micropulsations has a pronounced maximum in the auroral zone and decreases steeply toward both higher and lower latitudes (JACOBS and SINNO, 1960). Figure 115 shows the latitudinal dependence of the amplitude; JACOBS and SINNO (1960) classified Pi-2 micropulsations into two types, one which is accompanied by a negative bay (upper diagram) and the other accompanied by a positive bay (lower diagram).

The period of Pi-2 pulsations depends on the planetary index K_p; for small K_p values (~ 1), the period is of order 120 sec, but it decreases to 60 sec or even less for K_p values greater than 3 (TROITSKAYA and GUL'ELMI, 1967; ROSTOKER, 1967b).

B. PULSATIONS IN HIGH LATITUDES

Characteristics of micropulsations in higher latitudes are much more complicated than those in lower latitudes where Pi-2 type pulsations are dominant. Figure 116

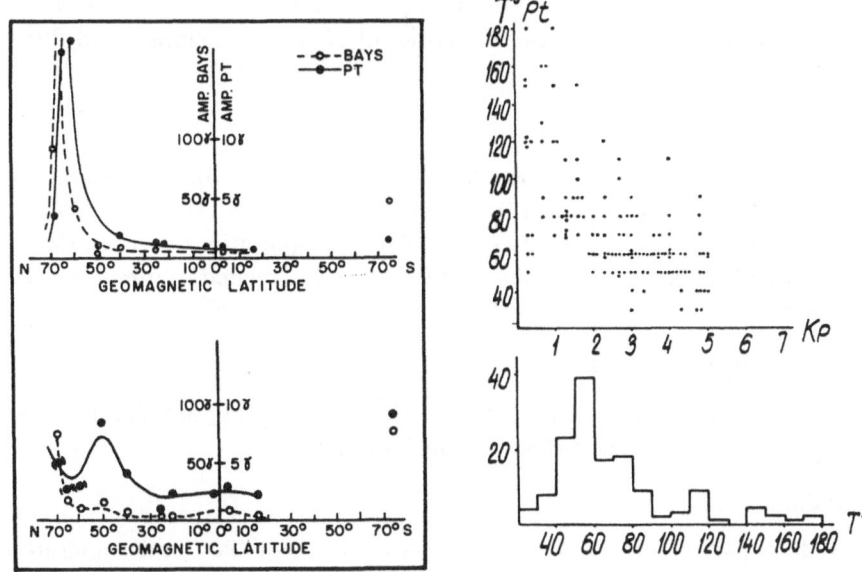

Fig. 115.   Left diagram: Amplitude of Pi-2 micropulsations and bays as a function of dipole latitude; the upper diagram for negative bays and the lower one for positive bays (Jacobs, J. A. and K. Sinno: *Geophys. J.* **3**, 1960). Right diagram: The period of Pi-2 micropulsations as a function of K_p and the occurrence frequency of Pi-2 micropulsations of different periods (Troitskaya, V. A. and A. V. Gul'elmi: *Space Sci. Rev.* **7**, 689, 1967).

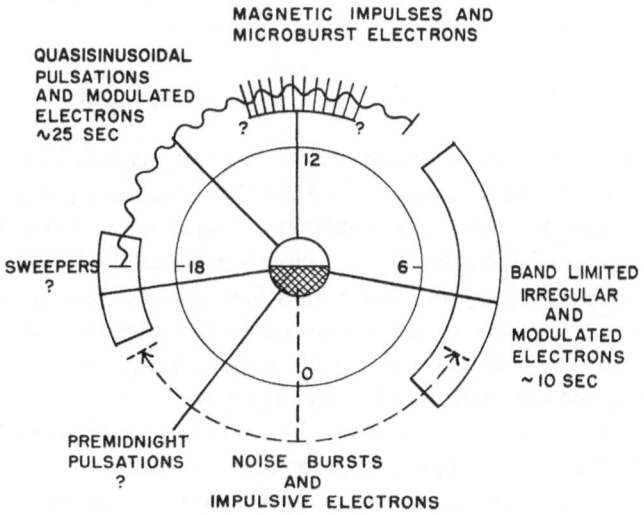

Fig. 116.   Daily variation of the characteristics of micropulsations and X-rays (McPherron, R. L., G. K. Parks, F. V. Coroniti, and S. H. Ward: *J. Geophys. Res.* **73**, 1697, 1968).

shows how characteristics of micropulsations in the auroral zone varies in the course of a day, together with the daily characteristics of X-ray precipitations (McPHERRON et al., 1968).

1. Local Time: 2000–0200

The occurrence of micropulsations are closely associated with the onset of the auroral substorm. CAMPBELL and MATSUSHITA (1962), YANAGIHARA (1963), VICTOR (1965), and WILHELM (1967) have shown that irregular micropulsations or noise bursts are characteristics of the micropulsations, and that they are related to impulsive electron precipitations.

2. Local Time: 0200–1000 LT

In these morning hours, Pi-1 of periods 5 to 10 sec occur; they are closely associated with the 5 to 10 sec modulated electron precipitation.

3. Local Time: 1000–1500 LT

In the midday part of the auroral zone, quasi-sinusoidal (Pc-3) micropulsations of 15 to 40 sec appear; the electron precipitation has a similar modulation.

4. Local Time: 1100–1300 LT

MILTON et al., (1967) have shown that magnetic impulses occur in this particular period; electron microbursts are associated with them (Sections 5.6 and 5.7).

5. Local Time: 1500–2200 LT

TROITSKAYA and MELNIKOVA (1959) and HEACOCK (1967a) reported the occurrence of I.P.D.P. type micropulsations (or the sweeper) during this period. The so-called 'pearl' pulsations have their peak occurrence at about 15 LT.

8.4. Micropulsations in the Auroral Bulge

Figure 117 shows simultaneous records of X-ray bursts, cosmic radio noise absorption micropulsations (5–30 sec periods), and the horizontal component magnetometer records from College and Sitka (CAMPBELL and MATSUSHITA, 1962). It is quite clear that the sudden onset of micropulsations is closely associated with the polar substorm which is manifested by X-ray bursts (Section 5.4), N type absorption (Section 4.4) and negative bays with sharp onsets (Section 2.2). A comparison of the onset times of micropulsations (ELF emissions), cosmic noise absorption, the λ 5577 intensity and other phenomena was made by MOROZUMI (1965) who noted that micropulsations precede the cosmic radio noise absorption by about 100 sec (Section 7.5); this is the case when active auroras are rapidly approaching the station.

There occur at least two types of micropulsations at the time of the onset of negative bays (or of the auroral substorm). At the very beginning of a negative bay, there appears a brief impulsive burst which lasts for about 15 min. Figure 118 shows

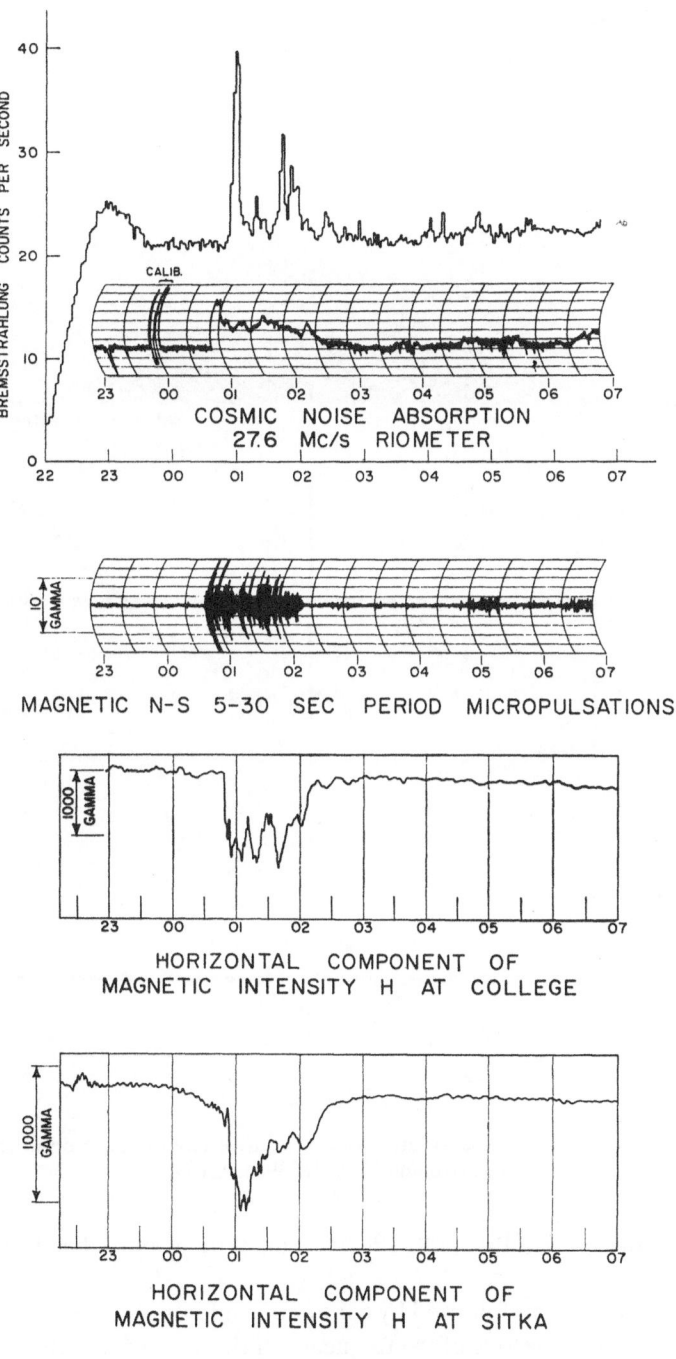

Fig. 117. Simultaneous occurrence of X-ray bursts; cosmic radio noise absorption, micropulsations and a negative bay (Campbell, W. H. and S. Matsushita: *J. Geophys. Res.* **67**, 555, 1962).

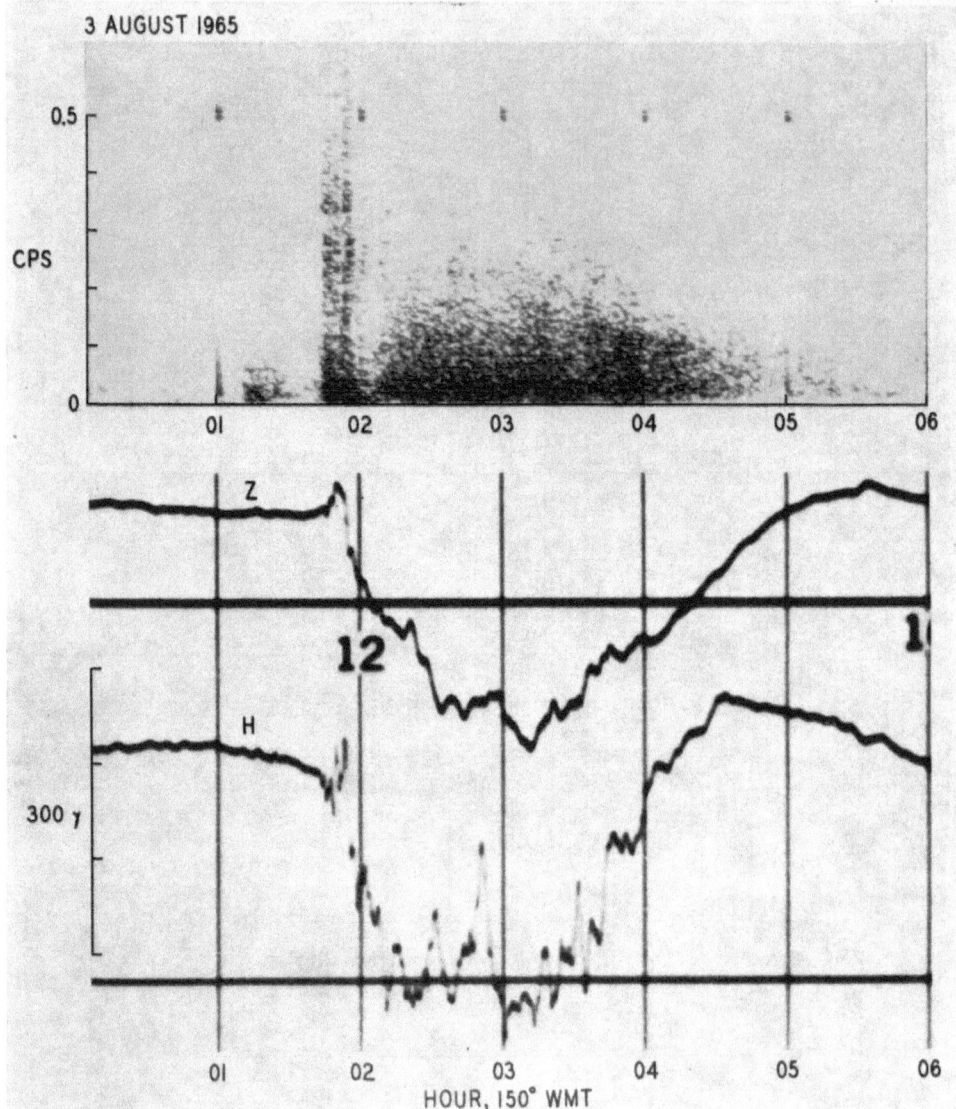

Fig. 118.   Bursts of Pi pulsations associated with a negative bay of sharp onset (Heacock, R. R.: *J. Geophys. Res.* **72**, 3905, 1967).

an example of the burst (HEACOCK, 1967b). This type of burst has been called S.I.P. by some workers.

   This short burst is then followed by continuous micropulsations. HEACOCK (1967b) noted that the upper envelope of the frequency of this type of pulsation closely follows the magnetic trace. This type of micropulsation has been called A.I.P. (auroral irregular pulsations) by GENDRIN and LACOURLY (1968). WILHELM (1967) showed that A.I.P. is associated with X-ray bursts.

8.5. Micropulsations and Positive Bays

HEACOCK (1966) noted that there occurs a particular type of micropulsation with rising mid-frequencies during positive bays; the mid-frequency rises from 0.1 to 0.5 c/s over one hour. We note that positive bays are the most common feature of the auroral zone of the fields of polar magnetic substorms in the evening hours of the auroral zone; they occur when a westward traveling surge passes near the poleward horizon (Section 2.3 and Figure 28). HEACOCK (1966) identified this particular type of micropulsation as I.P.D.P. discovered by TROITSKAYA (1961). It appears that Troitskaya's study was based on mid-latitude records, so that I.P.D.P. occurs only during an intense geomagnetic storm there. However, at an auroral zone station, it is a rather common phenomenon. Figure 119 shows an example of I.P.D.P., together with the corresponding College magnetogram. The magnetic changes associated with the I.P.D.P. suggest that College (with respect to the westward traveling surge) corresponds to Station B in Figure 28.

Figure 120 shows also an interesting example of I.P.D.P., together with a number

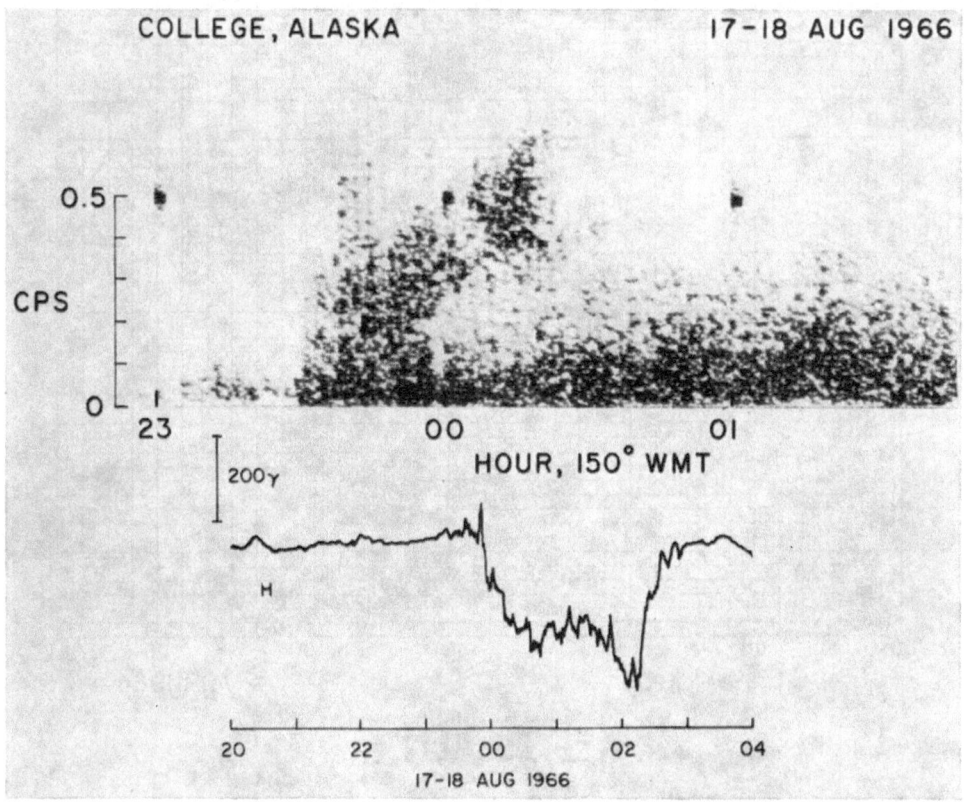

Fig. 119. Example of I.P.D.P. and the associated magnetic variations (Heacock, R. R.: *J. Geophys. Res.* **72**, 399, 1966).

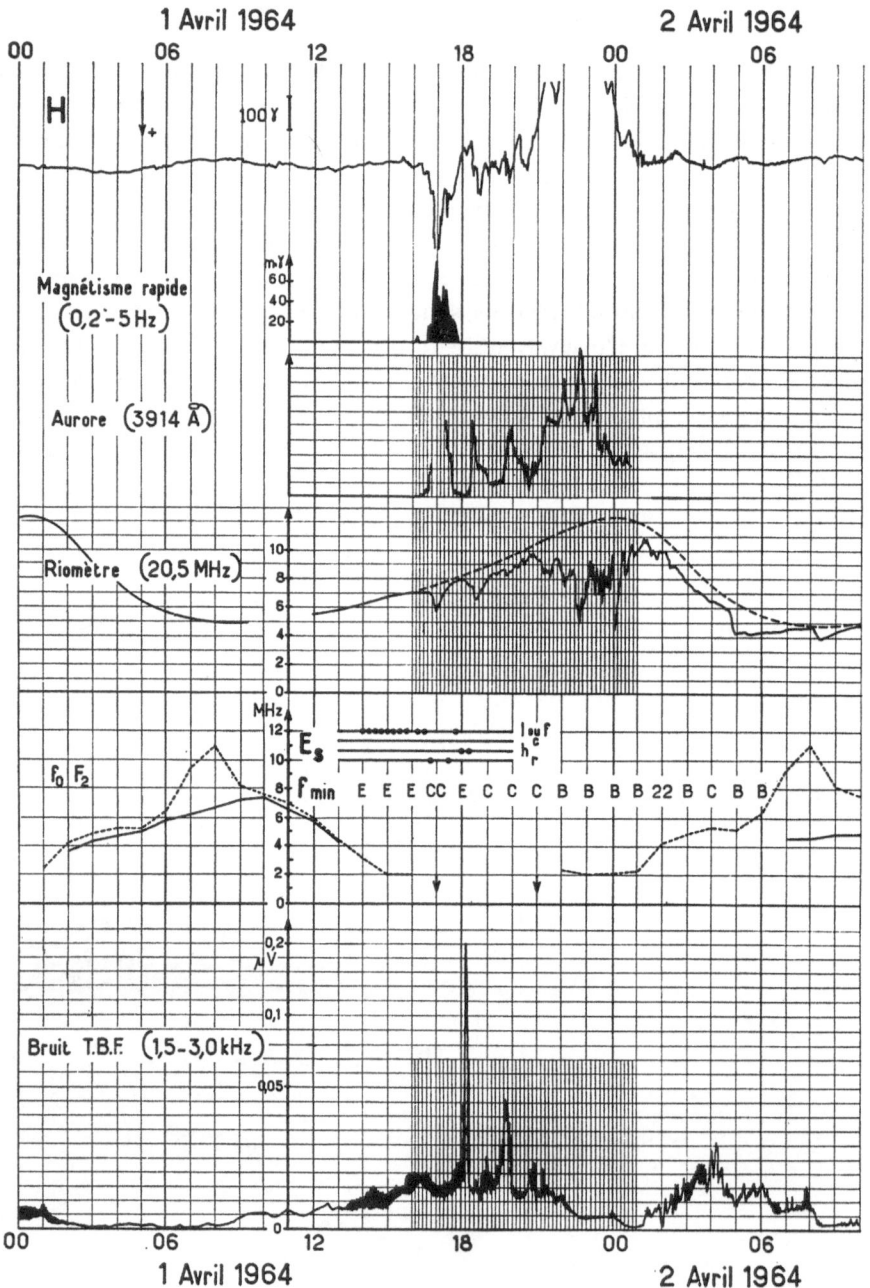

KERGUELEN

Fig. 120.   Example of I.P.D.P. and the associated geophysical phenomena (Gendrin, R., S. Lacourly, V. A. Troitskaya, M. Gokhberg, and R. V. Shepetnov: *Planetary Space Sci.* **15**, 1239, 1967).

of other polar geophysical phenomena; (GENDRIN *et al.*, 1967); from the top, the magnetic record, the amplitude of I.P.D.P., the intensity of the λ 3914 line, cosmic noise absorption, *Es*, f_{min} and VLF ($=$TBF). From an inspection of the records we can infer that the I.P.D.P. occurred when a westward traveling surge passed near (a little poleward) the zenith of the station; there occurred a large positive bay when the I.P.D.P. was recorded. GENDRIN *et al.* (1967) suggested that the waves are amplified by a wave-particle interaction with low energy protons (10 $\sim$ 100 keV). In Section 1.3, we noted that polar substorms are associated with the formation of a partial ring current.

8.6. Micropulsations and Pulsating Patches in the Morning Sector

An intense micropulsation activity occurs in the morning sector during polar substorms. In the fourth diagram in Figure 121, both the amplitude and frequency of micro-

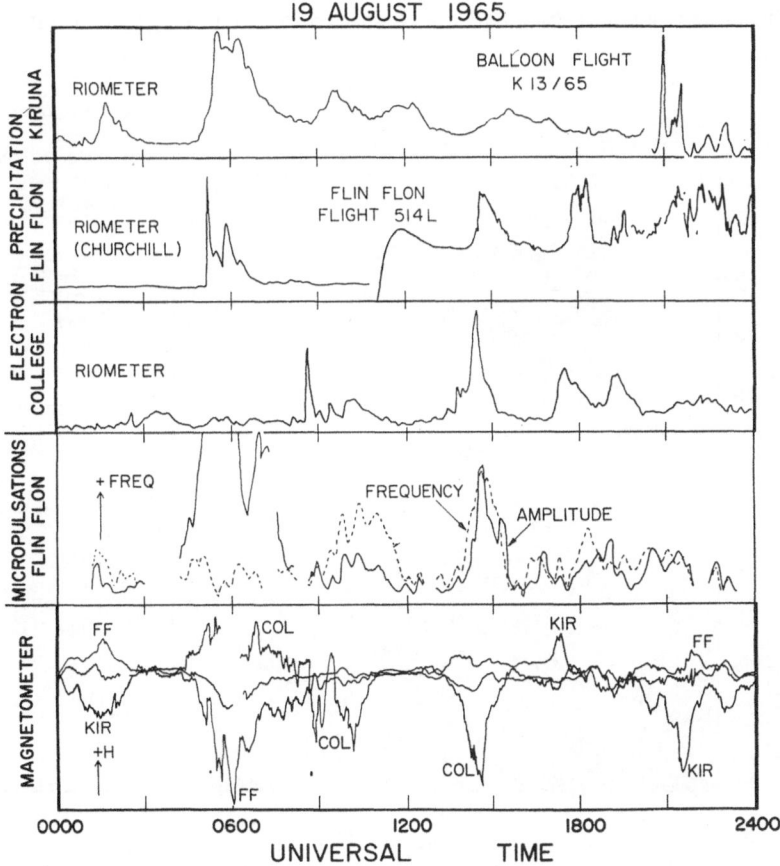

Fig. 121.   Micropulsation activity observed at Flin Flon, Canada, and the simultaneous magnetic, riometer and X-ray records from College, Flin Flon, and Kiruna (F. V. Coroniti, R. L. McPherron, and G. K. Parks: *J. Geophys. Res.* **73**, 1715, 1968).

pulsations observed at Flin Flon, Canada (06 U = 00 LT) are compared with other types of polar substorm activity, by superposing Kiruna, Flin Flon and College magnetic records (the bottom diagram) and showing the simultaneous variations of the cosmic radio noise (supplemented by X-ray observations by balloons) from Kiruna, Churchill and College (the upper three diagrams).

The first intense polar storm which began at about 03 UT caused an intense negative bay at Flin Flon, a less intense negative bay at Kiruna, and a positive bay at College, the N type absorption at Flin Flon and the M type absorption at Kiruna, and no absorption at College (since College was located well equatorward of the oval at that time). During this substorm, there occurred a considerable increase of the amplitude of micropulsations recorded at Flin Flon.

The second substorm, though weak, was registered as a negative bay with the E type absorption at College at about 08 UT, and an M type absorption at Kiruna. However, there was little magnetic or absorption activity at Flin Flon, although micropulsation activity was appreciably increased there. The third substorm was registered as a well-defined negative bay and an N absorption at College at about 13 UT, an

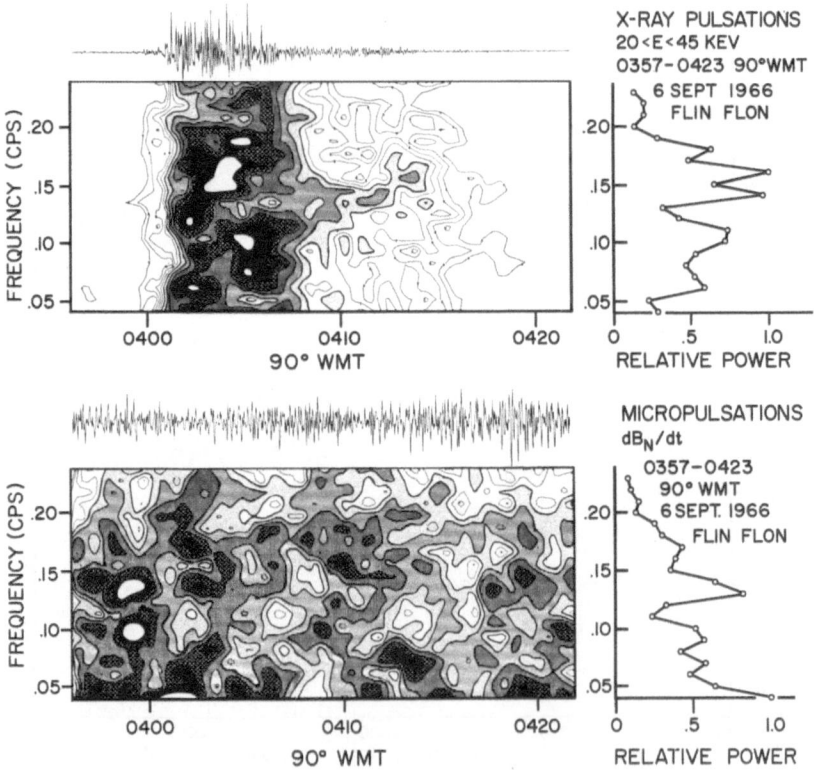

Fig. 122.   Micropulsations and the associated X-ray bursts in the morning sector; both analog and sonagraph records are shown. The power spectra are also shown on the right-hand side. (McPherron, R. L., G. K. Parks, F. V. Coroniti and S. H. Ward: *J. Geophys. Res.* **73**, 1697, 1968).

X-ray burst at Flin Flon and a weak cosmic radio noise absorption at Kiruna. Corresponding to this polar substorm, there was a clear increase of micropulsations at Flin Flon, which was located in the midmorning sector (07 LT).

Characteristics of micropulsations in the morning sector have been studied in great detail. Figure 122 shows a typical morning micropulsation and the corresponding X-ray bursts (both amplitude records and 'computed sonagrams'). X-ray bursts tend to have a sharper onset and a faster decay than the micropulsation activity. Micropulsations have a definite peak in their power spectrum at 7.1 sec, and X-rays at 6.2 sec; their power spectra are not similar. McPHERRON *et al.* (1968) noted that this type of micropulsations and the corresponding X-ray bursts occur most frequently at 05 LT and are generally associated with diffuse patchy aurora accompanying negative bays.

8.7. Micropulsations in the Daytime (10–15 LT)

Quasi-sinusoidal micropulsations are the most characteristic feature during the period between 10 and 15 LT; X-ray bursts are also modulated in a quasi-sinusoidal way.

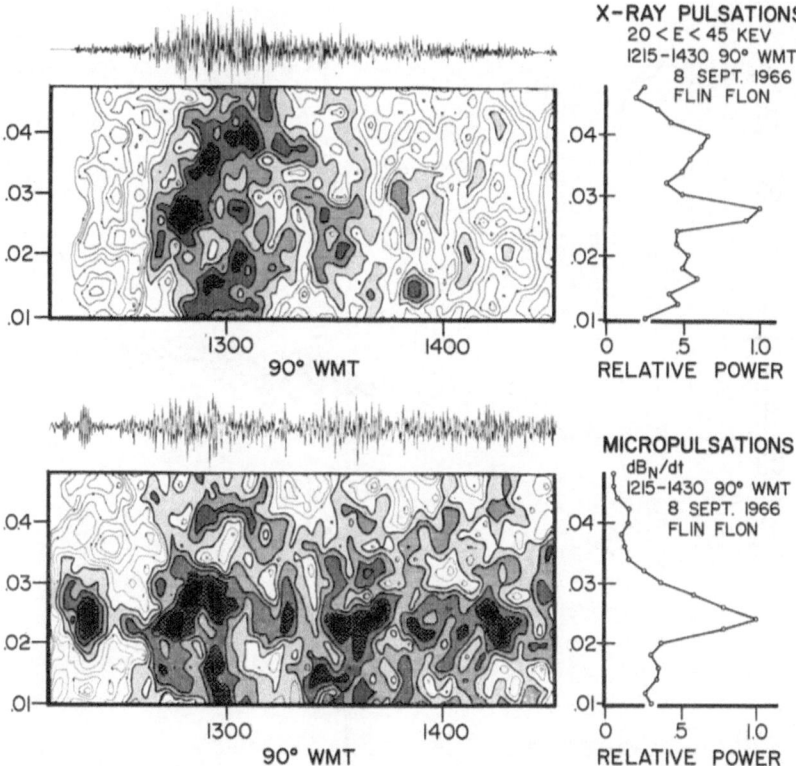

Fig. 123.   Micropulsations and the associated X-ray bursts in the daytime; both analog and sonagraph records are shown. The power spectra are also shown on the right-hand side (McPherron, R. L., G. K. Parks, F. V. Coroniti, and S. H. Ward: *J. Geophys. Res.* **73**, 1697, 1968).

Figure 123 shows an example. Again, the corresponding X-ray bursts tend to have a sharper onset and a faster decay than do micropulsations. In the power spectrum of micropulsations, a clear single peak is seen at a 40 sec period. MILTON *et al.* (1967) showed that X-ray microbursts (Sections 5.6 and 5.7) in the noon hours are associated with impulsive micropulsations.

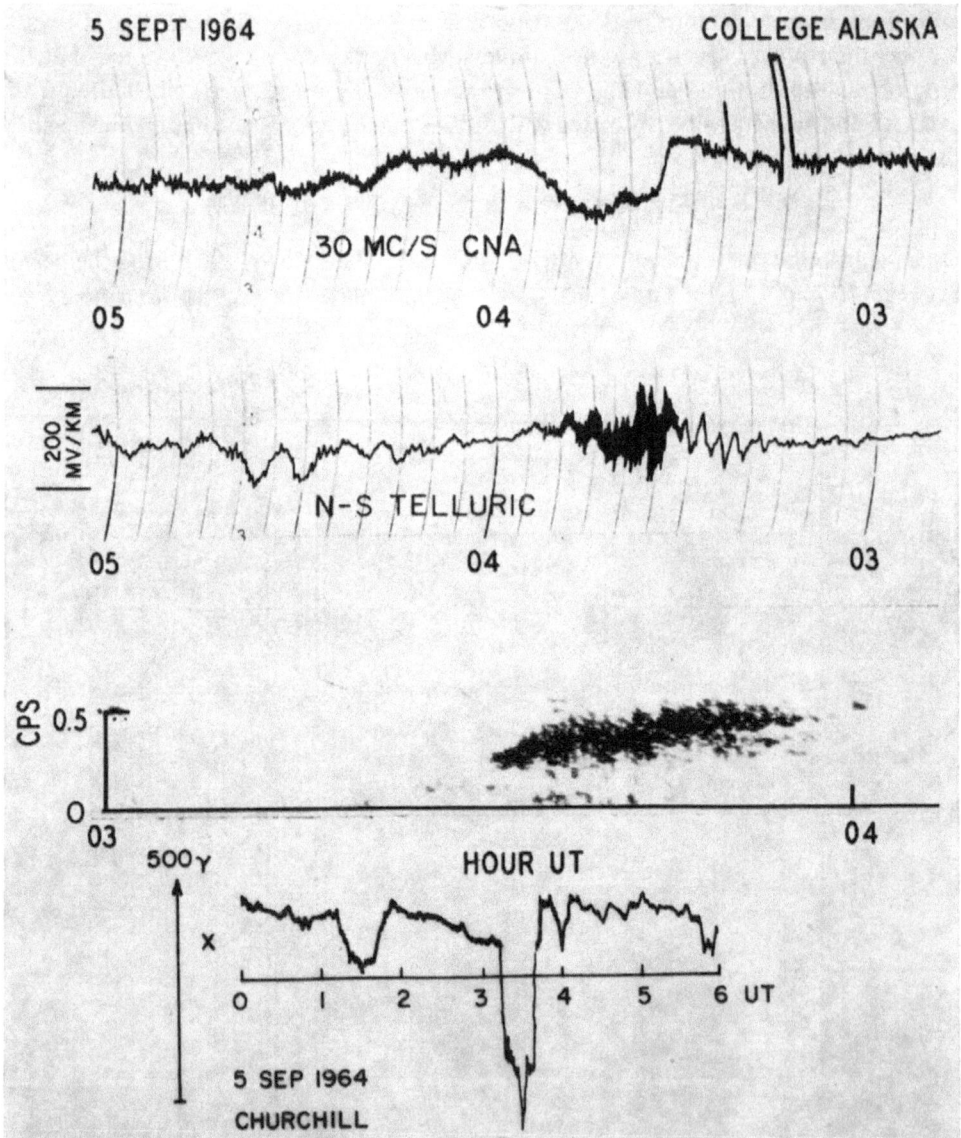

Fig. 124.   Intense Pc-1 event, together with the associated cosmic radio noise absorption (Heacock, R. R.: *J. Geomag. Geoelec.* 1968, in press).

8.8. Micropulsations in the Afternoon Hours

It has been known that Pc-1 type pulsations (or the so-called 'pearl' type pulsations) appear most frequently in the afternoon hours in the auroral zone (HEACOCK, 1963; HESSLER and HEACOCK, 1967). Since the nature of this particular type of pulsation has been discussed in detail in review articles (see General References), we shall not repeat it here.

HEACOCK (1963, 1968) noted that large-amplitude Pc-1 events typically have a measurable amount of simultaneous ionospheric cosmic noise absorption. Figure 124 shows the riometer record, the middle diagram, the analog Pi-1 record, and the bottom diagram, its sonagraph record. Figure 124 shows also the corresponding magnetic record from Churchill which was located in the mid-evening sector during the event. It is quite clear that this particular event was associated with the polar magnetic substorm, since the Churchill magnetic record shows an intense negative bay. It may be noted that the cosmic noise absorption occurred a little later than the onset time of the negative bay. This and other characteristics of the absorption are quite similar to those of the M type absorption.

8.9. Pi-2 Pulsations

There have been attempts to try to locate the source of Pi-2 micropulsations. SAITO (1961) made an extensive study of magnetic vectors during an initial phase of Pi-2 pulsations and found that the vectors tend to converge toward the auroral latitude in the late evening or midnight sector.

By using a set of stations, HERRON (1966) examined the phase lag of Pi-2 pulsations and suggested that the energy associated with Pi-2 pulsations moves away from the 10 ~ 11 pm meridian; however he noted that this is not a unique interpretation of his results.

ROSTOKER (1967) examined the polarization of Pi-2 micropulsations by using data from an East-West chain of six stations in Canada (from Victoria (dp long 292.7°) to Montreal (dp long 354.3°)) and found that the polarization is counterclockwise across the chain of the stations, regardless of local times. By taking into account the ionospheric screening, he deduced that magnetohydrodynamic waves propagate from the auroral zone to lower latitudes through the ionosphere.

8.10. Development of the Micropulsation Substorm: Figure 125

During an early phase of the micropulsation substorm (corresponding to the explosive phase of the auroral substorm), I.P.D.P. is observed a little equatorward of the auroral oval; this coincides with the region in which a positive (magnetic) bay is observed. Pi bursts are observed along a narrow region a little poleward of the expanding bulge. Pi-1 type pulsations are observed within the auroral bulge and along the auroral zone in the morning sector. Pi-2 pulsations are observed most clearly in middle latitudes in the late evening sector.

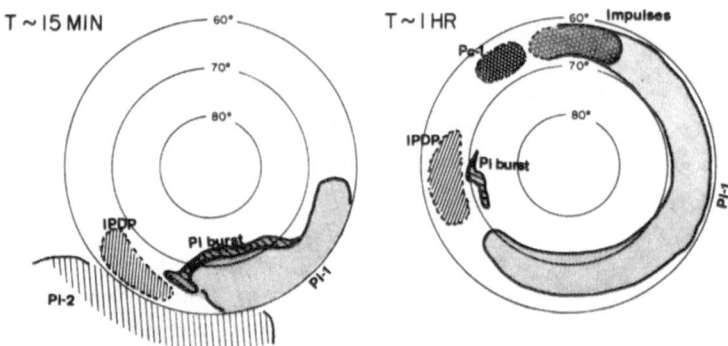

Fig. 125.   Development of the micropulsation substorm.

During a later phase of the micropulsation substorm (corresponding to the recovery phase of the auroral substorm), I.P.D.P. may be observed a little equatorward of the surge; the surge may be associated with Pi bursts. The area in which Pi-1 pulsations are observed expands considerably, to as far as the noon sector. This expansion is closely associated with that of the M type absorption, X-ray bursts and D type VLF emissions. In the noon sector, impulsive micropulsations are superposed on Pi-1 pulsations. Further, when the M type cosmic noise absorption extends as far as the afternoon sector, Pc-1 type pulsations occur there. The Pi-2 type pulsations may subside by this time.

References

GENERAL

GENDRIN, R.: 1967, 'Progrès récents dans l'étude des ondes T.B.F. et E.B.F.', *Space Sci. Rev.* **7**, 314–395.
HULTQVIST, B.: 1966, 'Plasma waves in the frequency range 0.001–10 cps in the earth's magnetosphere and ionosphere', *Space Sci. Rev.* **5**, 599–695.
TROITSKAYA, V. A. and GUL'ELMI, A. V.: 1967, 'Geomagnetic micropulsations and diagnostics of the magnetosphere', *Space Sci. Rev.* **7**, 689–768.

REFERRED TO IN TEXT

CAMPBELL, W. H. and MATSUSHITA, S.: 1962, 'Auroral-zone geomagnetic micropulsations with periods of 5 to 30 seconds', *J. Geophys. Res.* **67**, 555–573.
CORONITI, F. V., McPHERRON, R. L., and PARKS, G. K.: 1968, 'Studies of the magnetospheric substorm. 3. Concept of the magnetospheric substorm and its relation to electron precipitation and micropulsations', *J. Geophys. Res.* **73**, 1715–1722.
DOWDEN, R. L.: 1965, '"Micropulsation mode" propagation in the magnetosphere', *Planetary Space Sci.* **13**, 761–772.
DUNCAN, R. A.: 1961, 'Some studies of geomagnetic micropulsations', *J. Geophys. Res.* **66**, 2087–2094.
DUNGEY, J. W.: 1958, *Cosmic electrodynamics*, Cambridge Univ. Press, England.
GENDRIN, R. and LACOURLY, S.: 1968, Irregular micropulsations and their relations with the far magnetospheric perturbations, *Ann. Geophys.* **24**, 267–273.
GENDRIN, R., LACOURLY, S., TROITSKAYA, V. A., GOKHBERG, M., and SHEPETNOV, R. V.: 1967, 'Caractéristiques des pulsations irrégulières de période décroissante (I.P.D.P.) et leurs relations avec les variations du flux des particules piégées dans la magnétosphère', *Planetary Space. Sci.* **15**, 1239–1259.
HEACOCK, R. R.: 1963, 'Auroral-zone telluric-current micropulsations, $T < 20$ seconds', *J. Geophys. Res.* **68**, 1871–1884.

HEACOCK, R. R.: 1966, 'The 4-second summertime micropulsation band at College', *J. Geophys. Res.* **71**, 2763–2775.
HEACOCK, R. R.: 1967a, 'Evening micropulsation events with a rising midfrequency characteristic', *J. Geophys. Res.* **72**, 399–408.
HEACOCK, R. R.: 1967b, 'Two subtypes of type Pi micropulsations', *J. Geophys. Res.* **72**, 3905–3917.
HEACOCK, R. R.: 1968, 'Large amplitude Pc-1 events at College', *J. Geomag. Geoelec.* **20** (in press).
HERRON, T. J.: 1966, 'Phase characteristics of geomagnetic micropulsations', *J. Geophys. Res.* **71**, 871–890.
HESSLER, V. P. and HEACOCK, R. R.: 1967, 'Telluric current micropulsations at the auroral zone', Final Rep. AF 19(628)-1695, Geophys. Inst., Univ. of Alaska, UAG R-202.
JACOBS, J. A. and SINNO, K.: 1960, 'World-wide characteristics of geomagnetic micropulsations', *Geophys. J.* **3**, 333–353.
JACOBS, J. A. and WATANABE, T.: 1964, 'Micropulsation whistlers', *J. Atmospheric Terrest. Phys.* **26**, 825–829.
LIEMOHN, H. B.: 1967, 'Cyclotron-resonance amplification of VLF and ULF whistlers', *J. Geophys. Res.* **72**, 39–55.
MCPHERRON, R. L., PARKS, G. K., CORONITI, F. V., and WARD, S. H.: 1968, 'Studies of the magnetospheric substorm. 2. Correlated magnetic micropulsations and electron precipitation occurring during auroral substorms', *J. Geophys. Res.* **73**, 1697–1713.
MILTON, D. W., MCPHERRON, R. L., ANDERSON, K. A., and WARD, S. H.: 1967, 'Direct correspondence between X-ray microbursts and impulsive micropulsations', *J. Geophys. Res.* **72**, 414–417.
MOROZUMI, H. M.: 1965, 'Diurnal variation of aurora zone geophysical disturbances', *Rep. Ionos. Space Res. Japan* **19**, 286–298.
ROSTOKER, G.: 1967a, 'The polarization characteristics of Pi-2 micropulsations and their relation to the determination of possible source mechanisms for the production of nighttime impulsive micropulsation activity', *Canadian J. Phys.* **45**, 1319–1335.
ROSTOKER, G.: 1967b, 'The frequency spectrum of Pi-2 micropulsation activity and its relationship to planetary magnetic activity', *J. Geophys. Res.* **72**, 2032–2039.
SAITO, T.: 1961, 'Oscillations of geomagnetic field with progress of pt type pulsation', *Sci. Rept. Tohoku Univ., Ser. 5 Geophys.* **13**, 53–61.
TROITSKAYA, V. A.: 1961, 'Pulsation of the earth's electromagnetic field with periods of 1 to 15 seconds and their connection with phenomena in the high atmosphere', *J. Geophys. Res.* **66**, 5–18.
TROITSKAYA, V. A. and GUL'ELMI, A. V.: 1967, 'Geomagnetic micropulsations and diagnostics of the magnetosphere', *Space Sci. Rev.* **7**, 689–768.
TROITSKAYA, V. A. and MELNIKOVA, M. W.: 1959, 'On characteristic intervals of pulsations diminishing in period and their connection with phenomena in the high atmosphere', *Doklady Akad. Nauk USSR* **128**, No. 5.
VICTOR, L. J.: 1965, 'Correlated auroral and geomagnetic micropulsations in the period range 5 to 40 seconds', *J. Geophys. Res.* **70**, 3123–3130.
WATANABE, T.: 1962, 'Law of electric conduction for waves in the ionosphere', *J. Atmospheric Terrest. Phys.* **24**, 117–125.
WILHELM, K.: 1967, 'Geomagnetic micropulsations and electron bremsstrahlung in the northern auroral zone', *J. Geophys. Res.* **72**, 1995–1999.
YANAGIHARA, K.: 1960, 'Geomagnetic pulsations in middle latitudes – morphology and its interpretation', *Mem. Kakioka Mag. Obs.* **9**, 15–74.

CHAPTER 9

SATELLITE OBSERVATIONS DURING POLAR SUBSTORMS

9.1. Introduction

In this chapter, we combine available observations by satellite-borne instruments of the variations of the magnetic field and of the flux of both electrons and protons in and near the equatorial plane during auroral (or polar magnetic) substorms. The location at a point in the equatorial plane in and near the magnetosphere is specified by the radial distance r_e.

For the purpose of the presentation, we suppose in this section that four magnetospheric substorms, A, B, C, and D, happen to occur during the course of a particular day, say, at 04:00, 07:00, 10:00, and 13:00 UT, respectively. These are the onset times of the four magnetospheric substorms; that is to say, $T=0$ for each of the magnetospheric substorms. It is assumed also that each substorm lasts for two hours. On the basis of practicality in the analysis, each time is assumed to be the onset time of the polar magnetic substorm near the equatorward edge of the auroral oval in the midnight sector, where the onset can be most clearly determined. This assumption can be justified, because so far there is no definite evidence which suggests magnetospheric changes prior to the onset of the polar magnetic substorm, so that the onset of a magnetospheric substorm is immediately manifested by the onset of a polar magnetic substorm with an accuracy of 30 sec $\sim$ 1 min.

Again for the purpose of the presentation, we construct a diagram which shows typical variations of the magnetic field and of the flux of electrons and protons at several distances along an equatorial radial line passing through the Central Pacific sector. In this longitude sector, the four substorms are observed at about 18:00, 21:00, 00:00, and 03:00 local time, respectively. Such frequency of occurrence of the substorms is common during the period of a medium magnetospheric storm. This sector in particular was chosen because a typical auroral zone station, College (dp lat 64.7° N), and a typical low latitude station, Honolulu (dp lat 21° N), lie in this sector and because a synchronous satellite (ATS) which has been placed above this sector has provided extensive data.

Such an arrangement of the presentation implies that the occurrence of magnetospheric substorms does not depend on the location of a particular point on the earth with respect to the sun. This has been well justified in an extensive work by DAVIS and SUGIURA (1966) who showed that the AE index (a measure of the intensity of the polar electrojet) has no obvious UT dependence (at least as a first approximation). However, the nature of the variations of the magnetic field and of the flux of charged particles during the substorms are strongly controlled by the local time of the point.

We shall review here very briefly characteristics of typical magnetic variations and precipitation of electrons in the auroral zone and corresponding magnetic variations in low latitudes during the four substorms discussed above; the details have already been discussed in earlier chapters. Figure 126 shows typical magnetic variations at an

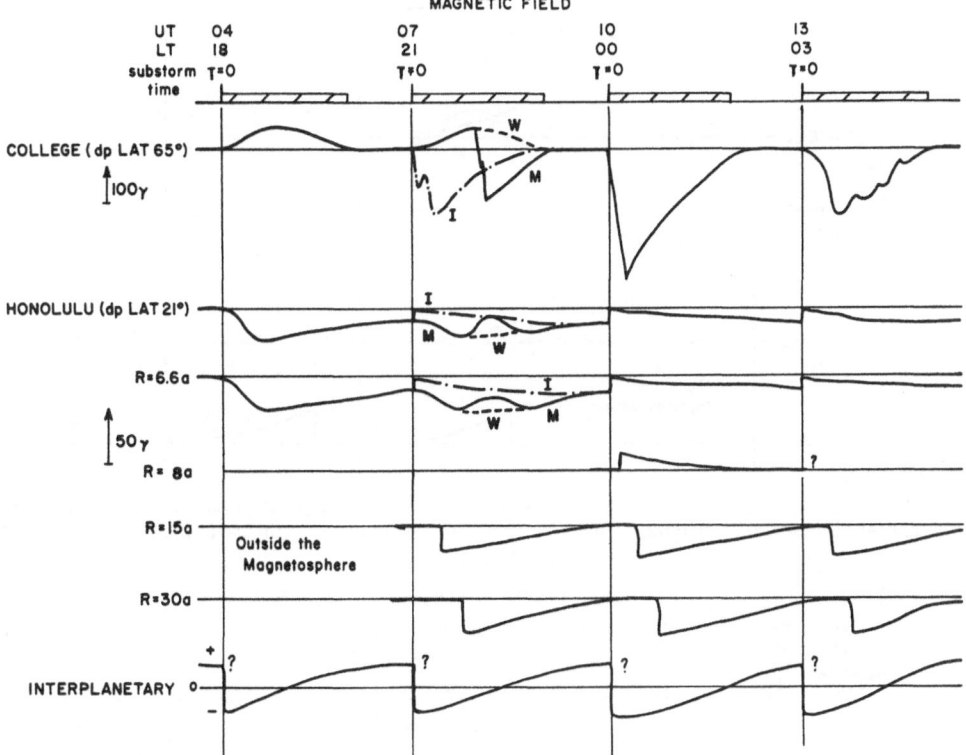

Fig. 126. Typical magnetic (the *H* component) variations during polar substorms in the auroral zone, low latitudes, and at several radial distances; the substorms are assumed to occur at 18, 21, 00, and 03 local times.

auroral zone station (College) and at a low latitude station (Honolulu) for the situation assumed above. One of the purposes of this chapter is to relate the variations observed at ground-based stations to the corresponding typical variations at $r_e = 6.6a$, $8a \sim 10a$, $15a$ and beyond. Figures 127 and 128 show the precipitation of electrons at an auroral zone station for the corresponding situation; Figure 127 for electrons of energies of order $1 \sim 10$ keV and Figure 128 for electrons of energies of order 50 keV. Figure 129 shows the precipitation of protons.

1. Auroral Zone

A positive bay is the most common feature in the afternoon and early evening hours

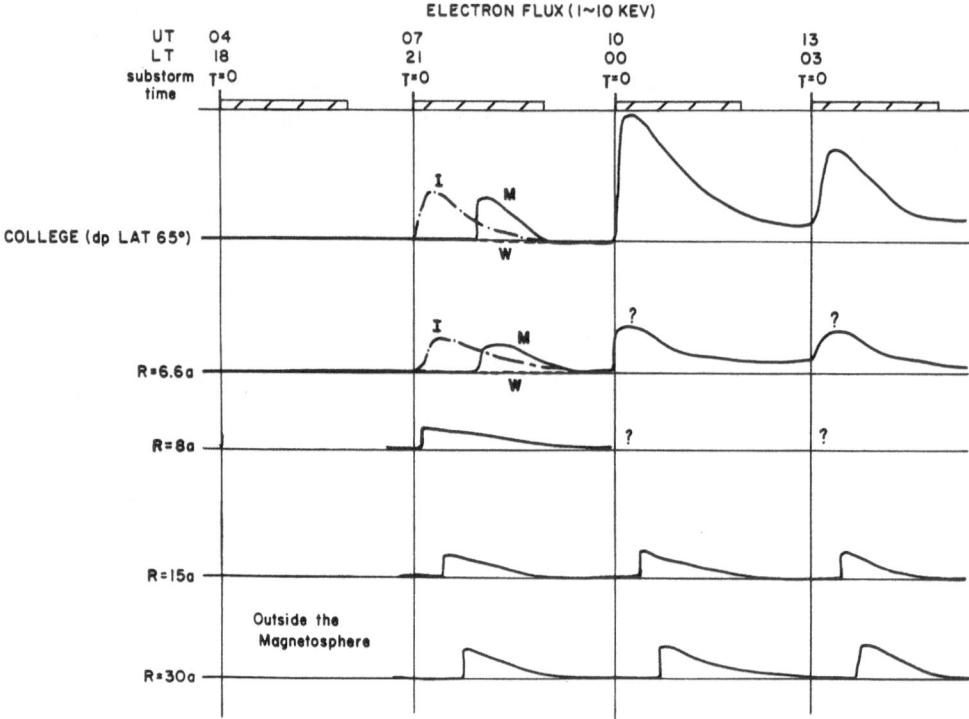

Fig. 127. Typical variations of electron (1 ∼ 10 keV) flux during polar substorms in the auroral zone and at several radial distances; the substorms are assumed to occur at 18, 21, 00, and 03 local times.

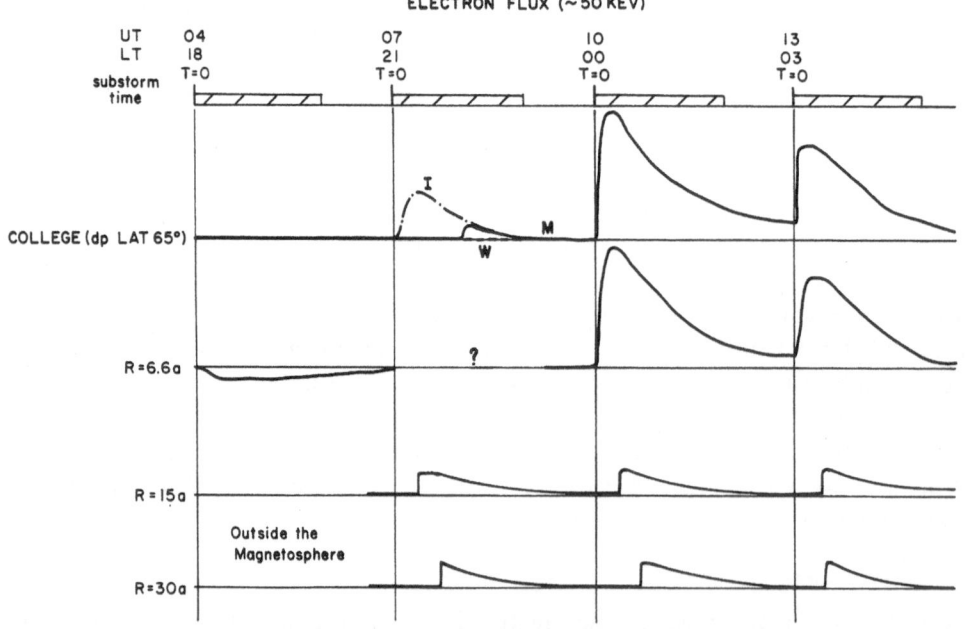

Fig. 128. Typical variations of electron (50 keV) flux during polar substorms in the auroral zone and at several radial distances; the substorms are assumed to occur at 18, 21, 00, and 03 local times.

of the auroral zone at the time of substorms, namely for the first substorm which begins at 04 UT or 18 LT in the Alaska-Hawaii sector. Westward traveling surges travel along the auroral oval, well poleward of an auroral zone station, so that particle precipitation (both proton and electron) does not occur during the first substorm over an auroral zone station which is College, Alaska in this particular study (Section 2.3, Figure 28).

In the evening hours (for the second substorm, 07 UT = 21 LT), the magnetic variations may be considered to be a transition type between a simple positive bay in the afternoon hours and a simple negative bay in the midnight hours (Section 2.3 and Figure 28). If a substorm is weak, there will be a positive bay which is similar to that seen in the afternoon hours; a westward surge travels well poleward of the station, so that there will be no precipitation of electrons. However, the proton aurora will be located over the auroral zone and its luminosity will be enhanced at this time. This case will hereafter be referred to as the W case.

If a substorm is of medium intensity, there will be a positive bay which is then followed by a negative bay (Section 2.3 and Figure 28). A westward surge may be considered to be the western leading edge of the expanding auroral bulge in this case, and a negative bay begins at the time the surge passes above the station. The precipitation of electrons also begins at this time, causing an E type cosmic radio noise absorption. A proton aurora is located over the auroral zone at the time of the initial positive bay and shows an enhancement of luminosity; however, as soon as visible auroras expand equatorward, the proton aurora shifts equatorward. This case will be called the M case.

If a substorm is intense or an intense geomagnetic storm is in progress, a negative bay will occur in the evening hours. The auroral bulge has an extensive East-West extent in such a case, so that the precipitation of electrons occurs at the time when the negative bay begins. The precipitation of electrons causes the N type cosmic radio noise absorption. Since the proton aurora is located well equatorward of the expanded oval, the proton precipitation cannot be seen in this case at an auroral zone station.

In the midnight hours (for the third substorm, 10 UT = 00 LT), a sharp negative bay is the most common feature. The precipitation of both soft (1 ~ 10 keV) and hard (50 keV) electrons begins when the poleward expanding bulge passes directly over the station (Sections 2.2, 4.4, 5.4). The protons precipitate in the same area (Section 6.4).

In the morning hours, the onset of negative bays (for the fourth substorm, 13 UT = 03 LT) is less sharp than that of midnight bays. This is also the case for the onset of the precipitation of electrons; the precipitation is associated with the M type absorption (Sections 4.6 and 5.6). Eastward drifting bands and patches are the most common feature associated with morning negative bays (Section 2.4). The protons precipitate fairly uniformly in a wide area where the patches are seen (Section 6.4).

2. Low Latitude

A negative bay is the most common feature in the afternoon and early evening hours

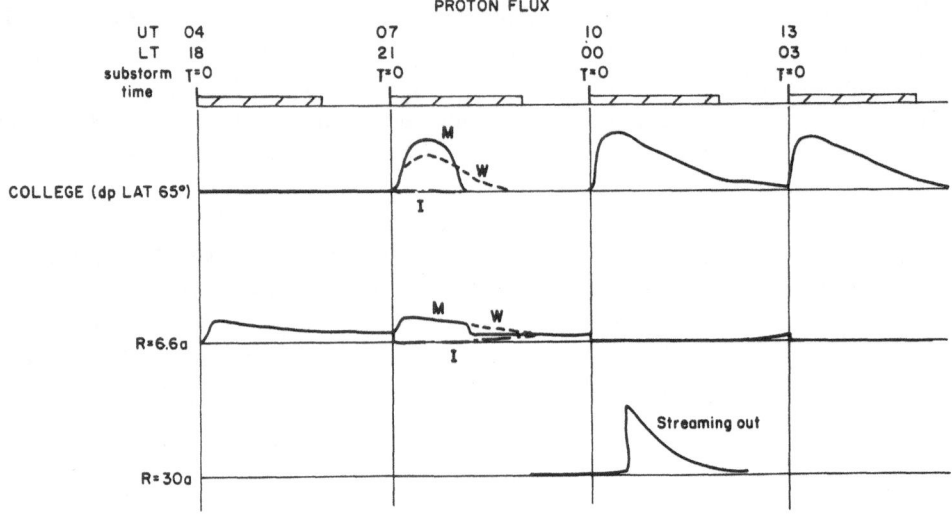

Fig. 129. Typical variations of proton (10 keV) flux during polar substorms in the auroral zone and at several radial distances; the substorms are assumed to occur at 18, 21, 00, and 03 local times.

in low latitudes at the time of the first substorm (Chapter 3). In the evening hours (the second substorm), the substorm variations may also be considered a transition type between a simple negative bay in the afternoon hours and a simple positive bay in the midnight hours. If a substorm is weak, there will be a negative bay which is similar to that seen in the afternoon hours. If a substorm is intense or an intense geomagnetic storm is in progress, a positive bay is observed. The most common case is a negative bay which is followed by a positive bay. In the midnight and early morning hours (the third and fourth substorms, respectively) a positive bay is the most common feature.

9.2. Magnetic Field and Particle Flux Variations at $r_e = 6.6a$: Synchronous Satellite (ATS)

A. MAGNETIC FIELD

1. Midnight Sector

A typical change in the midnight sector is a sharply defined positive change. A typical event occurred a little after 10 UT (= 00 LT in the Alaska-Hawaii-ATS sector) on December 25, 1966. Figure 130a shows the horizontal component traces from the ATS and low latitude stations widely separated in longitude, Honolulu, Kakioka, Tashkent, M'Bour and San Juan. We can see that an *evening* station (Kakioka, 18 LT time), also recorded a positive change, corresponding to the I case in Figure 126; at an *afternoon* station (Tashkent, 14 LT time), the simultaneous change consisted of a weak negative and then positive variation corresponding also to the I case.

DEC 25, 1966

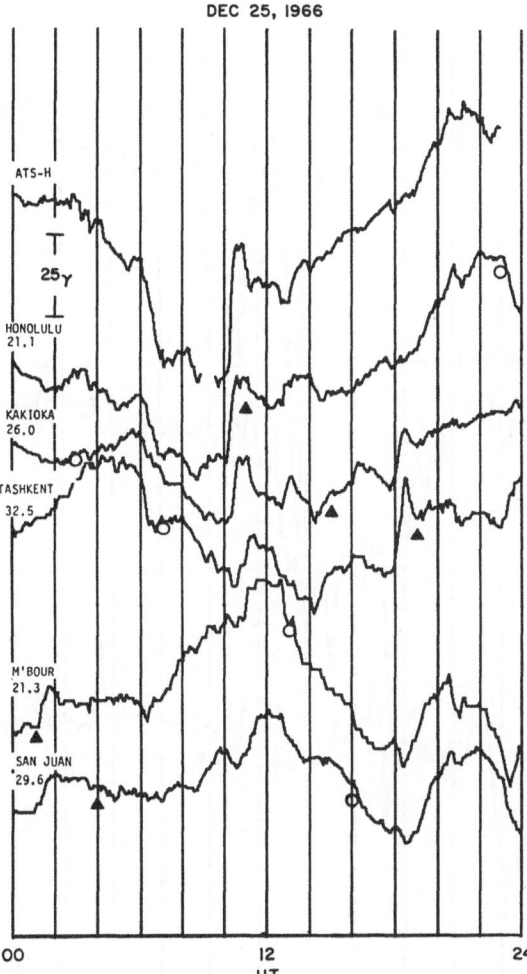

Fig. 130a. Horizontal component magnetic records from the synchronous satellite and low latitude stations widely separated in longitude, Honolulu, Kakioka, Tashkent, M'Bour, and San Juan; December 25, 1966: the ATS record was provided by Cummings, W. D. and P. J. Coleman Jr. (1968).

2. Evening Sector

Figure 130a provides an example of an evening event in the College-Honolulu-ATS sector. In the midnight sector of the auroral zone, a negative bay began at 0550 UT on December 25 (see the Great Whale River record in Figure 130b). At about 0600 UT (= 20 LT), both the ATS and Honolulu records show a sharp negative change. This corresponds to the W case.

The negative change appeared not only in the College-Honolulu-ATS sector, but also in an extensive local time range, extending to as far as the late morning sector, namely the Middle East sector. In the midmorning sector, a slight negative change was followed by a positive change.

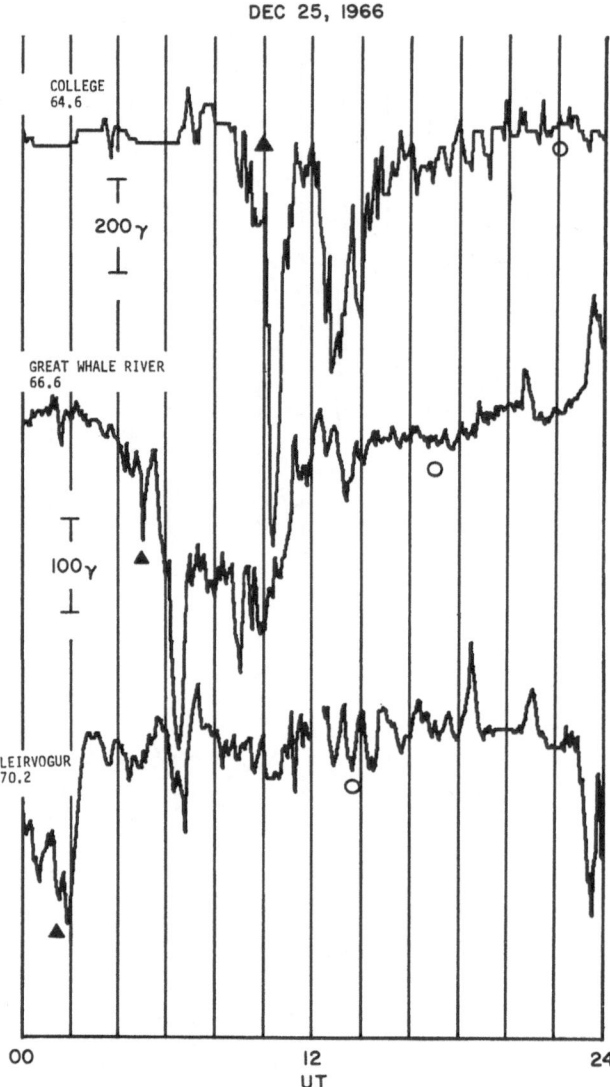

Fig. 130b.   Magnetic records (the *H* component) from auroral zone stations (College, Great Whale River, and Leirvogur) on December 25, 1966.

When polar substorms are intense, the corresponding changes in the evening sector of the auroral zone become very complicated. However, as mentioned earlier, they can be considered to be transitional changes between a simple negative bay in the midnight sector and a simple positive bay in the afternoon sector. A typical example of the M case (namely, the changes consisting of positive and then negative changes) occurred on January 21, 1967; Figure 131. In the midnight sector of the auroral zone (Churchill) there was an indication of the onset of a weak substorm at about 03 UT, but a distinct negative bay began at 0520 UT. The College records show a

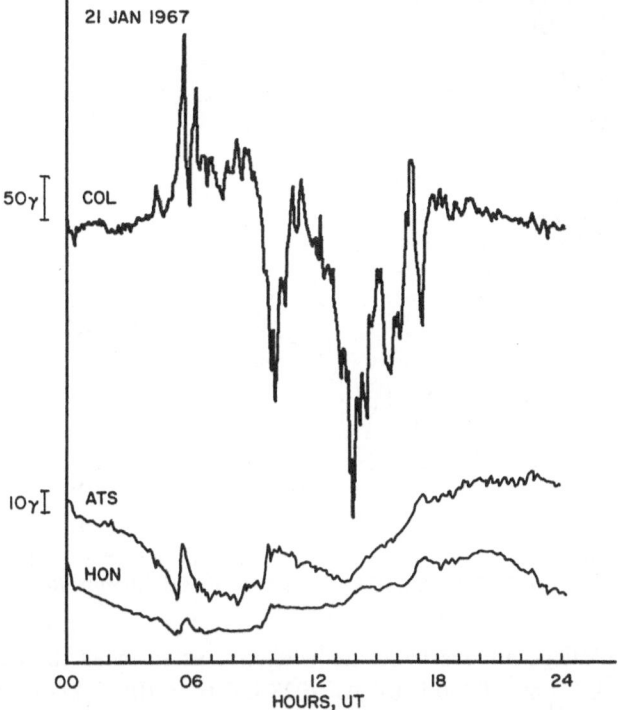

Fig. 131. Simultaneous magnetic records (the *H* component) from College, Honolulu, and the synchronous satellite on January 21, 1967: the ATS record was provided by Cummings, W. D. and P. J. Coleman Jr. (1968).

gradual increase from about 04 UT to about 0540 UT (= 18 LT to 1940 LT), when suddenly a negative excursion began. Both the ATS and Honolulu records showed first a gradual decrease and then a sudden increase, at about 0530 UT (= 1930 LT). This rapid positive change was then followed by a slow decrease, to about the pre-substorm level or below. At Whitehorse (dp lat 63.4°) Canada, which is closest to the longitude of the ATS, a westward traveling surge appeared near the northeastern horizon at 0520 UT (1920 Alaska time). Unfortunately, the Alaskan sky was cloudy on that day, but on the basis of Figure 28 it may be inferred that the surge seen at Whitehorse at about 0520 UT reached College at about 0530 UT (= 1930 LT); the speed of the surge is then estimated to be 1.3 km/sec which is typical. Note that the January 21 examples also provide a good example of the midnight event which began at about 0930 UT (= 2330 LT); a negative bay was observed at College, while a positive change was recorded by the ATS and Honolulu.

When an intense storm is in progress, a negative bay tends to appear earlier in the evening along the auroral zone; this is partly due to the equatorward expansion of the auroral oval along which the electrojet flows.

A typical example of such an anomalous appearance of a negative bay occurred at about 06 UT (= 20 LT) on January 8, 1967 (Figure 132); see the I case in Figure

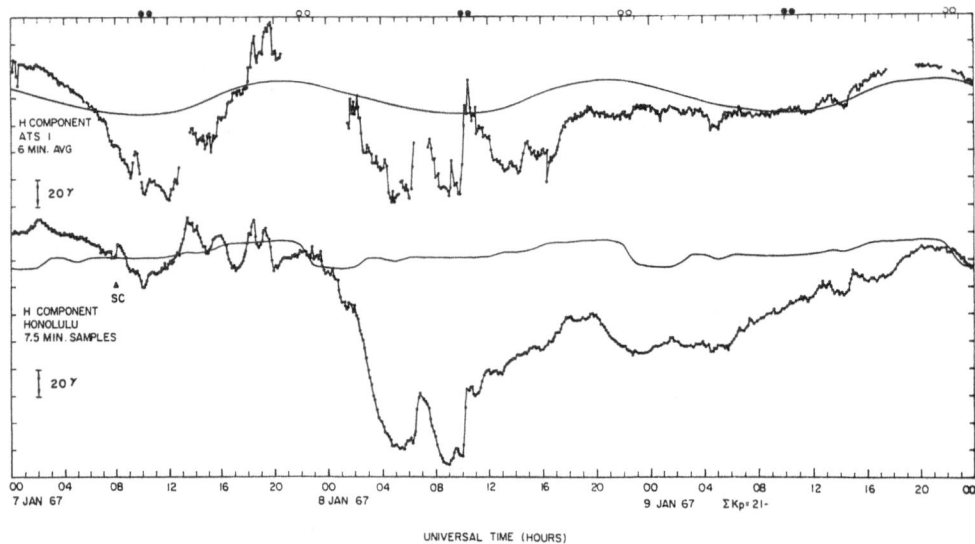

UNIVERSAL TIME (HOURS)

Fig. 132. Simultaneous magnetic records (the *H* component) from the synchronous satellite and Honolulu on January 8, 1967 (Coleman, P. J., Jr. and W. D. Cummings: *Radio Sci.* **3**, 762, 1968).

126. In fact, at College an overhead arc was seen as soon as the sky became dark enough, at 1630 LT time (0230 UT). At 2028 LT time (0623 UT), an intense surge appeared from the eastern horizon and covered the entire sky in a few minutes; the intense display lasted until 2100 LT. A negative bay was observed at College during this period. Figure 132 also provides a good example of the midnight case; intense auroral activity began at 0007 LT (1007 UT).

3. *Midmorning Sector*

Figure 130a also provides an example of a midmorning substorm. Although records from Siberia are not available at this time, it is certain that a polar substorm began at about 1750 UT (=0750 LT): a positive bay was recorded at Kakioka and Tashkent, which were located in the early morning and late evening sector, respectively. The corresponding change in the 18 local time sector (European Sector) was a slight negative change which was followed by a larger positive change. Both the ATS and Honolulu were located in the midmorning sector and recorded a slight negative change. Based on the above study, Figure 126 has been supplemented with the magnetic variations at $r_e = 6.6a$.

B. ELECTRONS (50 KEV)

Unfortunately, the synchronous satellite did not carry a detector for electrons of energies $1 \sim 10$ keV, so that it is not possible to supplement Figure 127. As we shall see in the next section, there is some indication that such electrons appear at $r_e = 6.6a$ in the midnight sector during substorms.

The observations of electrons of energies 50 keV $< E < 150$ keV have been examined in detail by PARKS *et al.* (1968) and PARKS and WINCKLER (1968). They showed that cosmic radio noise absorption observed at College is well correlated with an enhanced flux of electrons in the equatorial plane. Figure 133 gives an example of this good correlation. The first enhancement in the equatorial plane (middle curve) at about 0010 LT was associated with a negative bay (top curve) and N type absorption

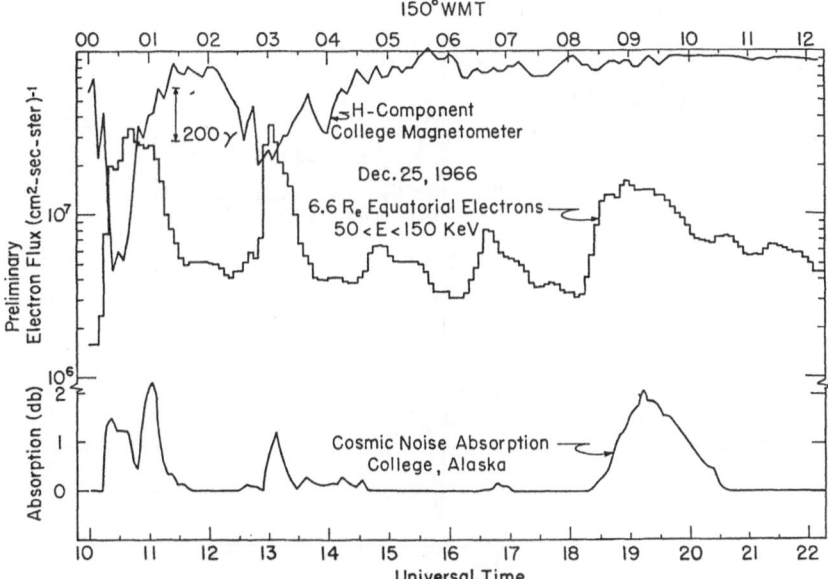

Fig. 133.   Variations of electron (50 keV) flux and the corresponding College magnetic (the *H* component) and cosmic radio noise absorption records; December 25, 1966 (Parks, G. K. and J. R. Winckler: School of Physics and Astronomy, University of Minnesota, Report 1968).

(bottom curve); the second enhancement at about 0230 LT was associated with a weaker negative bay and M type absorption (see Section 4.5). The third enhancement which began at about 0430 LT was not correlated in any obvious manner with events on the ground, and the fourth one (0615 LT) had only a slight corresponding indication in the riometer record. Then, a new, large enhancement began at about 0815 LT. It was associated with an intense M type cosmic radio noise absorption at College; since College was well outside the auroral oval at this local time, there was only slight indication of a negative bay at that time (Sections 4.6 and 5.6). We noted earlier that the polar magnetic substorm which corresponds to this enhancement began at about 1750 UT (0750 LT at College). Therefore, there was a time difference of about 25 min between the two onsets. This is in agreement with the time difference deduced in Section 4.6.

There is some uncertainty as to the variation of the electron flux in the afternoon sector. Figure 134 shows the variation of the flux for the December 25, 1966 case. The substorm which began at 0550 UT appeared to be related to a decrease of flux. Figure 127 has been supplemented on the basis of the above discussion.

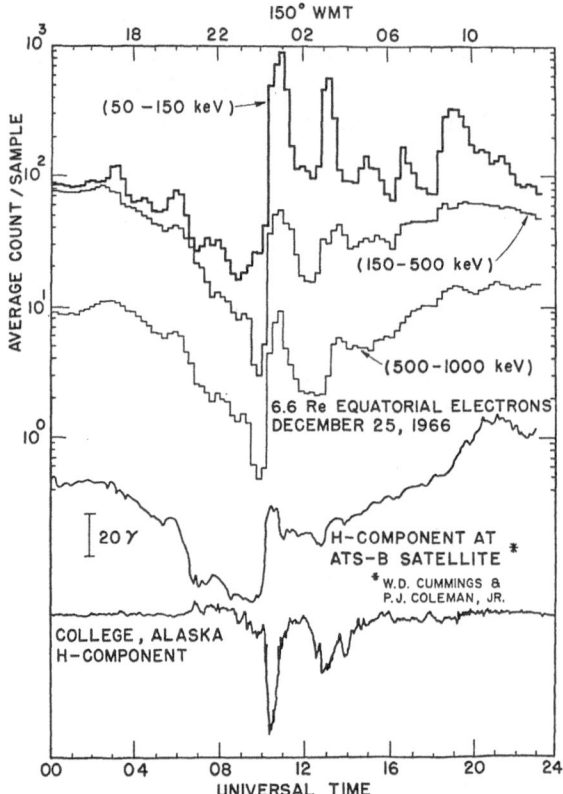

Fig. 134. Variations of electron flux of different energies observed by the synchronous satellite, together with the corresponding magnetic records from the satellite and College (Lezniak, T. W., R. L. Arnoldy, G. K. Parks, and J. R. Winckler: *Radio Sci.* 3, 710, 1968).

C. ELECTRONS (0.4–2 MEV)

LANZEROTTI *et al.* (1967) found that the electron flux sometimes oscillates with 7–25 min periods and the period is a function of electron energy. They showed also that the computed periods and the observed periods agree fairly well. On these bases, they suggested that the oscillating flux observed at the ATS is due to the longitudinal drifts around the earth of a group of electrons that has been injected into the magnetosphere or locally accelerated. It is interesting to note that in this connection KONRADI (1967) reported satellite observations which suggest that protons of energies of a few hundred kilovolts drift to the dayside after being injected on the night side.

9.3. Magnetic Field and Particle Flux Variations in the Range $r_e = 4a \sim 15a$

A. MAGNETIC FIELD

The radial range between $r_e = 4a$ and $15a$ has been extensively surveyed by the OGO satellites. HEPPNER *et al.* (1967) divided the quiet-time magnetosphere into several regions, depending on the difference between the observed field and the computed field (based on the spherical harmonic coefficients for the earth's main field). Figure 135 shows their diagram. In the equatorial radial range between $r_e = 4.0a$ and $4.5a$, the field is approximately the same as the computed field. The range between $r_e = 4.7a$ and $r_e = 11a$ is characterized by a weaker field than what we expect for the computed field ($\Delta B < 0$), while beyond $r_e = 11a$ the field is greater than the computed value ($\Delta B > 0$).

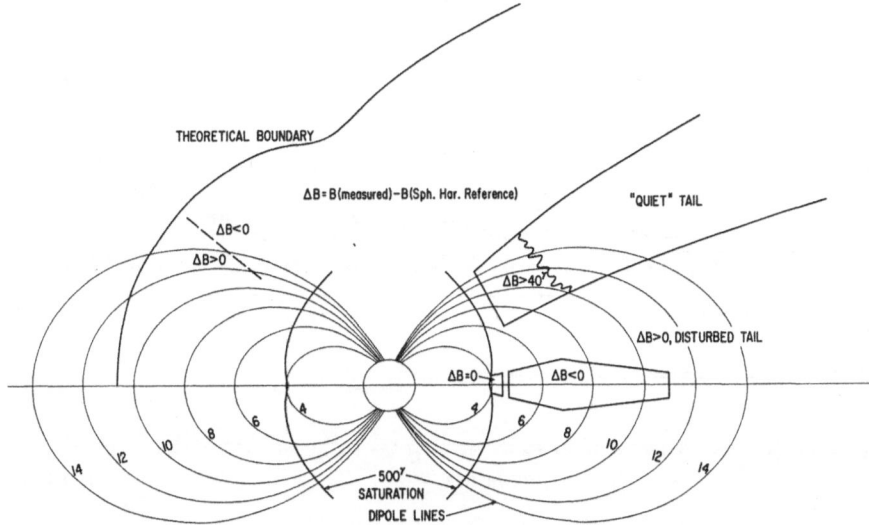

Fig. 135. Distortion of the magnetic field in the magnetosphere (Heppner, J. P., M. Sugiura, T. L. Skillman, B. G. Ledley, and M. Campbell: *J. Geophys. Res.* **72**, 5417, 1967).

Superposed on such a permanent distortion of the geomagnetic field, various types of temporal changes occur. Here we are particularly interested in changes during substorms. HEPPNER *et al.* (1967) showed that the sign of the changes during polar magnetic substorms is opposite to the sign of ΔB. At a point beyond $r_e = 11a$ (where $\Delta B < 0$), the sign of the changes is positive. Figure 136a, b show examples in the two regions: In Figure 136a, a negative bay, which began at 1922 UT at Kiruna, had a corresponding negative change at 1937 UT at $r_e \simeq 12.5a$. Another negative bay, which began at 2125 UT, was associated with a negative change at 2139 UT at $r_e \simeq 10a$. Both changes occurred while the satellite was in the region of $\Delta B > 0$.

On the other hand, in Figure 136b, we see that a negative bay, which was recorded at Cape Chelyuskin at 1409 UT, was associated with a positive change at the satellite location at 1410.8 UT at $r_e \simeq 9.5a$; the satellite was in the region of $\Delta B < 0$. The latter result is in agreement with that obtained by the synchronous satellite, except for the fact that there is a definite delay of the onset of the corresponding event at the satellite location. Therefore it appears that in the midnight sector the positive change occurs not only at $r_e = 6.6a$, but also in an extensive radial range, at least to $r_e = 10a$, with a delay of order a few minutes. These limited observations have been incorporated in the fourth and fifth rows of Figure 126.

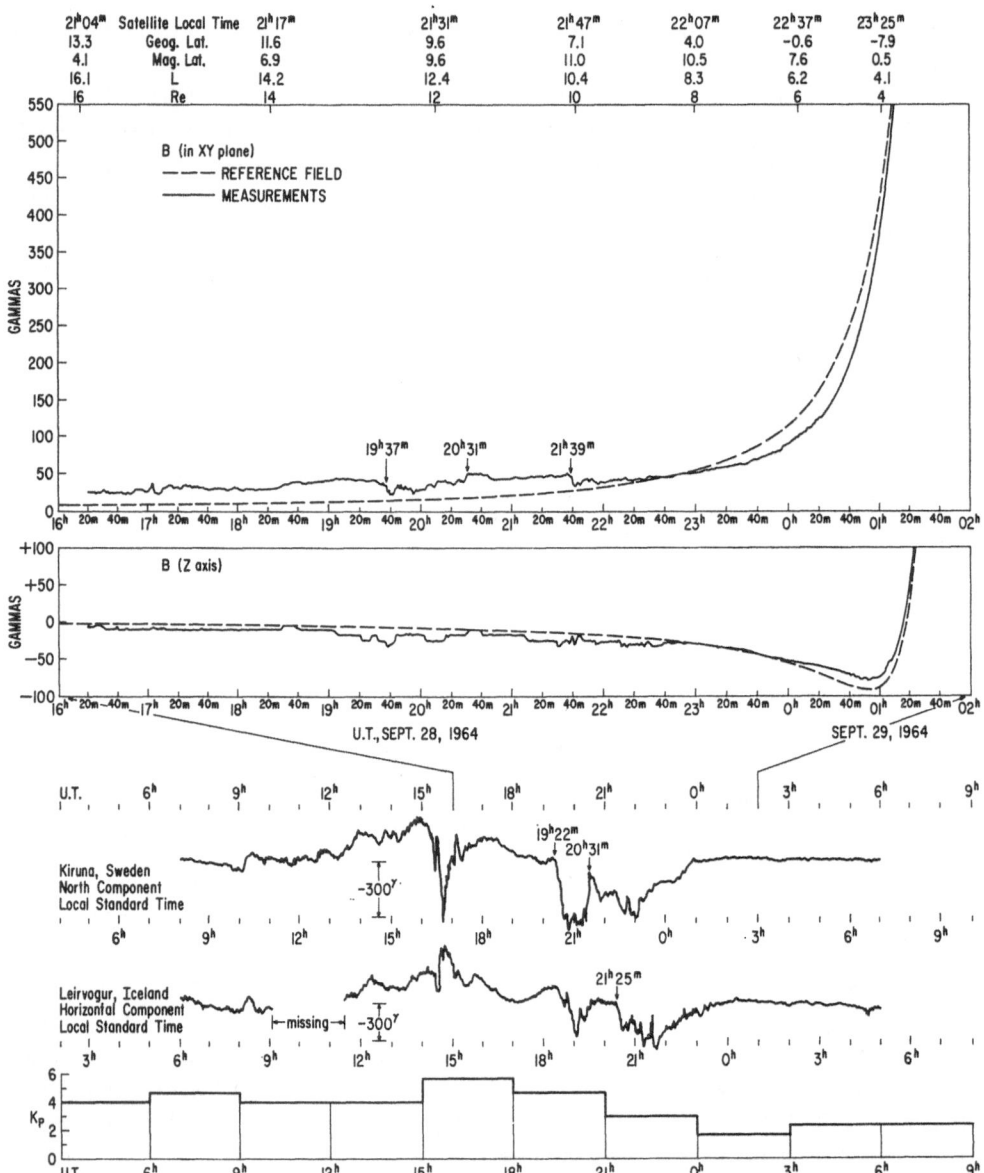

Fig. 136a, b. Simultaneous magnetic records from the OGO-1 satellite and a number of auroral zone stations (Heppner, J. P., M. Sugiura, T. L. Skillman, B. G. Ledley, and M. Campbell: *J. Geophys. Res.* **72**, 5417, 1967).

B. PARTICLE FLUX

Electrons of energies 125 eV to 2 keV have been studied by VASYLIUNAS (1968). Figures 137 and 138 show the distribution of an intense low energy electron flux in the late evening meridian and equatorial cross-section, respectively. The inner boundary

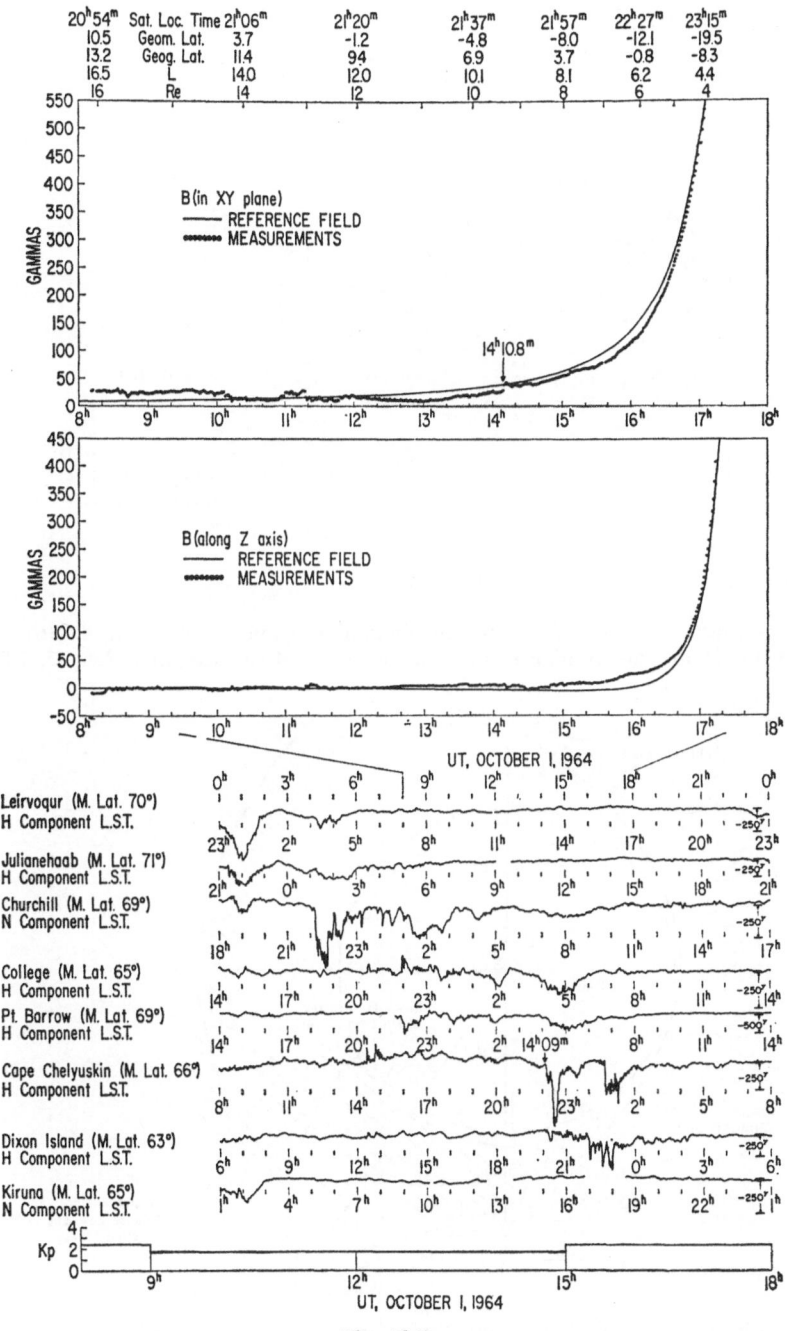

Fig. 136b.

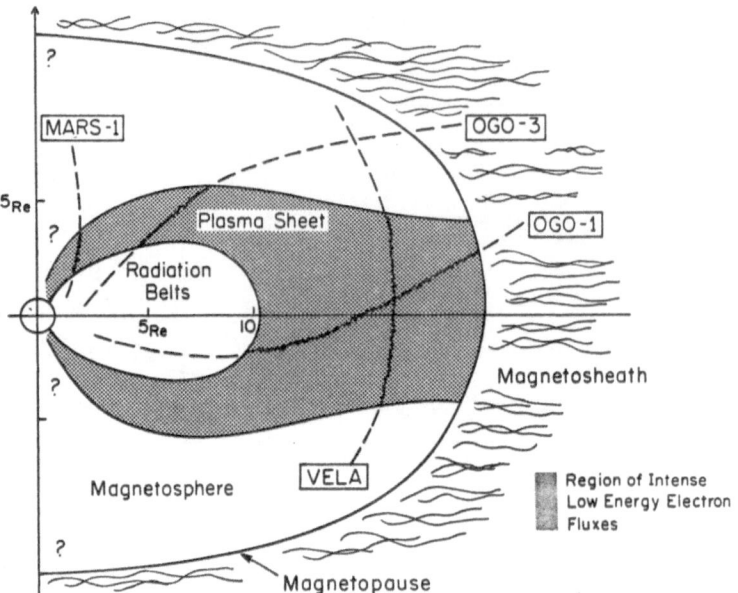

Fig. 137. Schematic diagram showing the distribution of an intense low energy electron flux in the meridian plane in the late evening sector (Vasyliunas, V. M.: *J. Geophys. Res.* **73**, 2839, 1968).

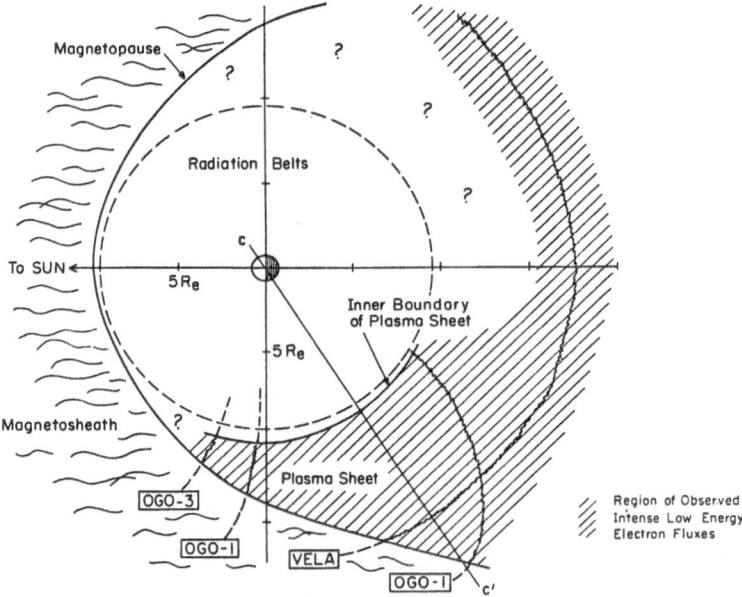

Fig. 138. Schematic diagram showing the distribution of an intense low energy electron flux in the equatorial plane (Vasyliunas, V. M.: *J. Geophys. Res.* **73**, 2839, 1968).

of the plasma sheet is sharply bounded by the outer boundary of the trapping region. VASYLIUNAS (1968) showed that during polar substorms the plasma sheet extends closer to the earth from the tail region than during quiet periods. He interpreted

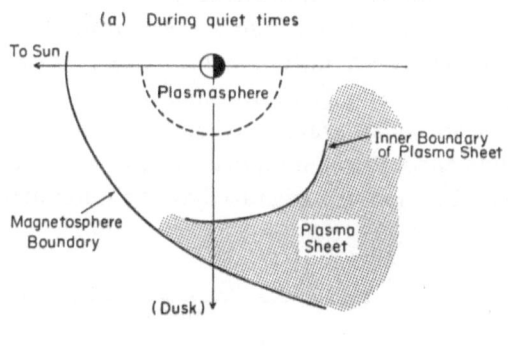

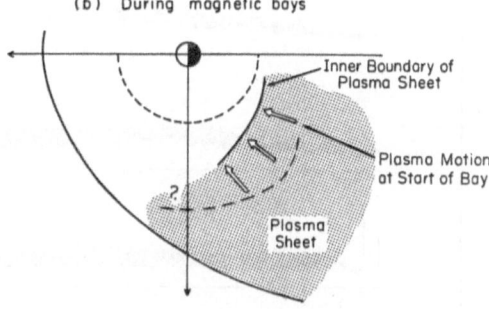

Fig. 139a.   Schematic diagram showing the changes of the distribution of an intense flux of low energy electrons (Vasyliunas, V. M.: *J. Geophys. Res.* **73**, 2839, 1968).

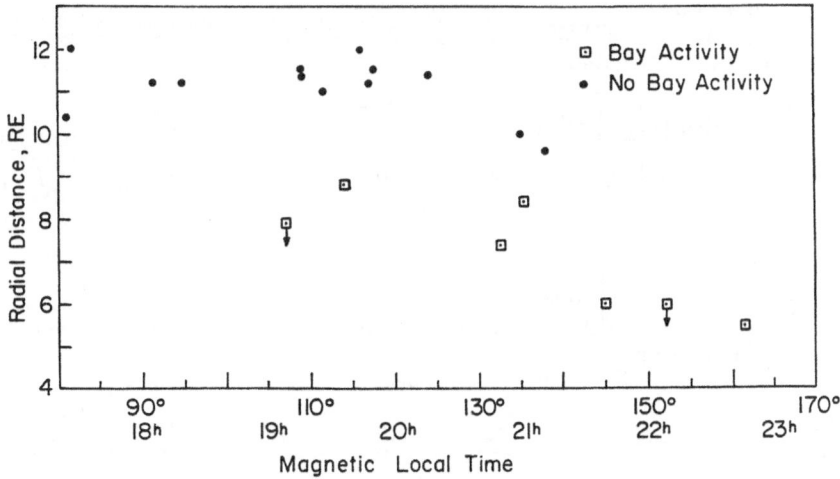

Fig. 139b.   Radial distance of the inner edge of the intense flux of low energy electrons as a function of the K_p index (Vasyliunas, V. M.: *J. Geophys. Res.* **73**, 2839, 1968).

that this results from an inward motion of the plasma during polar substorms. In one case, the estimated speed of the inward motion was 12 km/sec; supposing then that the motion is caused by the $(E \times B)$ drift, the magnitude of the potential difference across the magnetospheric tail is 48 kilovolts (see also FREEMAN, 1968). Figure 139a shows this situation schematically and Figure 139b shows the radial distance of the inner edge of the plasma sheet (as a function of local time) for two different conditions: for periods without polar magnetic substorms and for periods during polar magnetic substorms. Figure 139c shows an example of observations on October 11–12, 1964. In this case, the satellite encountered an intense flux of low energy electrons at 0645 UT, three minutes after the onset of a negative bay at Churchill (0642 UT), which was

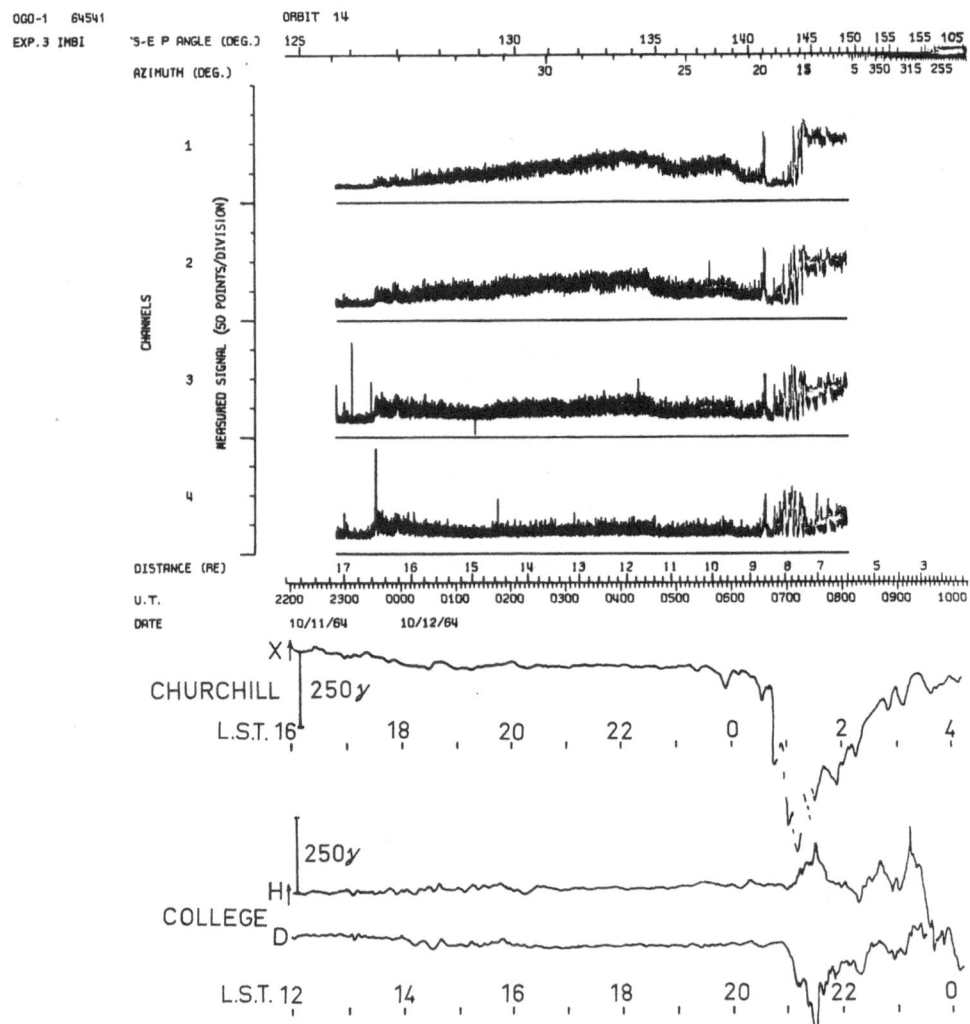

Fig. 139c. Example of the observation of the intense flux of low energy electrons during a polar magnetic substorm (Vasyliunas, V. M.: *J. Geophys. Res.* **73**, 2839, 1968).

located in the midnight sector. This particular substorm was simultaneously observed by a VELA satellite which was located at $r_e = 17a$; the VELA observations will be discussed in the next section.

An enhancement of the flux occurs for more energetic electrons ($E > 45$ keV). ANDERSON (1965) and SERLEMITSOS (1966) showed that the electron fluxes just outside the trapping region undergo rapid changes which are characterized by a rapid increase of electron flux in a time of the order of a minute, followed by a much slower decay.

REID and PARTHASARATHY (1966) and PARTHASARATHY and REID (1967) showed that such bursts of the electrons are associated with cosmic radio noise absorption. Figure 140a shows two bursts of electrons at about 0704 UT and 1012 UT, respectively,

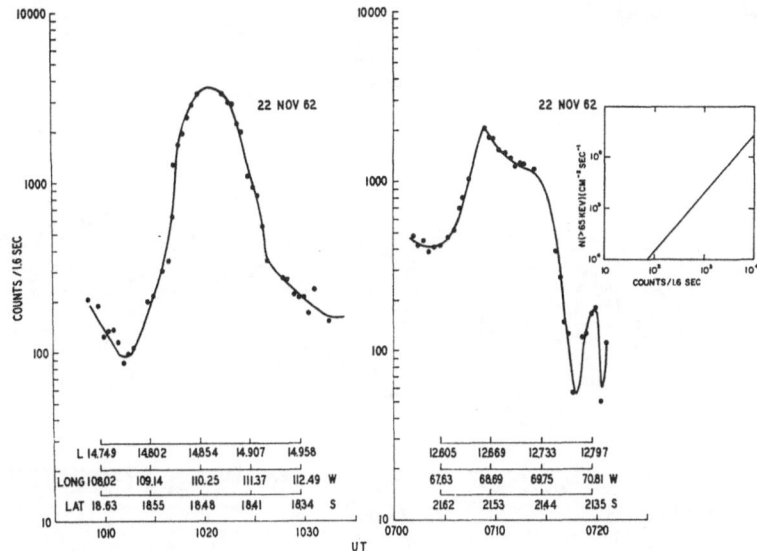

Fig. 140a.   Details of the two electron bursts observed in the tail region of the magnetosphere by the Explorer 14 satellite on November 22, 1962 (Reid, G. C. and R. Parthasarathy: *J. Geophys. Res.* **71**, 3267, 1966).

on November 22, 1962, observed by Explorer 14; Figure 140b the corresponding riometer records from Thule, Fort Yukon, College, and Healy, and Figure 140c the corresponding all-sky photographs from College.

This particular day was a rather disturbed one so that the southern boundary of the auroral oval was located well south of the zenith of College as early as 2100 LT (0700 UT).

A new arc was formed just poleward of the southernmost arc at 0702 UT. A westward traveling surge appeared along this newly formed arc at 0709 UT, which covered the College sky at 0715 UT. In the riometer record, we can see that the E type absorption began at 0710 UT at Healy and 0712 UT at College; the earlier onset at Healy is due to the fact that the initial location of the activated arc was closer to Healy than to College. The satellite ($r_e = 12.6a$) and Central Canada were located

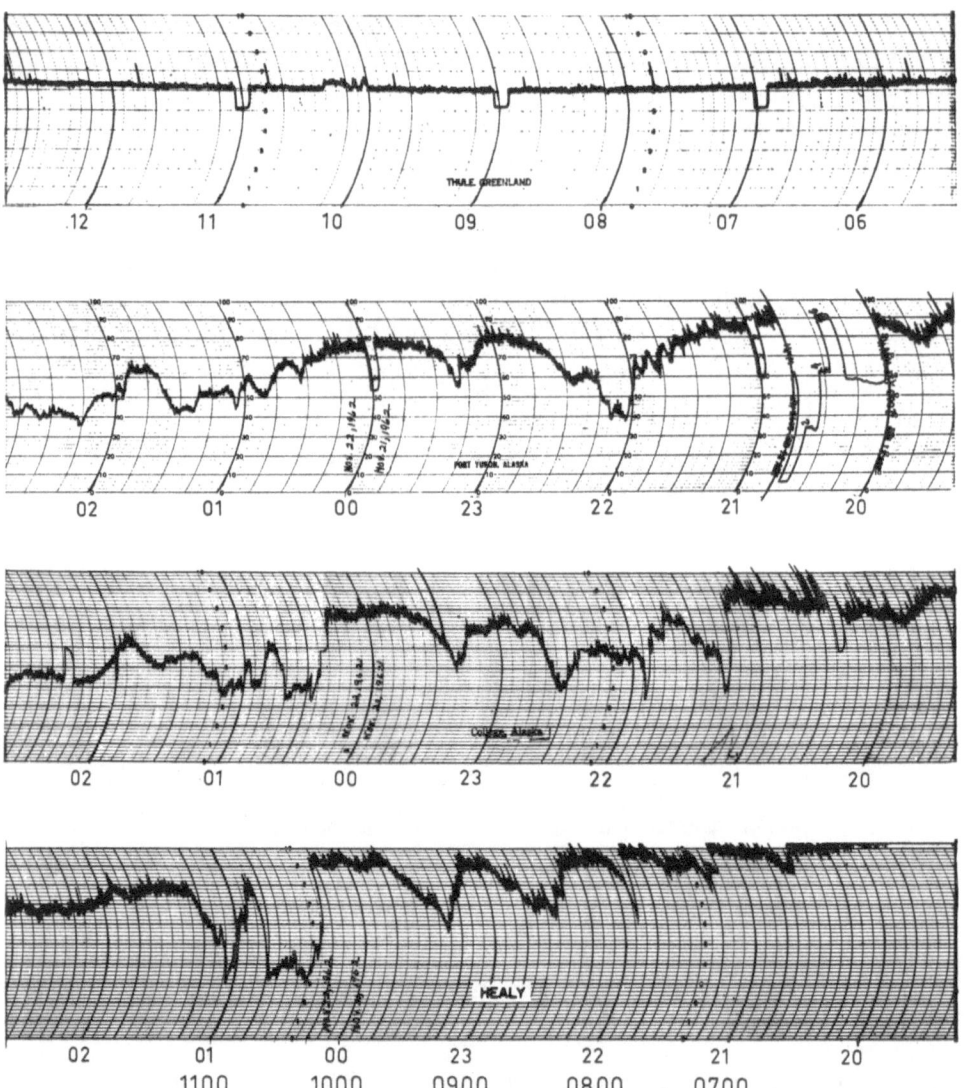

Fig. 140b.   Riometer records from Thule, Fort Yukon, College, and Healy on November 22, 1962
(Reid, G. C. and R. Parthasarathy: *J. Geophys. Res.* **71**, 3267, 1966).

in the midnight sector during this event, but Alaska was located in the evening sector.
Thus, we may infer that the substorm began in Canada at about 0702 UT or earlier.
We note also that the expanding bulge did not cover Fort Yukon ($L=6$) during this
event, so that the region of intense flux of electrons appeared to be confined along a
rather narrow band across the southern part of Alaska.

Fig. 140c.   College all-sky photographs, corresponding to the two electron bursts in Figure 140a.

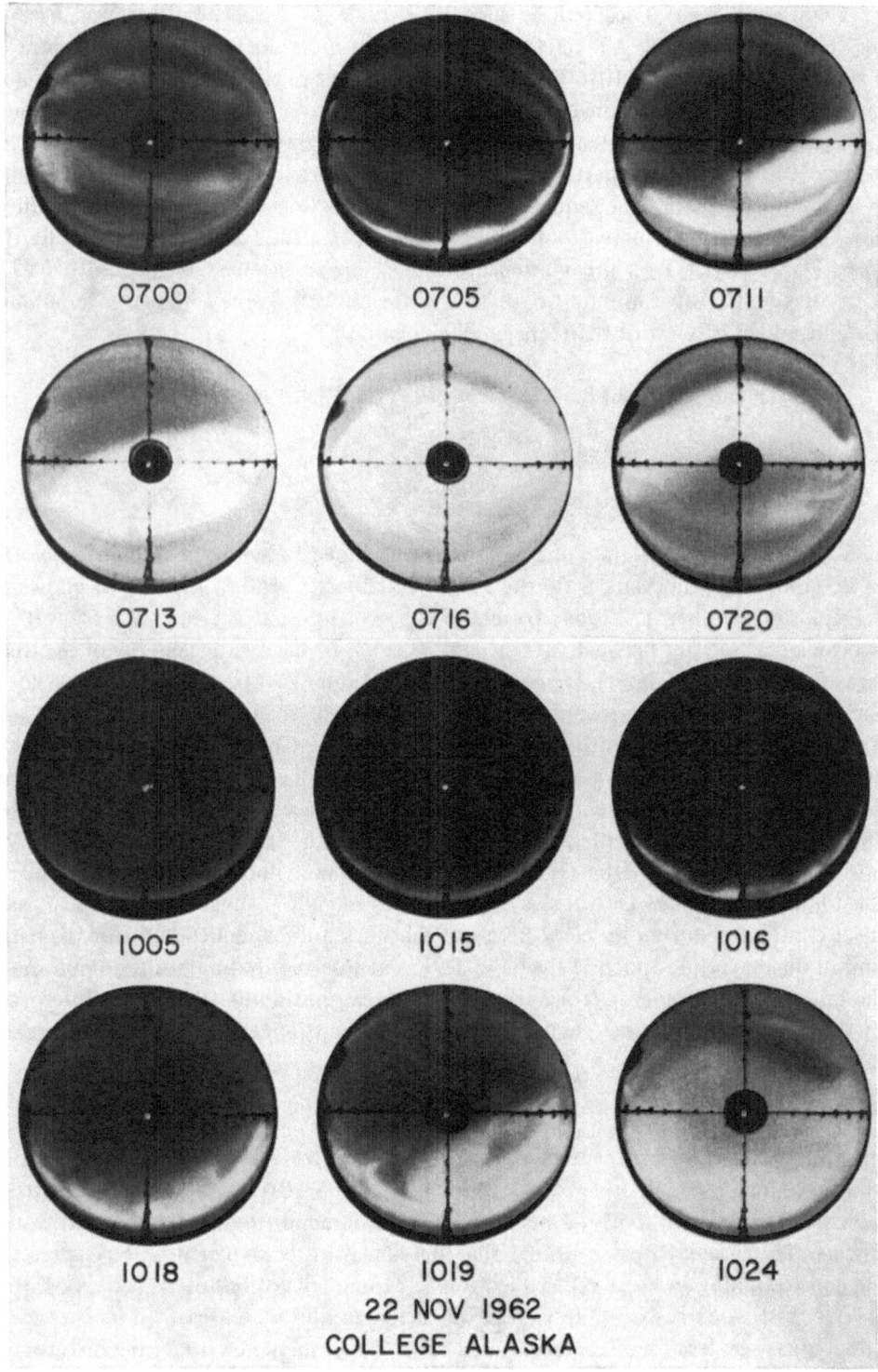

There were then at least three successive weak substorms, but auroras became very quiet by 0950 UT. AT 1004 UT, the southernmost arc increased in brightness; it became very bright at 1016 UT and began to move rapidly poleward. This is also clearly seen by the sharp onset of the N type absorption which began almost simultaneously at 1017 UT at Healy and College. The satellite began to see an enhanced flux at 1013 UT. In this particular case both the satellite and Alaska were located in the midnight sector. The sequence of auroral events was as follows: the first indication of the event (the increase of the brightness of an arc) occurred near Healy (dp lat 63.5°) at 1004 UT; a great intensification of the brightness began at 1016 UT; then, in a matter of a minute or so, an intense auroral display spread over almost the entire field of view of the College all-sky camera.

9.4. Tail Region of the Magnetosphere: the VELA and IMP Satellites

A. PARTICLE FLUX

1. Low Energy Electrons

Low energy electrons in the tail during polar substorms have been studied by Hones *et al.* (1967) by using data from the VELA satellites ($r_e = 17a$). Figure 141 shows an example on October 12, 1964; from the top, F (the analyzer flux; $F \times 5.2 \times 10^6 =$ electrons/cm^2 sec ster between 300 eV and 20 keV), E (the average energy of electrons measured with the analyzer), Geiger counter count rate (3×10^3 electrons ($E > 45$ keV)/ cm^2 sec ster), and the corresponding magnetic records from Churchill, Barrow, and College. Two distinct negative bays were recorded at Churchill (at about 0643 UT and 1050 UT). During an early phase of the negative bays, the Geiger counter rate decreased to background, and the analyzer flux decreased. However, both increased abruptly near the time of maximum development of the bays at about 0730 UT. We saw in Section 9.3, that the OGO satellite encountered the inward moving plasma sheet boundary at 0645 UT on this day. Hones *et al.* (1967) suggested that the plasma sheet contracted during an early phase of the bays and expanded suddenly near the time of the maximum epoch of the bays. They also inferred that bursts of hot plasma in the tail derive their energy from processes occurring inside the VELA satellite orbit. In the next subsection, we shall study the behavior of electrons of energies greater than 45 keV.

2. Electrons (>45 keV)

Electrons of energies greater than 45 keV show a characteristic variation which consists of a sharp increase and a slow decay (Anderson, 1965). Rothwell and Wallington (1967) examined relationships between this phenomenon and polar magnetic substorms. They clearly demonstrated that the sharp onset of negative bays precedes the corresponding increase of the electrons by some tens of minutes. The time difference is least when the satellite is close to the earth and increases with its increasing radial distance from the earth (Figure 142). They suggested that the disturbance

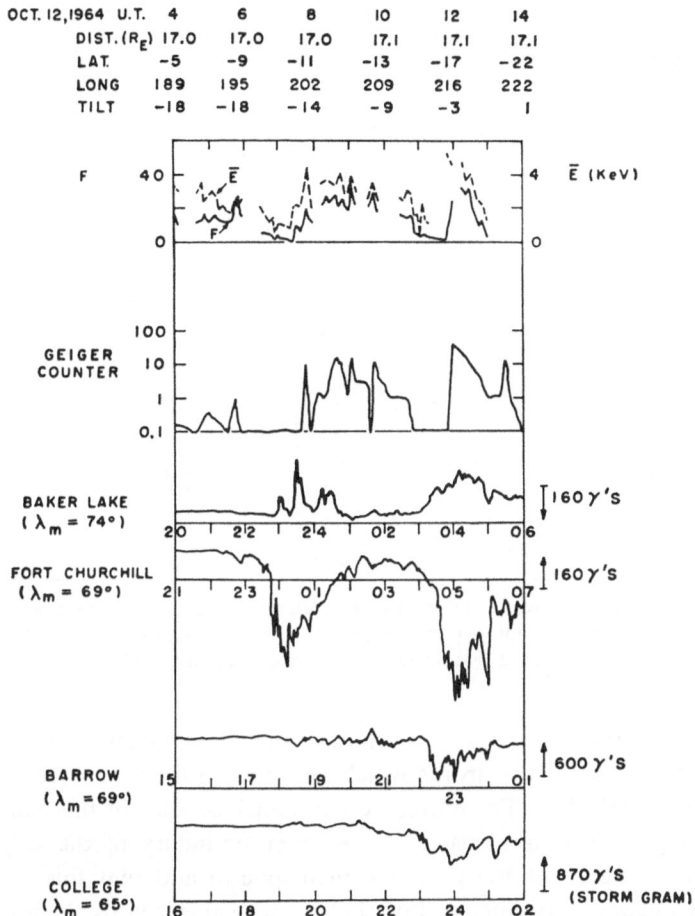

Fig. 141.   Electron fluxes observed by the VELA satellite and the simultaneous magnetic records from Churchill, Barrow, and College (Hones, E. W., Jr., J. R. Asbridge, S. J. Bame, and I. B. Strong: *J. Geophys. Res.* **72**, 5879, 1967).

producing the magnetic bays and associated particle acceleration originates fairly deep in the magnetosphere and propagates outward to higher L values.

PARTHASARATHY and REID (1967) showed that bursts of electrons in the tail region are also associated with cosmic radio noise absorption. Figure 143 shows an example of an electron burst observed at 2210 UT on May 28, 1964, by the IMP-I satellite which was located at $r_e = 28a$. Figure 143 also shows cosmic radio noise absorption observed at the South Pole and Kiruna. The absorption at the South Pole (dp lat 78.5°) began at about the time when the electron burst was observed in the tail. On the other hand, the corresponding magnetic record from Halley Bay (located in the auroral zone) shows that the polar substorm began at about 2205 UT (Figure 143). This important example suggests that the auroral substorm also began in the auroral zone at about 2205 UT and that the expanding auroral bulge reached the

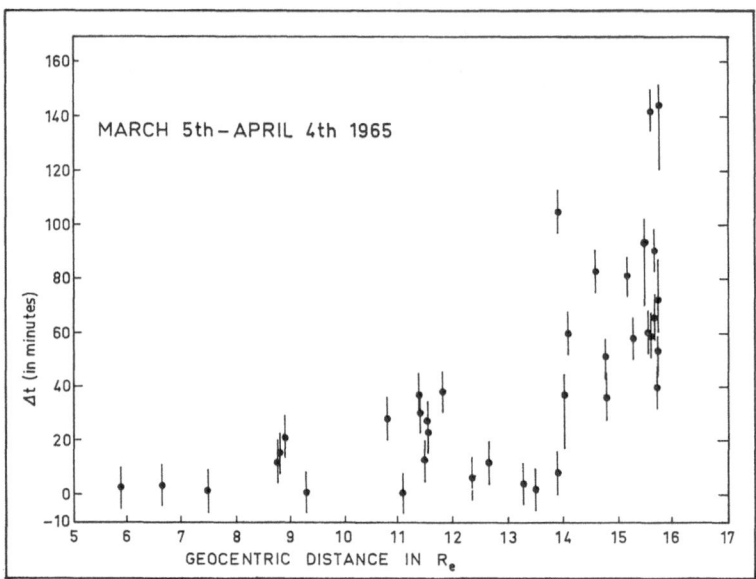

Fig. 142. Time difference between the onset of a sudden increase of electron flux in the tail region and the onset of the corresponding polar magnetic substorm (Rothwell, P. and V. Wallington: Birkeland Symposium, Sandfjord, Norway, 1967).

South Pole at 2210 UT. Indeed, the simultaneous all-sky camera photographs (not illustrated here) show active auroras which approached the zenith from the equatorward horizon at 2210 UT. Therefore, we can conclude that in this particular event the substorm process began first near the outer boundary of the trapping region (or the inner boundary of the outer magnetosphere) and that this process moved progressively outward. An intense flux of electrons generated by the process reached a distance of 28 earth radii in the tail region and over South Pole at about the same time. This event provides an important indication that the poleward explosive motion of the auroral system in the midnight sector is related to the outward progression of the substorm process. However, a number of examples should be carefully examined before we reach a definite conclusion on this very important point. A similar example on May 15, 1964, was examined by BRICE (1967). Comparing one of the bursts which began at about 0250 UT with the simultaneous riometer records from Canadian stations, he concluded that the electron burst and the cosmic radio noise absorption began at about the same time. However, the earliest onset time of this particular polar magnetic substorm was registered as a negative bay at Murmansk (dp lat 63.5°) at about 0240 UT, about 10 min earlier than the onset time of the burst.

The correlation between the electron burst and the corresponding cosmic radio noise absorption was recently examined in great detail by HARGREAVES *et al.* (1968) on the basis of extensive riometer and VELA satellite data. Their conclusions are:

(1) The association between particle events in the magnetospheric tail and absorp-

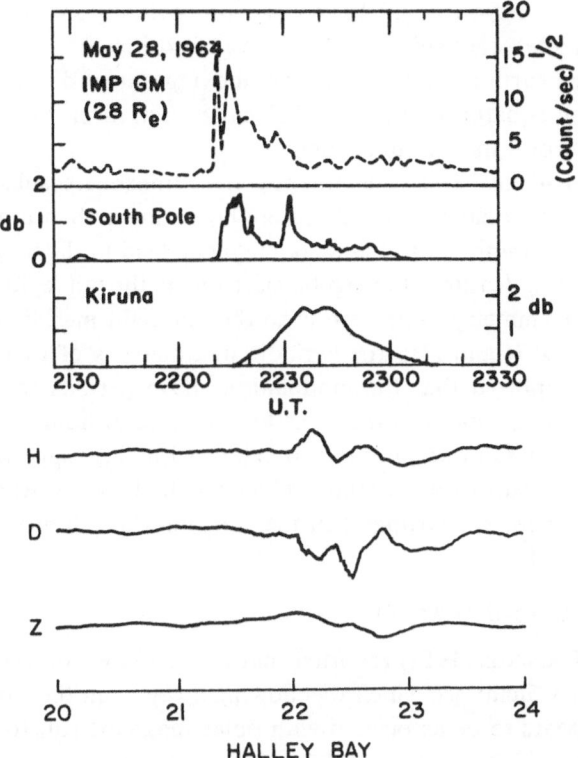

Fig. 143.   Electron bursts in the tail regions and the corresponding riometer records from South Pole and Kiruna. The simultaneous Halley Bay magnetic record shows that the substorm associated with the electron burst began at 2205 UT: the upper diagram was provided by Reid, G. C. and R. Parthasarathy (1966).

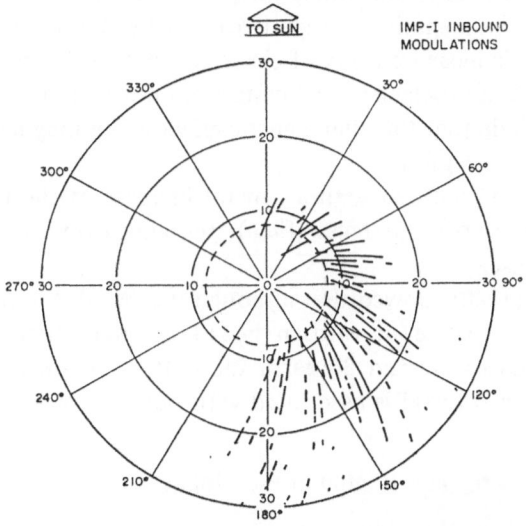

Fig. 144.   Location in which periodically modulated electron fluxes are observed in the magnetosphere (Lin, R. P. and K. A. Anderson: *J. Geophys. Res.* **71**, 1827, 1966).

tion events in the D region could be verified for virtually every event provided a satellite and a riometer were suitably placed to record it.

(2) At seventeen earth radii, the events are moving outward, away from the earth. The field lines at an equatorial crossing of 17 earth radii meet the earth considerably poleward of the auroral maximum at night.

(3) The flux of 45 keV electrons entering the atmosphere substantially exceeds the associated flux at 17 earth radii. This may be no more than a geometrical effect due to the bursts spreading out in the weak magnetic field in the magnetospheric tail.

Another interesting feature of energetic electrons in the tail region of the magnetosphere is a great asymmetry (with respect to the midnight meridian) of the distribution (BAME et al., 1967) and also the periodic modulations of their fluxes. LIN and ANDERSON (1966) reported that the modulations have periods ranging from a few minutes to half an hour, and a duration as long as several hours. They showed that the modulations increase in amplitude and extent with K_p, suggesting that they are associated with polar substorms. Figure 144 shows the location where such modulations are detected. They are strongest in the early morning hours and are bounded rather sharply at $L=8$.

3. Energetic Protons (>0.31 MeV)

ARMSTRONG and KRIMIGIS (1968) reported that the Explorer 33 satellite encountered a small flux of order $5/cm^2$ sec which was flowing away from the earth along the tail. Such an event appears to be associated with polar magnetic substorms.

B. MAGNETIC FIELD

The magnetic field variation during polar magnetic substorms is a rather simple negative change. ANDERSON and NESS (1966) showed that such a negative change is well correlated with the electron bursts mentioned in Section A2 (see Figure 145). They suggested that the negative change is caused by the diamagnetic effect of the plasma, although their measured flux of electrons account for only about 1% of the total diamagnetism. Since such electron bursts occur during polar magnetic substorms, it is natural to conclude that this diamagnetic effect is the magnetic manifestation of the substorm in the tail region.

This should not be confused with a general increase of the tail field during disturbed periods (BEHANNON and NESS, 1966); the substorm effect is superposed on such a general increase.

MIHALOV et al. (1968) showed that although the North-South component of the tail field is, in general, positive (upward) in the tail region, its direction is often reversed (downward) at distances beyond 30 earth radii. It is of great interest to examine whether or not such a reversal is associated with polar substorms.

9.5. Trapping Region of the Magnetosphere ($r_e < 6a$)

Since the discovery of the Van Allen radiation belts, variations of the flux and

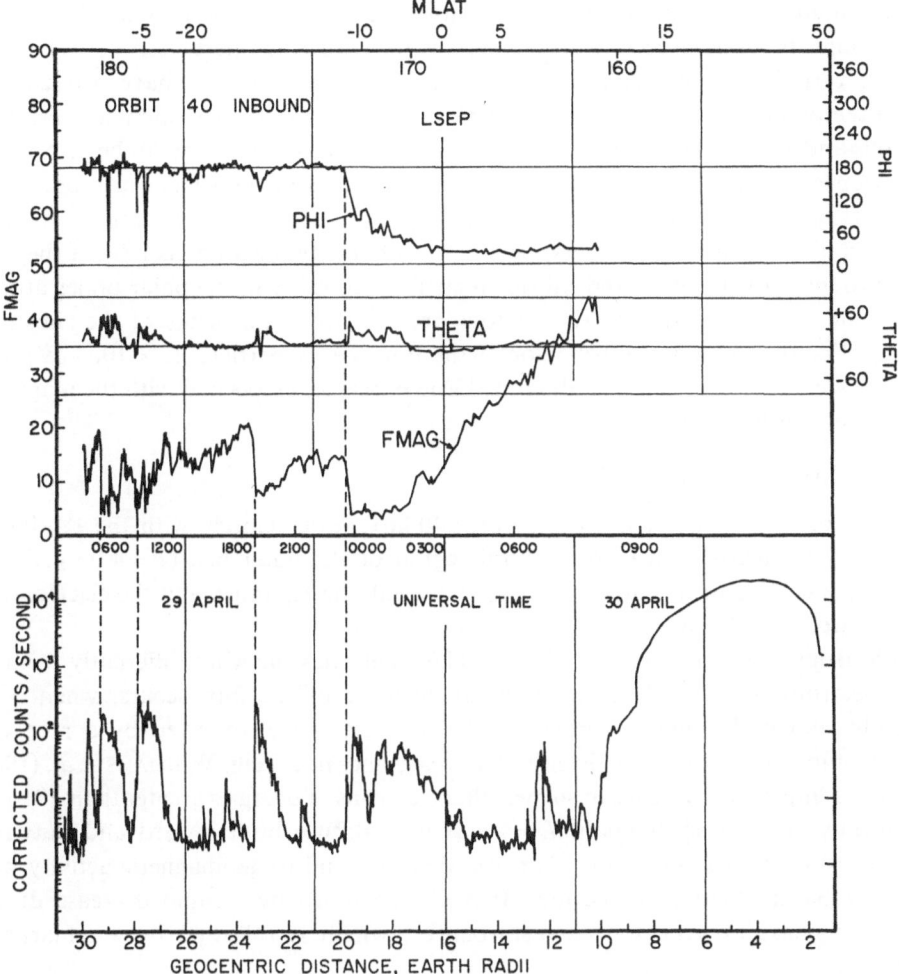

Fig. 145.   Correlation between the electron bursts and negative changes of the magnetic field intensity in the tail region (Anderson, K. A. and N. F. Ness: *J. Geophys. Res.* **71**, 3705, 1966).

spectrum of energetic particles in the trapping region have been investigated by a number of workers. Since these studies are well summarized by HESS (1968) and ROEDERER (1968), they are not repeated here. Unfortunately, most of those studies have reported changes of the radiation belts as a function of the storm time, so that the time scale is too large to examine the variations during individual substorms. Undoubtedly, the magnetospheric substorm is a very energetic phenomenon, so that it must play an important role in controlling the behavior of the energetic particles in the trapping region. In Section 9.2, we saw that electrons of energies greater than 50 keV increase considerably during individual substorms, so that a large fraction of 50 keV electrons must be generated inside the trapping region or supplied from outside impulsively and sporadically by processes associated with substorms. Therefore, without examining the variation of the flux with substorm time, we may not be able to understand the basic process which generates the Van Allen particles.

The variations of the Van Allen belts during magnetic storms have two aspects. From the point of view of the total kinetic energy involved, energetic particles, both protons and electrons of energies greater than 100 keV, are not likely to be of primary importance; some of them may simply be by-products of the substorm process. On the other hand, particles of energies less than 100 keV, in particular protons of energies less than 50 keV, carry a considerable amount of energy (Section 10.4 A). If they are a by-product of substorms, the substorm energy deposited in the polar upper atmosphere and that transferred to the protons are the two major sinks of the substorm energy. For this reason, the behavior of the low energy particles ($E < 100$ keV) may provide a clue toward understanding the basic processes associated with the magnetospheric substorm.

A. ENERGETIC ELECTRONS

The flux of electrons of energies of order 40 keV varies closely with the K_p index. This is most clearly seen in the central region of the outer belt ($L = 4.0 \sim 4.5$), but becomes less clear for smaller L values; at $L = 3.0$, the K_p index and the electron flux are not well correlated.

Electrons of energies greater than 280 keV behave somewhat differently from 40 keV electrons. At $L > 4.5$, a rapid initial decrease of the flux occurs, when $\sum K_p$ (= daily sum of K_p indices) exceeds 20; this decrease is followed by a recovery. On the other hand, at $L < 4.0$, the initial decrease does not occur. WILLIAMS *et al.* (1968) presented important evidence to suggest that electrons of energies greater than 300 keV are locally accelerated at about $L = 4$ and then diffuse both inward and outward.

Electrons of energies greater than 1.6 MeV respond to geomagnetic activity with considerable fluctuation in the flux. It is characterized by a rapid decrease during the early phase of geomagnetic disturbances, which is followed by a remarkable recovery.

CRAVEN (1966) examined the above-mentioned behaviors using the best time resolution (~2 hours) attained by the Injun 3 satellite for two periods, between March 28 and April 12, 1963, and between June 4 and June 10, 1967. Figure 146 shows the 3-hour averages of the horizontal component of the College magnetic record, the variations of the flux for electrons ($E > 40$ keV, $E > 230$ keV, $E > 1.6$ MeV) and K_p index. We can see from Figure 146 that the above mentioned changes of the electron fluxes can take place in several hours, much shorter than the lifetime of geomagnetic *storms*.

B. PROTONS (1 ~ 50 KEV)

Unfortunately, the variation in the flux of protons (1 ~ 50 keV) during polar substorms has not been determined. However, FRANK (1967a, b, c, d) has shown that protons appear deep in the trapping region even during rather weak magnetic storms and that the ratio of the kinetic energy density of the protons to the magnetic field energy density sometimes exceeds unity.

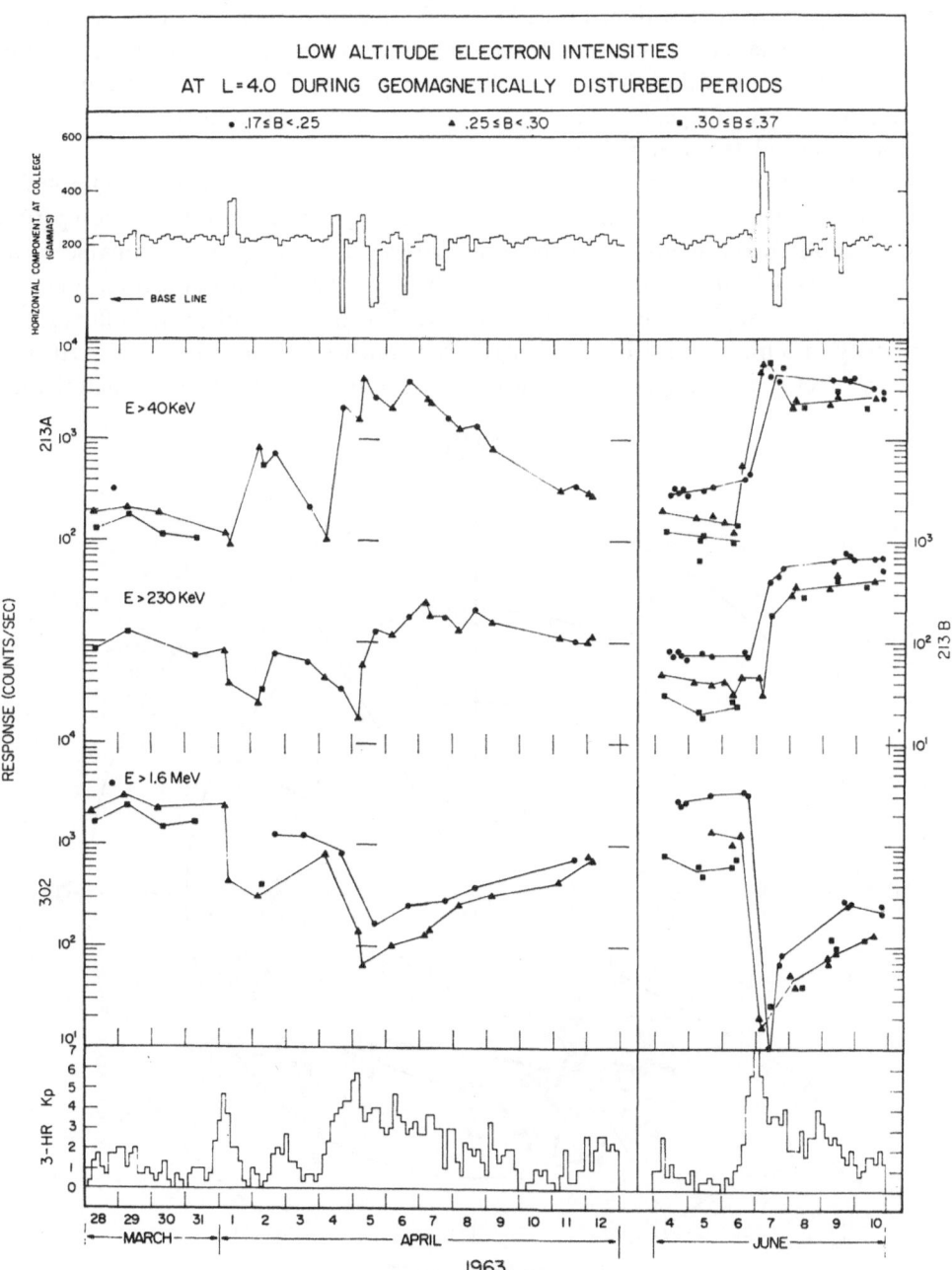

Fig. 146.   College magnetic record (3-hour average) and the corresponding variations of electrons for different energies ($E > 40$ keV, $E > 230$ keV, $E > 1.6$ MeV) (Craven, J. D.: *J. Geophys. Res.* **71**, 5643, 1966).

9.6. Observations by Low Altitude Satellites

An extensive study of variations of the fluxes of energetic particles has also been made by low latitude satellites; some of the results have been reported in the preceding sections. Since most of the studies are well summarized by O'BRIEN (1967) and others, we shall not review the general outcome of such studies but confine our attention to substorms and related phenomena.

Orbital characteristics of low altitude polar orbiting satellites are not suitable for observing the development of individual polar substorms, since their revolution period is comparable to the lifetime of substorms. However, they are very valuable in at least two respects. First, since they move rapidly in a North-South direction, they can be used to give rough estimates of the latitudinal distribution of particle fluxes and luminosity of auroral emissions at a particular epoch of a substorm in a particular local time sector. Secondly, they can provide a vast amount of statistical material far

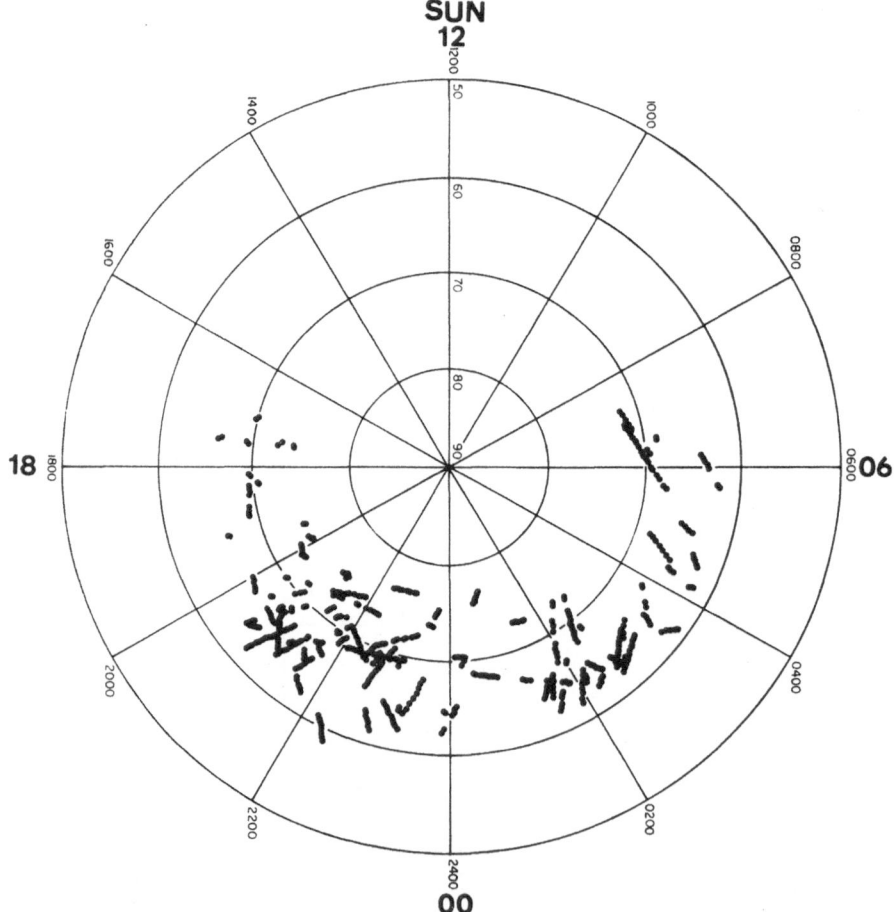

Fig. 147.   Distribution of the locations where an intense flux of electrons of energies greater than 10 keV was observed (Fritz, J. A. and D. A. Gurnett: *J. Geophys. Res.* **70**, 2485, 1965).

more efficiently than a North-South chain of observatories on the ground. It is mainly for this latter reason that some of the results of satellite observations have already been mentioned in preceding chapters and sections.

A. ELECTRONS (E ~ 10 KEV)

FRITZ and GURNETT (1965) constructed a polar map to indicate the location where the Injun 3 satellite observed an intense flux of electrons of energies greater than 10 keV. Their diagram is shown in Figure 147. Their distribution of the precipitation is well confined to the dark hemisphere, and the equatorward boundary of the precipitation suggests that the precipitation occurs along the auroral oval. Further, the satellite encountered intense fluxes in the region where the westward traveling surges appear most frequently.

An extensive study of the particle flux along the midnight-noon meridian was made by EVANS *et al.* (1966), SHARP *et al.* (1967), and SHARP and JOHNSON (1968). They found that in the midnight meridian, there is a single narrow band in which both electrons and protons precipitate, while in the noon meridian two distinct bands appear (Figure 148); soft electrons precipitate in a band centered at dp lat 77°, and hard electrons in a band centered at dp lat 71°; note that colatitude is used in Figure 148. It is likely that the higher latitude corresponds to the auroral oval and the lower band to the precipitation region associated with the M type absorption.

B. ELECTRONS (E ~ 50 KEV)

FRANK *et al.* (1964) and McDIARMID and BURROWS (1968) obtained the intersection line between the ionosphere and the outer boundary of the trapping region (that is, the projection of the outer boundary of the trapping region onto the ionosphere along

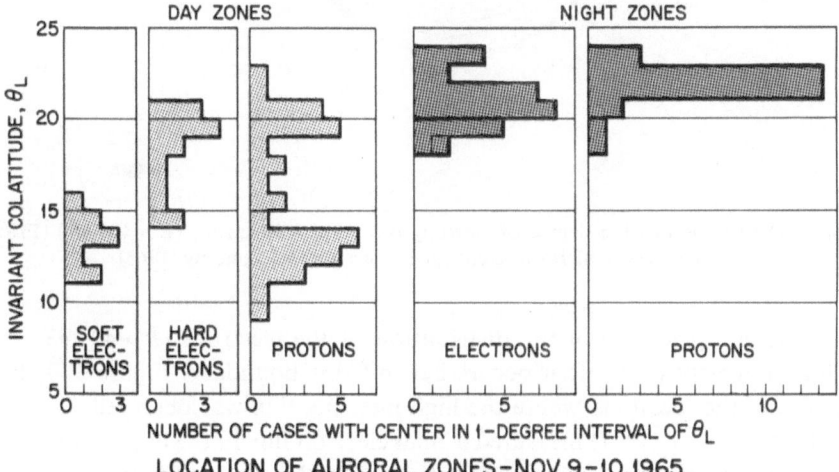

Fig. 148. Latitudinal distribution of the precipitation of electrons and protons in the noon-midnight meridian (Sharp, R. D. and R. G. Johnson: 'Earth's Particles and Fields', *Proc. NATO Advanced Study Institute, July 31-August 11*, 1967).

the geomagnetic field lines). In general, the flux of electrons of energies greater than 40 keV with their pitch-angle at 90° sharply decreases beyond a certain latitude. They showed that this boundary is eccentric with respect to the dipole pole. In Section 1.1 we noted that this intersection line between the ionosphere and the outer boundary of the trapping region coincides approximately with the auroral oval.

Recently, FRITZ (1968) examined this 'boundary' in more detail by introducing a parameter which is the ratio of the flux of electrons with $E > 40$ keV being precipitated into the atmosphere to the flux of electrons with $E > 40$ keV mirroring near the position of the Injun 3 satellites. He showed that in the midnight sector the parameter ϕ increases very rapidly near the boundary toward higher latitudes from about 10^{-2} to unity. Beyond this point, the flux remained isotropic ($\phi = 1$). However, the tendency toward isotropy near the boundary becomes less with increasing local time; at 18 LT, isotropy was seldom observed at the boundary (Figure 149).

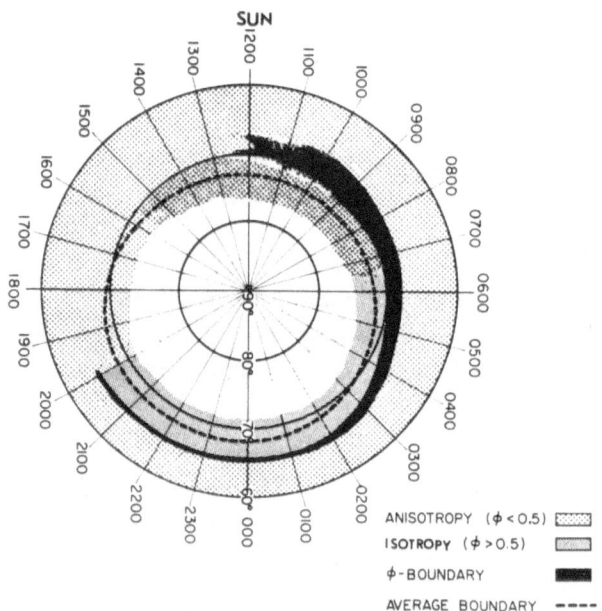

Fig. 149.   Distribution of the degree of isotropy of energetic electrons ($E > 40$ keV) (Fritz, T. A.: University of Iowa, Dept. of Physics and Astronomy, 1968).

An interesting feature of the distribution of the electrons ($E > 40$ keV) is that a very high intensity peak often occurs beyond this boundary. Figure 150 shows the polar plot of the locations where the high intensity flux was observed.

JELLY and BRICE (1967) measured a high electron flux in the day sector when polar magnetic substorms were occurring in the midnight sector. Figure 151 shows that during successive paths of a satellite over Canada, a marked increase of the electron flux was observed when a sharp negative bay began in Siberia (Dixon, Tixie Bay) which was located in the night sector at that time.

C. PROTONS

As mentioned earlier, SHARP *et al.* (1967) found that along the midnight-noon meridian there is a single peak of the proton precipitation near the auroral zone in the night sector and a double peak in the day sector. The energy spectrum of these peaks appears to be similar. They showed also that the flux in the evening sector is more intense than in the morning sector. ROMICK and SHARP (1967) observed the Hα emission from a hydrogen aurora by both a satellite-borne detector and ground instruments (Figure 152). During one particular observation, the hydrogen aurora had a latitudinal width of about 3° (between 65.7° and 68.3°). Within the hydrogen aurora, there was one narrow region in which hard electrons precipitated; λ 5577 luminosity was seen in this region. There were also three regions in which soft electrons precipitated; one of them was within the hydrogen aurora, the second near the poleward boundary of the hydrogen aurora and the third poleward of the second.

D. MAGNETIC FIELD

ZMUDA *et al.* (1966) and ZMUDA *et al.* (1967) found that transverse (to *B*) magnetic disturbances occur over the auroral oval, but not elsewhere. When the oval shifts equatorward during geomagnetic storms, there is also a corresponding shift of the region of transverse magnetic disturbance. They have interpreted that the transverse disturbances are produced by electric currents flowing along geomagnetic field lines which originate in the oval.

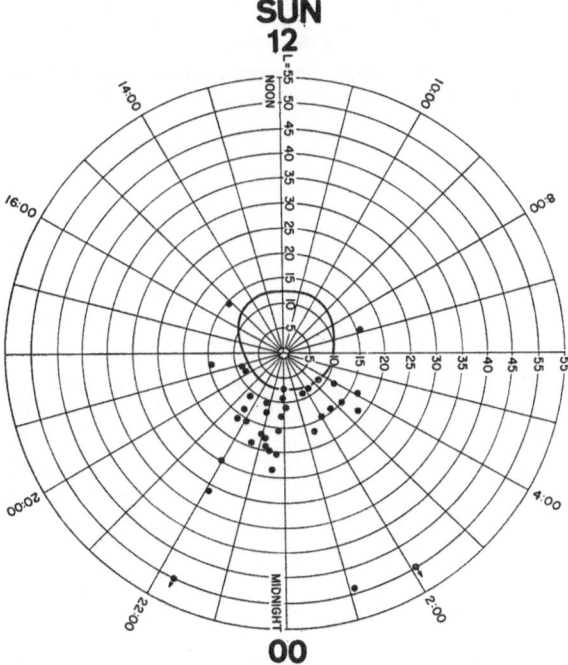

Fig. 150.   Distribution of the locations where an intense flux of electrons (40 keV) was observed (McDiarmid, I. B. and J. R. Burrows: *Canadian J. Phys.* **46**, 49, 1968).

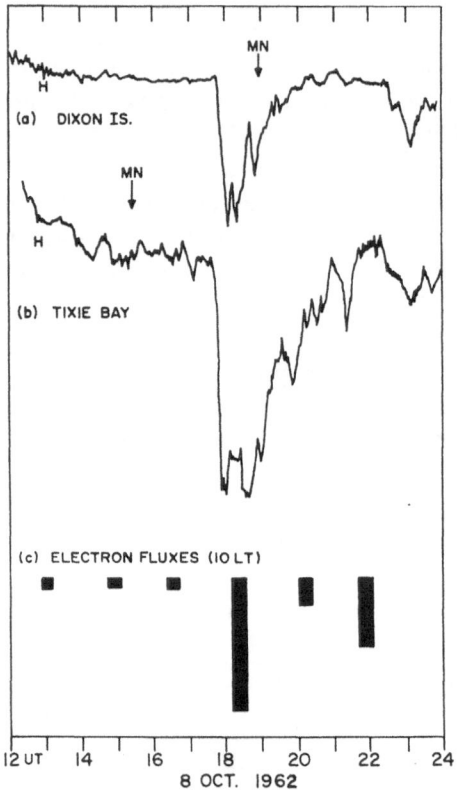

Fig. 151. Sudden increase of electron fluxes observed over Canada, when an intense negative bay occurred in Siberia (Jelly, D. and N. Brice: *J. Geophys. Res.* **72**, 5919, 1967).

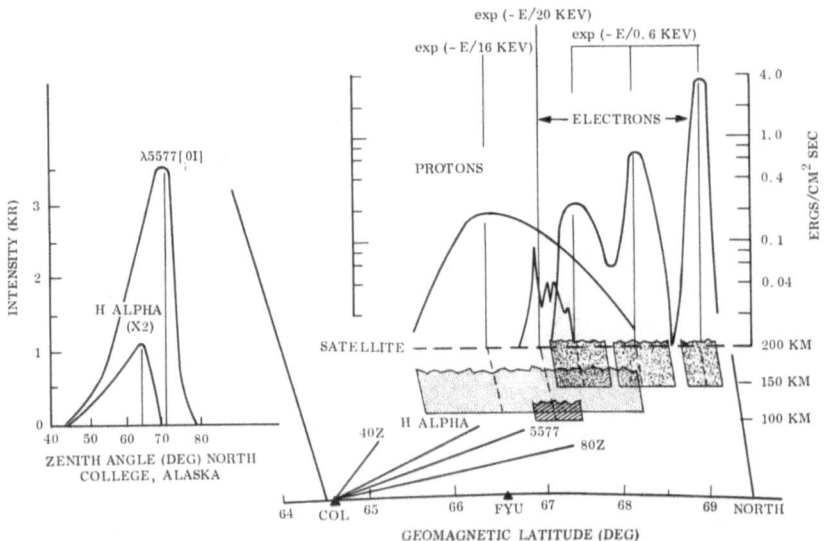

Fig. 152. Simultaneous observation of the proton aurora by both a satellite-borne detector and ground instruments (Romick, G. J. and R. D. Sharp: *J. Geophys. Res.* **72**, 4791, 1967).

LANGEL and CAIN (1967) examined OGO-2 (polar orbiting) satellite data during the geomagnetic storm of March 13–14, 1966. Although the satellite measured only the (scalar) magnitude of the disturbance field, they were able to infer a reasonable current distribution across the path of the satellite. They indicated that the two cell model current system (the *SD* current system) fits better than the one cell model current system; for details of the model current system, see Chapter 3 (see Figures 31a, b, c).

9.7. Solar Wind and Polar Magnetic Substorms

A. SOLAR WIND PARTICLES

So far there have been only a few studies of the possible relationship between solar wind quantities, such as the number density, velocity, particle flux, energy flux and momentum flux and the onset of individual polar substorms. However, by using the geomagnetic planetary index K_p, SNYDER *et al.* (1963), WILCOX *et al.* (1967) and SCHATTEN and WILCOX (1967) showed that among the quantities mentioned above, the solar wind velocity is the only quantity which has an obvious relation with K_p. There is little correlation between K_p and solar wind density.

This conclusion is supported by observation of the solar wind during the geomagnetic storm of April 17/18, 1964, by the VELA satellite as noted by GOSLING *et al.* (1967). They showed that during an early phase of the storm when intense polar substorms occurred frequently there was no obvious change of the solar wind quantities.

B. INTERPLANETARY MAGNETIC FIELD

FAIRFIELD and CAHILL (1966) and FAIRFIELD (1967) found that a southward directed component of the interplanetary magnetic field tends to be more closely associated with polar magnetic substorms than a northward directed field. Figure 153 shows such an example on November 5/6, 1964. The top three traces represent the field magnitude *F*, solar ecliptic latitude and longitude angles θ and ϕ. The fourth trace in Figure 153 represents the superposition of six Arctic records whose envelope is defined as the AE index. The bottom traces represent the disturbance magnitudes at the three polar cap stations, Resolute Bay, Mould Bay, and Alert.

The interval from 0900 UT to 1100 UT was quiet. At 1105 UT, the θ value became negative, indicating that the North-South component changed in sign and was directed southward. The AE index became large soon after this change. Then, at about 1400 UT, the θ value became positive; it became negative at 1430 and 1740 and 2300 UT, respectively, with a positive value in between. Corresponding to these changes, there was some indication of polar magnetic disturbance. A northward directed field is associated with a decrease in polar magnetic disturbances. Figure 154 constructed by FAIRFIELD (1967), shows statistical results of this study. The greatest disturbance is associated with large southward (negative) fields and the quietest conditions with weak northward (positive) fields. This study suggests that the magnitude of the fields, as well as the direction, plays an important role in the geomagnetic activity.

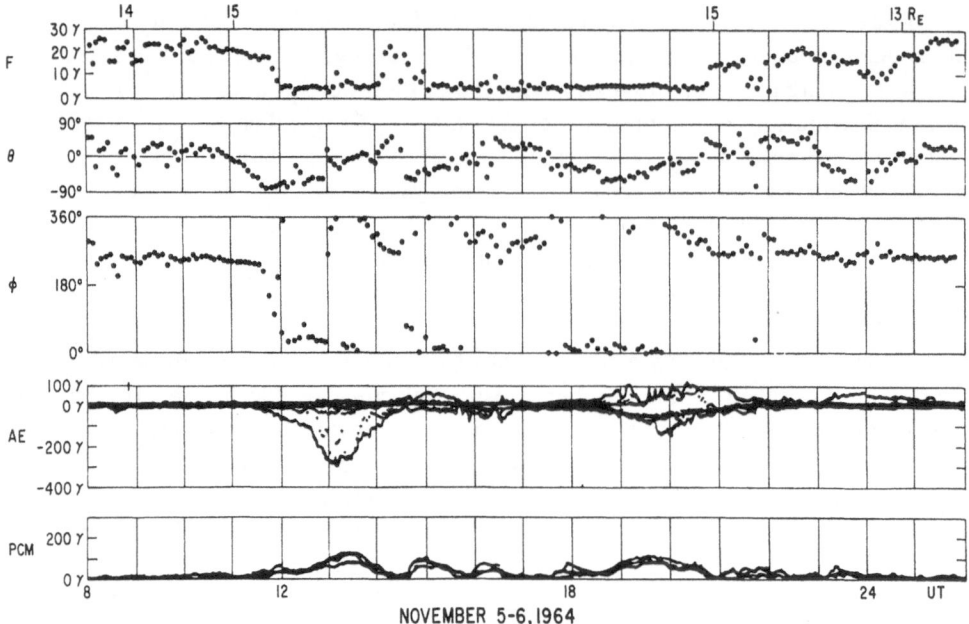

Fig. 153. Example suggesting a possible relationship between changes of interplanetary magnetic fields and polar magnetic substorms (Fairfield, D. H.: Goddard Space Flight Center Pub. X-612-67-338, July, 1967).

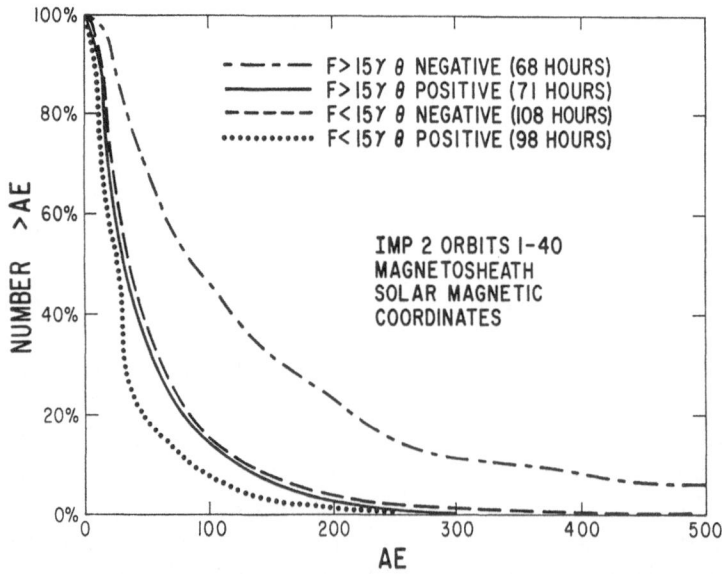

Fig. 154. Number of cases with the AE index greater than the abcissa for four groups depending on direction and magnitude (Fairfield, D. H.: Goddard Space Flight Center Pub. X-612-67-338, July, 1967).

On the other hand, WILCOX *et al.* (1967) and SCHATTEN and WILCOX (1967) indicated that the magnitude of the field is the only quantity which has an obvious relationship with K_p.

C. ELECTRIC FIELD

SCARF *et al.* (1968) observed VLF electric fields in the frequency range 0.1 to 100 Kc/s in interplanetary space by Pioneer 8 satellite. Figure 155 shows an example on December 29/30, 1967. On that day, Pioneer 8 was located at a distance corresponding to a solar wind travel time of about 55–65 min for the normal quiet flow speed (380 ~ 450 km/sec). The simultaneous College magnetogram is also shown. SCARF *et al.* (1968) suggested that an intense negative bay which began at 1344 UT at College was associated with VLF electric field noise bursts observed by Pioneer 8 at 1450 UT.

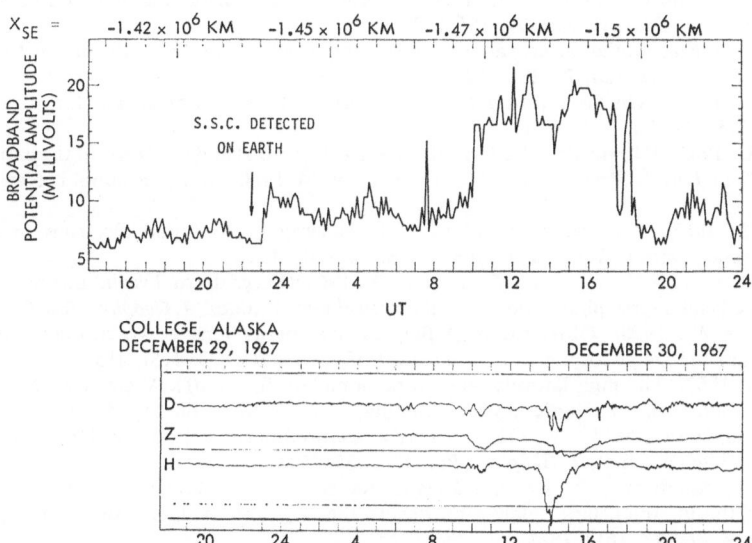

Fig. 155. Intense VLF electric fields observed in interplanetary space and the corresponding College magnetic record (Scarf, F. L., G. M. Crook, I. M. Green, and P. F. Virobik: *J. Geophys. Res.* **73**, 1968).

References

GENERAL

O'BRIEN, B. J.: 1962, 'Review of studies of trapped radiation with satellite-borne apparatus', *Space Sci. Rev.* **1**, 415–484.
O'BRIEN, B. J.: 1967, 'Interrelations of energetic charged particles in the magnetosphere', in *Solar-Terrestrial Physics* (ed. by J. W. King and W. S. Newman), Academic Press, N.Y., pp. 169–211.
BROWN, R. R.: 1966, 'Electron precipitation in the auroral zone', *Space Sci. Rev.* **5**, 311–387.
HESS, W. N.: 1968, *The radiation belt and magnetosphere*, Blaisdell Pub. Co., Waltham, Mass.

REFERRED TO IN TEXT

ANDERSON, K. A.: 1965, 'Energetic electron fluxes in the tail of the geomagnetic field', *J. Geophys. Res.* **70**, 4741–4763.

ANDERSON, K. A. and NESS, N. F.: 1966, 'Correlation of magnetic fields and energetic electrons on the IMP 1 satellite', *J. Geophys. Res.* **71**, 3705–3727.

ARMSTRONG, T. P. and KRIMIGIS, S. M.: 1968, 'Observations of protons in the magnetosphere and magnetotail with Explorer 33', *J. Geophys. Res.* **73**, 143–152.

BAME, S. J., ASBRIDGE, J. R., FELTHAUSER, J. E., HONES Jr., E. W., and STRONG, I. B.: 1967, 'Characteristics of the plasma sheet in the earth's magnetotail', *J. Geophys. Res.* **72**, 113–129.

BEHANNON, K. W. and NESS, N. F.: 1966, 'Magnetic storms in the earth's magnetic tail', *J. Geophys. Res.* **71**, 2327–2351.

BRICE, N.: 1967, Morphology of elementary magnetospheric substorms, Cornell-Sydney University Astronomy Center, CSUAC 94.

CRAVEN, J. D.: 1966, 'Temporal variations of electron intensities at low altitudes in the outer radiation zone as observed with satellite Injun 3', *J. Geophys. Res.* **71**, 5643–5663.

DAVIS, T. N. and SUGIURA, M.: 1966, 'Auroral electrojet activity index AE and its universal time variations', *J. Geophys. Res.* **71**, 785–801.

EVANS, J. E., JOKI, E. G., JOHNSON, R. G., and SHARP, R. D.: 1966, 'Austral and boreal zone precipitation patterns for low-energy protons', *Space Res.*, Vol. VI (ed. by R. L. Smith-Rose), Spartan Books, Washington, pp. 773–788.

FAIRFIELD, D. H.: 1967, 'Polar magnetic disturbances and the interplanetary magnetic field', Goddard Space Flight Center, Pub. X-612-67-338, July.

FAIRFIELD, D. H. and CAHILL Jr., L. J.: 1966, 'Transition region magnetic field and polar magnetic disturbances', *J. Geophys. Res.* **71**, 155–169.

FRANK, L. A.: 1967a, 'Initial observations of low-energy electrons in the earth's magnetosphere with OGO 3', *J. Geophys. Res.* **72**, 185–195.

FRANK, L. A.: 1967b, 'On the extraterrestrial ring current during geomagnetic storms', *J. Geophys. Res.* **72**, 3753–3767.

FRANK, L. A.: 1967c, 'On the distributions of low-energy protons and electrons in the earth's magnetosphere', in *Earth's Particles and Fields* (ed. by B. M. McCormac), Reinhold, New York, pp. 67–87.

FRANK, L. A.: 1967d, 'Several observations of low-energy protons and electrons in the earth's magnetosphere with OGO 3', *J. Geophys. Res.* **72**, 1905–1916.

FRANK, L. A., VAN ALLEN, J. A., and CRAVEN, J. D.: 1964, 'Large diurnal variations of geomagnetically trapped and of precipitated electrons observed at low altitudes', *J. Geophys. Res.* **69**, 3155–3167.

FREEMAN Jr., J. W.: 1968, 'Observation of flow of low-energy ions at synchronous altitude and implications for magnetospheric convection', *J. Geophys. Res.* **73**, 4151–4158.

FRITZ, T. A.: 1968, 'The high latitude outer zone boundary for > 40 keV electrons as observed by Satellite Injun 3', Dept. Physics and Astronomy, Univ. of Iowa, 68–21, March.

FRITZ, T. A. and GURNETT, D. A.: 1965, 'Diurnal and latitudinal effects observed for 10-keV electrons at low satellite altitudes', *J. Geophys. Res.* **70**, 2485–2502.

GOSLING, J. T., ASBRIDGE, J. R., BAME, S. J., HUNDHAUSEN, A. J., and STRONG, I. B.: 1965, 'Measurements of the interplanetary solar wind during the large geomagnetic storm of April 17–18', *J. Geophys. Res.* **72**, 1813–1821.

HARGREAVES, J. K., HONES Jr., E. W., and SINGER, S.: 1968, 'Relations between bursts of energetic electrons at 17 earth-radii in the magnetotail and radio absorption events in the ionospheric D-region', *Planetary Space Sci.* **16**, 567–580.

HEPPNER, J. P., SUGIURA, M., SKILLMAN, T. L., LEDLEY, B. G., and CAMPBELL, M.: 1967, 'OGO-A magnetic field observations', *J. Geophys. Res.* **72**, 5417–5471.

HESS, W. N.: 1968, *The radiation belt and magnetosphere*, Blaisdell Pub. Co., Waltham, Mass.

HONES Jr., E. W., ASBRIDGE, J. R., BAME, S. J., and STRONG, I. B.: 1967, 'Outward flow of plasma in the magnetotail following geomagnetic bays', *J. Geophys. Res.* **72**, 5879–5892.

JELLY, D. and BRICE, N.: 1967, 'Changes in the Van Allen radiation associated with polar substorms', *J. Geophys. Res.* **72**, 5919–5931.

KONRADI, A.: 1967, 'Proton events in the magnetosphere associated with magnetic bays', *J. Geophys. Res.* **72**, 3829–3841.

LANGEL, R. A. and CAIN, J. C.: 1967, 'OGO-2 magnetic field observations during the magnetic storm of March 13–15, 1966', Goddard Space Flight Center, Pub. X-612-68-88, October.

LANZEROTTI, L. J., ROBERTS, C. S., and BROWN, W. L.: 1967, 'Temporal variations in the electron flux at synchronous altitudes', *J. Geophys. Res.* **72**, 5893–5902.

LEZNIAK, T. W., ARNOLDY, R. L., PARKS, G. K., and WINCKLER, J. R.: 1968, 'Measurement and intensity of energetic electrons at the equator at 6.6 r_e, *Radio Sci.* **3**, 710–714.

LIN, R. P. and ANDERSON, K. A.: 1966, 'Periodic modulations of the energetic electron fluxes in the distant radiation zone', *J. Geophys. Res.* **71**, 1827–1835.

MCDIARMID, I. B. and BURROWS, J. R.: 1968, 'Local time asymmetries in the high-latitude boundary of the outer radiation zone for the different electron energies', *Canadian J. Phys.* **46**, 49–57.

MIHALOV, J. D., COLBURN, D. S., CURRIE, R. G., and SONETT, C. P.: 1968, 'Configuration and reconnection of the geomagnetic tail', *J. Geophys. Res.* **73**, 943–959.

O'BRIEN, B. J.: 1967, 'Interrelations of energetic charged particles in the magnetosphere', in *Solar-Terrestrial Physics* (ed. by J. W. King and W. S. Newman), Academic Press, N.Y., pp. 169–211.

PARKS, G. K. and WINCKLER, J. R.: 1968, 'Acceleration of energetic electrons observed at the 6.6 R_e equatorial plane during magnetospheric substorms', Cosmic Ray Group, School of Physics and Astronomy, Univ. of Minnesota Report.

PARKS, G. K., ARNOLDY, R. L., LEZNIAK, T. W., and WINCKLER, J. R.: 1968, 'Conjugate effects on energetic electrons between the equator at 6.6 R_e and the auroral zone', *Radio Sci.* **3**, 715–719.

PARTHASARATHY, R. and REID, G. C.: 1967, 'Magnetospheric activity and its consequences in the auroral zone', *Planetary Space Sci.* **15**, 917–929.

REID, G. C. and PARTHASARATHY, R.: 1966, 'Ionospheric effects of energetic electron bursts in the tail of the magnetosphere', *J. Geophys. Res.* **71**, 3267–3272.

ROEDERER, J. G.: 1968, 'Experimental evidence on radial diffusion of geomagnetically trapped particles', in *Earth's Particles and Fields* (ed. by B. M. McCormac), Reinhold, New York, pp. 143–155.

ROMICK, G. J. and SHARP, R. D.: 1967, 'Simultaneous measurements of an incident hydrogen flux and the resulting hydrogen Balmer alpha emission in an auroral hydrogen arc', *J. Geophys. Res.* **72**, 4791–4801.

ROTHWELL, P. and WALLINGTON, V.: 1967, 'The polar substorm and electron 'islands' in the earth's magnetic tail', paper presented at Birkeland Symposium, Sandfjord, Norway.

SCARF, F. L., CROOK, G. M., GREEN, I. M., and VIROBIK, P. F.: 1968, 'Initial results of the Pioneer 8 VLF electric field experiment', *J. Geophys. Res.* **73**, 6665–6686.

SCHATTEN, K. H. and WILCOX, J. M.: 1967, 'Response of the geomagnetic activity index Kp to the interplanetary magnetic field', *J. Geophys. Res.* **72**, 5185–5191.

SERLEMITSOS, P.: 1966, 'Low-energy electrons in the dark magnetosphere', *J. Geophys. Res.* **71**, 61–77.

SHARP, R. D. and JOHNSON, R. G.: 1967, 'Satellite measurements of auroral particle precipitation', in *Earth's Particles and Fields* (ed. by B. M. McCormac), Reinhold, New York, pp. 113–125.

SHARP, R. D. and JOHNSON, R. G.: 1968, 'Some average properties of auroral electron precipitation as determined from satellite observations', *J. Geophys. Res.* **73**, 969–990.

SHARP, R. D., JOHNSON, R. G., SHEA, M. F., and SHOOK, G. B.: 1967, 'Satellite measurements of precipitating protons in the auroral zone', *J. Geophys. Res.* **72**, 227–237.

SNYDER, C. W., NEUGEBAUER, M., and RAO, U. R.: 1963, 'The solar wind velocity and its correlation with cosmic-ray variations and with solar and geomagnetic activity', *J. Geophys. Res.* **68**, 6361–6370.

VASYLIUNAS, V. M.: 1968, 'A survey of low-energy electrons in the evening sector of the magnetosphere with OGO 1 and OGO 3', *J. Geophys. Res.* **73**, 2839–2884.

WILCOX, J. M., SCHATTEN, K. H., and NESS, N. F.: 1967, 'Influence of interplanetary magnetic field and plasma on geomagnetic activity during quiet-sun conditions', *J. Geophys. Res.* **72**, 19–26.

WILLIAMS, D. J., ARENS, J. F., and LANZEROTTI, L. J.: 1968, 'Observations of trapped electrons at low and high altitudes', Goddard Space Flight Center Report, X-611-68-63, February.

ZMUDA, A. J., HEURING, F. T., and MARTIN, J. H.: 1967, 'Dayside magnetic disturbances at 1100 kilometers in the auroral oval', *J. Geophys. Res.* **72**, 1115–1117.

ZMUDA, A. J., MARTIN, J. H., and HEURING, F. T.: 1966, 'Transverse magnetic disturbances at 1100 kilometers in the auroral region', *J. Geophys. Res.* **71**, 5033–5045.

CHAPTER 10

MAGNETOSPHERIC SUBSTORM

10.1. Introduction

We have learned in the preceding chapters of the various manifestations of the magnetospheric substorm in the polar upper atmosphere, as well as in the magnetosphere. As a first step toward understanding the basic processes involved in the magnetospheric substorm, these manifestations may be interpreted in terms of the changing distribution and intensity of electric fields and particle fluxes in the magnetosphere.

In Section 10.2, we shall examine the distribution of the particle fluxes in the polar upper atmosphere and in the magnetosphere on the basis of what we learned in the preceding chapters. In Section 10.3, we shall consider a plausible model of a three-dimensional current system and the associated electric fields. In the first part of Section 10.4, we shall list the basic requirements for *the* theory of the magnetospheric substorm and then review the various theories proposed to date.

10.2. Polar Substorm and the Distribution of the Particle Fluxes in the Polar Upper Atmosphere and in the Magnetosphere

A. ELECTRON FLUXES OVER THE POLAR REGION

In their synthetic study of polar upper atmospheric phenomena, HARTZ and BRICE (1967) proposed a model of the average daily precipitation pattern of electrons over the entire polar region. They classified the precipitation of electrons into two types, the 'splash-type' and 'drizzle-type'. The splash-type precipitation is associated with '*discrete*' phenomena that predominate at night, and the drizzle-type precipitation is associated with '*diffuse*' phenomena that maximize in the late morning hours.

They noted that the discrete phenomena include:

(1) discrete, localized, bright, often rapidly fluctuating auroral forms having characteristic heights greater than 100 km;

(2) abrupt auroral absorption events on riometer records;

(3) rapidly fading, impulsive type VHF signals scattered by field-aligned ionization at heights of about 105 km;

(4) intense sporadic E echoes – often termed auroral sporadic E – on ionograms at indicated heights of 100 km or more;

(5) spread F echoes on ionograms;

(6) bursts of relatively high frequency (>4 kc/sec) VLF emissions or auroral hiss;

(7) impulsive (Pi) micropulsations of the earth's magnetic field;

(8) balloon observations of bursts of bremsstrahlung X-rays of characteristically soft energies;

(9) a negative bay on a magnetometer, having a rapid or abrupt onset and a rather slower recovery; and

(10) relatively short duration bursts consisting of intense fluxes of soft electrons (energies of the order of a few keV).

By contrast, the diffuse phenomena – statistically at least – are associated with:

(1) steady, diffuse, weak to subvisual (mantle) aurora;

(2) strong slowly varying riometer absorption;

(3) a steady or slowly varying mean signal level for VHF forward scatter echoes from isotropic ionization irregularities at heights of the order of 85 km;

(4) sporadic E echoes from a height in the 80 to 90 km range;

(5) quasi-constant VLF emissions (polar chorus) at frequencies below 2 kc/sec;

(6) continuous (Pc) geomagnetic micropulsations;

(7) long duration, slowly varying, and characteristically hard X-ray events; and

(8) consistent, moderately intense fluxes of electrons of energies at least 40 keV.

On the basis of the above observations, HARTZ and BRICE (1967) proposed the

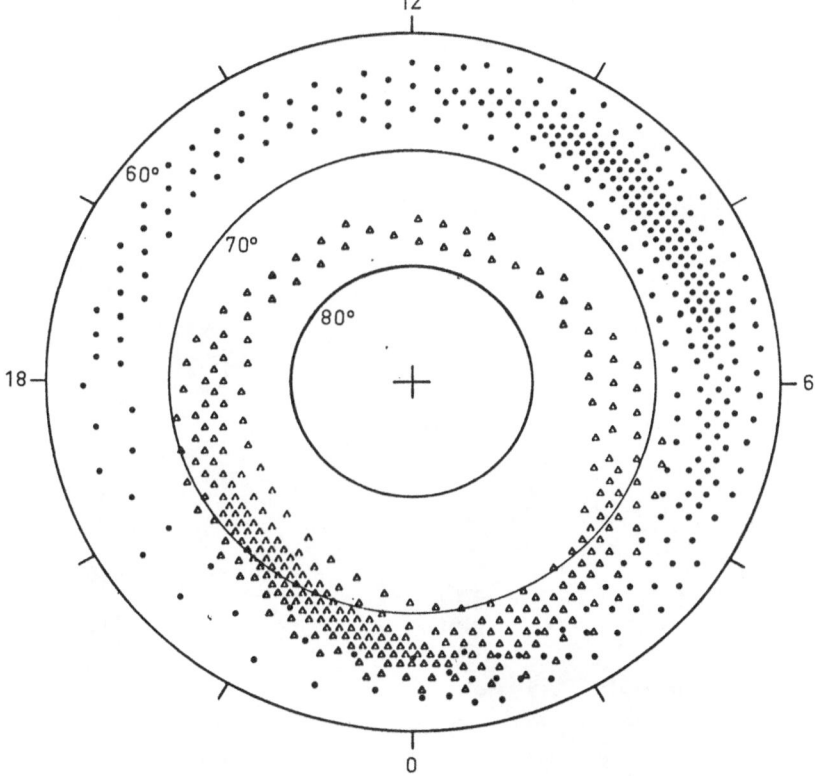

Fig. 156. Average daily precipitation pattern of electrons over the polar cap (Hartz, T. R. and N. M. Brice: *Planetary Space Sci.* **15**, 301, 1967).

average daily precipitation pattern of electrons over the entire polar region. It is reproduced here as Figure 156. The splash-type precipitation is represented by triangles and the drizzle-type precipitation by dots.

In the earlier chapters, we discussed in detail the daily variation pattern for each substorm phenomenon. We learned that the daily variation pattern tells us how characteristics of each phenomenon vary as a function of local time, but that it does not mean that such a pattern remains fixed throughout a day. A more proper way of describing each polar phenomenon is that such a pattern appears intermittently several times a day.

Now, from what we learned in the earlier chapters, it is not difficult to infer that the same argument can also be applied to the average daily precipitation pattern and that such a pattern appears intermittently with a lifetime of order $1 \sim 3$ hours. *The splash-type precipitation and the drizzle-type precipitation are not independent phenomena, but only different aspects of the precipitation during the magnetospheric substorm.*

We have already examined in some detail the development of each substorm over the entire polar cap. They provide important information on how characteristics of the electron precipitation or proton precipitation vary as a function of substorm time and local time. Figure 157 is constructed to illustrate schematically the develop-

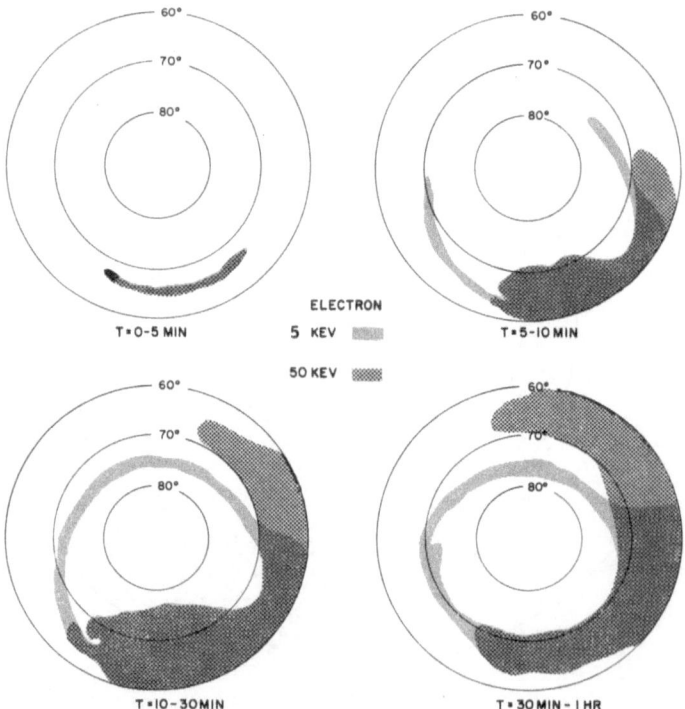

Fig. 157. Development of the electron precipitation over the polar region during the magnetospheric substorm.

ment of the precipitation of electrons at four stages ($T=0$–5 min, $T=5$–10 min, $T=10$–30 min, $T=30$ min–1 hour) on the basis of our study in Sections 2.1, 4.8, 5.9, and 7.7.

In order to single out the precipitation during a substorm, we have not included the precipitation at $T=0$, namely during a quiet period. During quiet periods between two substorms, 10 keV electrons precipitate along the auroral oval; the precipitation of 50 keV occurs along a narrow strip just equatorward of the oval in the midnight and the forenoon sector (Section 9.5.). In Figure 157, we choose two representative energies, 5 keV and 50 keV. Electrons of energy of order 5 keV are primarily responsible for exciting visible (structured) auroras, and electrons of energy 50 keV for the cosmic radio noise absorption and X-ray bursts.

B. ELECTRON FLUXES IN THE MAGNETOSPHERE

In this subsection, we attempt to construct the distribution of electrons at the three stages of the magnetospheric substorm, $T=5$–10 min, $T=10$–30 min, $T=30$ min–1 hour (Figure 158). Here again we have not included the distribution of the electrons during quiet time. The left hand side of the figure shows the noon-midnight meridian cross-section and the right hand side the equatorial cross-section of the magnetosphere. The circle on the equatorial cross-section represents an approximate boundary of the trapping region and the outer magnetosphere defined in Section 1.1.

In Chapter 9, we learned that the spatial extent of the electron flux increases outward outside the trapping boundary (Section 9.4). We noted important evidences to suggest that the increase outside the trapping boundary is closely related to the poleward advance of the precipitation during the explosive phase of a substorm (Section 9.4). Unfortunately, however, it is not certain at present whether this increase represents an outward *flow* of electrons or a local acceleration of electrons by some mechanism, such as a *propagation* of a shock wave (which leaves thermalized electrons behind it), or a propagation of the reconnection process of magnetic field lines (which converts the magnetic energy into thermal energy of plasma); this problem will be discussed in the next section. The difference between the above two possibilities should be manifested in the configuration of the magnetic field after the magnetospheric substorm. The shock wave would not drastically change the magnetic field configuration, although details depend on the nature of the shock wave (collisionless shock wave); we would expect, however, a significant diamagnetic effect of the heated plasma, if the shock wave leaves hot plasma behind it. On the other hand, the reconnection should result in a considerable change of the configuration after the substorm (see Section 10.4.B.9). Many of the field lines which constituted the magnetospheric tail before the onset of the substorm contract toward the earth during the substorm. However, since we do not have enough information on the magnetic field configuration, the field lines are not drawn in Figure 158.

The increase of the flux of 10 keV electrons inside the trapping boundary is likely to be related to the equatorward expansion of polar substorm phenomena. If the increase is caused by an inward flow of electrons, it is likely to be due to the $(E \times B)$

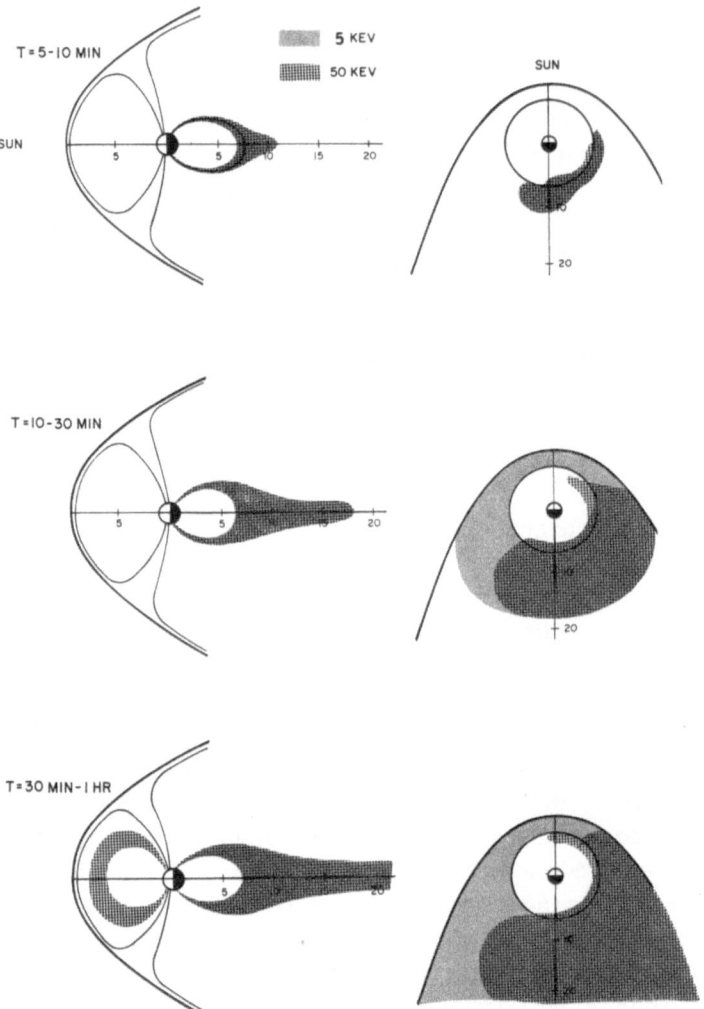

Fig. 158.   Distribution of the electron flux in the magnetosphere during the magnetospheric substorm.

drift motion, since the magnetic field configuration within the trapping region is not
far from a dipolar configuration and thus the electrons must move across the magnetic
field vector B. The electric field that can cause the required $(E \times B)$ motion should be
directed westward; the magnitude of the potential difference across the magnetosphere
is estimated to be 48 kilovolts (VASYLIUNAS, 1968).

If such an electric field is communicated to the ionosphere, we should expect a
westward electric field of order 10 volts/km. It is not difficult to determine the existence
of such an electric field by a rocket chemical release experiment; the electric field
should cause an equatorward drift motion of the ionized components of the chemicals
of order $E/B \simeq 200$ m/sec. HAERENDEL and LÜST (1968) reported that the drift motion

of the ionized component observed at Kiruna has an equatorward component of this order.

The population of 50 keV electrons increases in the morning sector of the trapping region (Sections 4.8, 5.9, 9.2 B), as well as in the tail region (Sections 4.8, 5.9, 9.3 B, 9.4 A). The increase of the flux in the morning sector could be due to either an actual drift motion of electrons from the midnight sector or a propagation of the acceleration mechanism of such electrons. An accurate measurement of the difference of the onset times of the cosmic radio noise absorption between the midnight sector and the morning sector should provide a clue in clarifying this question. In either case, there is no doubt that the magnetospheric substorm supplies a significant part of 50 keV electrons in the trapping region.

C. PROTON FLUX IN THE TRAPPING REGION

Since there is no direct observation of protons of energies $1 \sim 50$ keV during a magnetospheric substorm, we attempt to infer the changes of the flux mainly on the basis of ground magnetic observations in low altitudes (Section 3.4) and observations made by a synchronous satellite (Section 9.2 A). Figure 159 shows a schematic drawing of the changes of the proton population in the magnetosphere during a magnetic substorm.

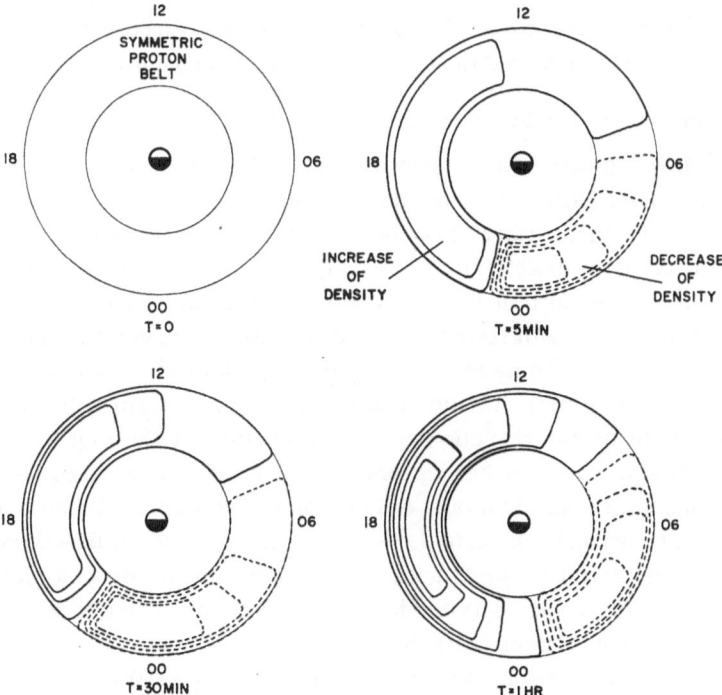

Fig. 159.   Changes of the distribution of ring current protons ($1 \sim 50$ keV) in the trapping region during the magnetospheric substorm.

We assume that there is a symmetric (with respect to the dipole axis) ring current at $T=0$. During an early phase of the substorm, a drastic reduction of the population of the protons occurs in the midnight and early morning hours. The area of the reduction appears to expand outward (Section 9.3. A), although it is not shown in Figure 159.

On the other hand, the population of the protons increases gradually and steadily in the 10–21 local hour range (Chapter 3, and Section 9.2. A). In the mid-evening sector (say 22 LT), there occurs an initial increase of the population which is followed by a decrease (Sections 9.1. and 9.2. A). This decrease may be considered to be due to a westward expansion of the area of the reduction from the midnight sector. The decrease is then followed by an increase of the proton population; this increase may be considered to be due to an expansion of the area of the increase from the day sector. After the maximum epoch of the substorm, the proton population decreases in the region where the increase took place. On the other hand, there is little change of the population of the protons in the early morning sector. Therefore, when substorms occur frequently, there will be few protons in the morning sector. Note that the above variations could be due to radial motions of the protons.

We should note that the above description refers to a substorm of medium intensity. If the substorm is very intense, the reduction of the proton population is greatly reduced in an extensive longitude range in the dark sector (20 LT–06 LT) during an early phase of the substorm (Section 9.2. A).

10.3. Polar Substorm and the Distribution of the Electric Fields in the Polar Upper Atmosphere and in the Magnetosphere

A. THREE-DIMENSIONAL CURRENT SYSTEM

In Section 3.4, we showed that a revised version of Kirkpatrick's model can reasonably well explain the distribution of magnetic disturbance fields in middle and low latitudes on the earth's surface. The model implies that new ring current particles (protons of energy $1 \sim 50$ keV) appear widely in the sunlit and the evening sectors (Section 10.2. C). The current flows into the auroral ionosphere from the morning end of this asymmetric (with respect to the dipole axis) ring current and flows out to the ring current in the evening sector, after flowing along the auroral oval.

As we learned in Section 9.2 and 10.2. C, the magnetic records from the synchronous satellite indicate the reduction of the population of the ring current particles in the midnight and early morning sectors. Therefore, the eastward equatorial current in the dark hemisphere in the model calculation in Section 3.4 should correspond to this reduction, rather than the actual eastward directed current. Based on the discussions above, Figure 160 is constructed to schematically illustrate a plausible three-dimensional current system. The model implies that the asymmetric growth of the proton belt (the ring current belt) discussed in Section 10.2. C is of fundamental importance among the various magnetospheric storm processes.

FEJER (1961) postulated that the ring current particles are injected into the trapping region in the sunlit sector and that the asymmetric ring current thus produced completes its circuit by generating Pedersen and Hall currents in the ionosphere. Since we now know that the ring current particles consist mainly of protons of a few kilovolts and that they form the belt well inside the trapping region (CAHILL, 1966; FRANK, 1967; Section 10.2. C), they cannot be directly injected there from outside. Nevertheless, the available evidence suggests that new ring current particles appear widely in the sunlit sector.

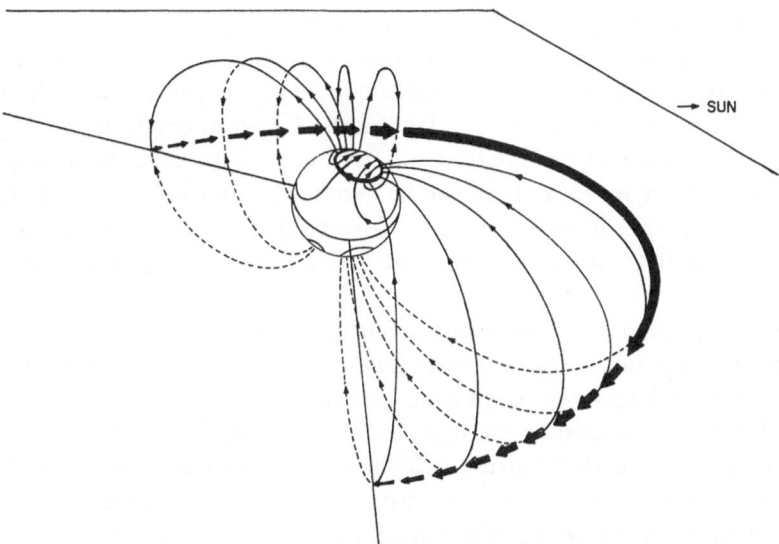

Fig. 160. Model, three-dimensional current system during the magnetospheric substorm (Akasofu, S.-I. and C.-I. Meng: *J. Geophys. Res.* **74**, 1969).

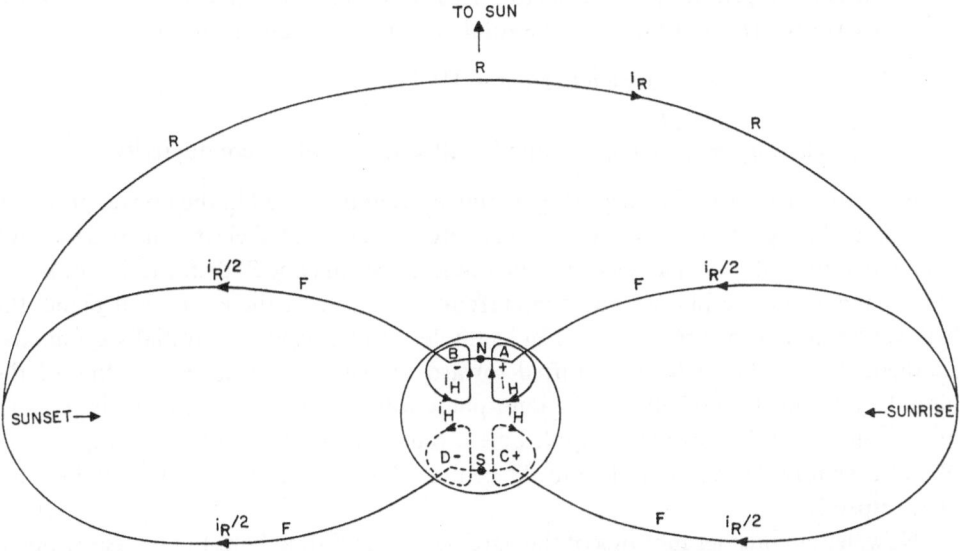

Fig. 161. Model current system proposed by Fejer (Fejer, J. A.: *Canadian J. Phys.* **39**, 1409, 1961).

Figure 161 shows Fejer's model current system. If the asymmetric ring current grows in the day sector, a current of intensity i_R flows from the sunset meridian to the sunrise meridian. This incomplete circuit generates space charges at both ends of the ring current, and consequently a potential difference across the magnetosphere. The resulting electric field is directed from the sunrise sector to the sunset sector. Such a system may complete the circuit by driving electric currents along the field lines and in the ionosphere. The ionospheric current thus generated flows from A to B in the Northern Hemisphere and C to D in the Southern Hemisphere.

The pattern of the currents in the ionosphere depends greatly on the distribution of the conductivity. Between A and B, the Pedersen current J_P flows along the direction of the electric field $J_P = \sigma_1 E$; here, σ_1 denotes the Pedersen conductivity. Figure 161 shows only the direct path from A to B and from C to D. Because the ionosphere has a high Hall conductivity (σ_2), there will be another current system (the Hall current J_H) which flows perpendicular to E; $J_H = -\sigma_2\, E \times B/B^2$, forming loops around the points A, B, C and D; for simplicity, only a single loop is shown at each point in Figure 161.

The current pattern would be greatly distorted from the simple one in Figure 161, if the conductivity is not uniform. For example, if the points A and B are located on the 06 and 18 LT meridian on the auroral oval respectively, the Pedersen current will be channeled along the oval in the dark hemisphere, since it is a highly conductive strip. Furthermore, the Hall current (which flows across the high conductive strip) generates space charges near the poleward and equatorial boundaries of the oval. In the Northern Hemisphere, a positive space charge appears on the poleward boundary of the oval and a negative space charge on the equatorial boundary; the electric field associated with the space charges is directed equatorward and drives an additional westward Hall current in the dark sector. This is often expressed as an increase of the conductivity (BOSTRÖM, 1967a, b); the total current J is then given by

$$E = (\text{Pedersen current} + \text{Hall current})$$
$$= \sigma_3 J,$$
where $\sigma_3 = \sigma_1 + \sigma_2^2/\sigma_1$ and is called the Cowling conductivity.

As mentioned earlier, the asymmetric ring current discussed in the above, produces a potential difference across the magnetosphere; the associated electric field is directed from dawn to dusk. Such an electric field could cause an $(E \times B)$ drift motion of large-scale magnetospheric plasma and the current belt itself. In the equatorial plane, the belt will be shifted toward the sun. For example, let us consider an initially symmetric energetic particle belt which is shifted toward the sun. The intersection line of the boundary of this shifted belt and the ionosphere will be a circle around the dipole pole, but its center is shifted from the dipole pole toward the dark hemisphere. In Figure 162, N and C signify the dipole pole and the center of the intersection line for the shifted belt, respectively.

Now, let us consider motions of the particles in the shifted belt. Let us also suppose that the ring current belt does not seriously change the magnetic field configuration.

Then, the particles tend to drift along circular paths (or along $|B|$ = constant) centered at the dipole axis from every point on the shifted belt. (Here, we suppose that the particles are very energetic, so that they will not be seriously affected by the electric field.) We shall be interested in motions of the particles whose initial points are on the inner and outer surfaces of the shifted belt. In the morning sector, positive particles on the outer boundary drift along $|B|$ = constant curves toward the midnight meridian; on the other hand, electrons on the inner boundary drift toward the noon meridian.

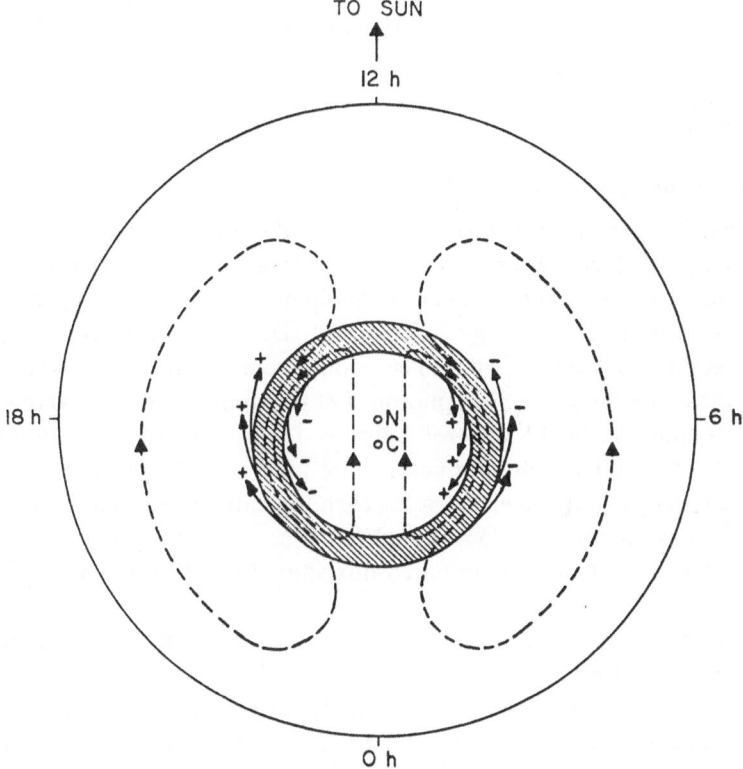

Fig. 162.   Secondary space charges generated in the model current system proposed by Fejer and the resulting Hall current in the ionosphere (Fejer, J. A.: *Canadian J. Phys.* **39**, 1409, 1961).

In the afternoon sector, electrons on the outer boundary drift toward the midnight meridian and positive particles drift toward the noon meridian from the outer boundary. The projection of this system onto the polar upper atmosphere is shown in Figure 162. FEJER (1961) suggests that the resulting Hall current system has a resemblance to the *SD* current system (Section 2.1).

   In this section, we examine the electric fields deduced from the study of the asymmetric growth of the ring current. They can account for a few important features of the magnetospheric substorm, and their further consequences will be discussed in Section 10.5. Other ideas on the origin of the electric field will be reviewed in the next section.

10.4. Review of Magnetospheric Substorm Theories

Most of the proposed theories on the magnetospheric substorm are very incomplete, because they attempted to explain only one (or at most two) of the polar upper atmospheric manifestations of the magnetospheric substorm without realizing that they are but one of the manifestations. Nevertheless, it is worthwhile to review these theories, because they certainly contain elements of truth to be seriously considered.

In the first part of this section, we shall examine the basic requirements to be fulfilled by the theories. The second part consists of brief reviews of the proposed theories to date.

A. BASIC REQUIREMENTS

1. Total Energy dissipated during a Single Substorm

Thus far, there has been no conclusive estimate of the total energy involved in a single magnetospheric substorm. There is no doubt that a significant portion of the total energy generated in the magnetosphere is dissipated in the polar upper atmosphere and that this portion of the energy is mainly carried by auroral electrons (of energy 5 keV). It is commonly assumed that in electron precipitation, about one ionization in 50 leads to the emission of a λ 3914 photon. Let us suppose that the average intensity of the λ 3919 emission in the auroral bulge is 1 kR. This requires an emission of 10^9 photons/cm^2 sec. The ion production rate must then be of order 5×10^{10} ions/cm^2 sec. The energy lost by an electron for each ionization collision is about (35 eV/ion) $(5 \times 10^{10}$ ions/cm^2 sec$) = 1.75 \times 10^{12}$ eV/cm^2 sec $= 2.8$ ergs/cm^2 sec; if this energy is carried by 5 keV electrons, the required flux should be of order 2.0×10^8 electrons/cm^2 sec.

Assuming that the auroral bulge has an area of 8×10^{16} cm^2 (North-South dimension $\simeq 15°$; East-West dimension $\simeq 90°$), the energy injection rate is of order 2×10^{17} ergs/sec. Taking the lifetime of the substorm to be of order 2 hours $= 7200$ sec, the energy deposited in the form of the precipitation of auroral electrons in the two hemispheres is of order $2 \times 1.4 \times 10^{21}$ ergs $= 2.8 \times 10^{21}$ ergs. HARGREAVES et al. (1968) observed that the electron flux observed in the ionosphere is significantly greater than that observed simultaneously by the VELA satellite, but noted that the difference may simply be due to the difference of the magnetic field configuration (namely, the convergence of the field line toward the earth). Therefore, it may not be too unreasonable to assume that one half of the electron flux appears in the polar upper atmosphere and the other half in the tail region, so that the total energy carried in the form of auroral electrons is $2 \times 2.8 \times 10^{21}$ ergs $= 5.6 \times 10^{21}$ ergs.

A rough estimate of the energy carried in the form of the ring current protons may be made by an equation derived by DESSLER and PARKER (1959). Taking the magnitude of the low latitude negative bay to be 50 γ (Chapter 3, Section 9.2), it is estimated to be 2×10^{22} ergs; the equation is given for a symmetric ring current, so that the estimated value is divided by a factor of 2 to get a rough estimate of the asymmetric partial ring current. The above value is, however, certainly an underestimation, since any loss of protons during the formation of the ring current is not taken into account. The total energy carried by auroral electrons and ring current protons will thus be of order 2×10^{22} ergs. The contribution of the energies carried by more energetic particles should not drastically alter this estimate.

AXFORD (1967a) suggested that energy is dissipated in the magnetosphere during magnetic *storms* in three main forms:

(1) production of aurora ($\approx 10^{18}$ ergs/sec);

(2) heating of the ionosphere by the electric currents causing magnetic substorms ($\approx 10^{18}$ ergs/sec); and

(3) inflation of the magnetosphere (ring current) ($\approx 10^{18}$–10^{19} ergs/sec).

2. Initial Indication of the Substorm

The first indication of the magnetospheric substorm may be most clearly observable in the first indication of the auroral substorm which can give us information on certain aspects of the precipitation of electrons of a few kilovolts. It suggests that the precipitation of electrons of a few kilovolts near the equatorward edge of the auroral oval in the midnight sector suddenly increases by one or even two orders of magnitude in a few minutes; this is observed either as a sudden increase in the brightness of the equatorwardmost arc or as a sudden formation of an arc near the poleward boundary of the proton aurora (SWIFT, 1967a). Therefore, the first indication of the magnetospheric substorm should be a sudden activation of the process that is responsible for the generation of auroral particles near the trapping boundary in the midnight sector.

3. Storage of the Substorm Energy

There is no visible indication that the auroral substorm is preceded by any auroral motion. If the magnetospheric substorm results from a direct injection of solar wind plasma across the boundary of the magnetosphere, some luminous effect should be observable in the polar upper atmosphere before the onset of an auroral substorm. In Section 9.7. A, we noted also that the K_p index (which is a rough measure of the intensity of polar magnetic substorms) is not related to the energy flux or momentum flux of the solar wind particles.

Furthermore, since the first indication of the magnetospheric substorm occurs near the trapping boundary and also since the increase of the electron fluxes at $r_e = 17a$ occurs as late as the time of the maximum epoch of the magnetospheric substorm ($T = 15 \sim 30$ min; Section 9.4. A), we may conclude that the magnetospheric substorm is essentially an internal process in the magnetosphere and that the energy for the substorm (2×10^{22} ergs) must be first stored rather deep in the magnetosphere, perhaps near the trapping boundary.

4. Explosive Phase and Recovery Phase

The polar substorm has two characteristic phases, the explosive phase and the recovery phase. This is best manifested by the auroral substorm in the midnight sector, but other substorms in the midnight sector also have similar characteristics. We can thus conclude that the magnetospheric substorm also has an explosive phase.

The explosive phase suggests that the energy stored near the trapping boundary in the midnight sector must be released in an explosive way. During the explosive phase, the volume occupied by auroral electrons expands rapidly in all directions.

5. *Asymmetric Proton Belt*

We have already noted that the formation of a proton belt requires as much or even greater energy than the total energy of auroral electrons. Therefore, the formation of the proton belt is one of the major aspects of the magnetospheric substorm. We note that the proton belt is formed deep in the trapping region in the sunlit sector.

B. REVIEW OF THEORIES OF THE MAGNETOSPHERIC SUBSTORM

Based on the discussions in Section 10.4. A, we can construct the following energy flow diagram chart.

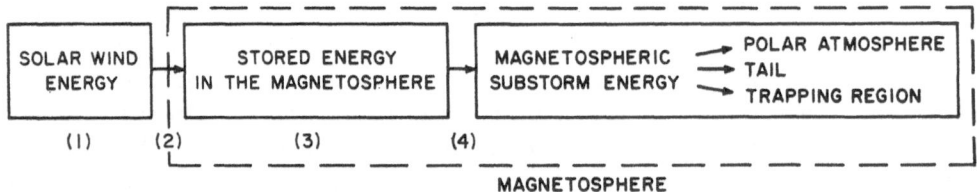

On the basis of the discussions in 10.4. A and the above energy flow diagram, a suitable theory of magnetospheric substorms must answer at least the following questions:

(1) What is the original form of the solar wind energy for the magnetospheric substorm?

(2) How is it converted into energy suitable for storage in the magnetosphere?

(3) What is the form of this energy during the storage?

(4) How is it converted into energy for the magnetospheric substorm?

We shall review in the following the various substorm theories that have been proposed. These may be classified into ten groups:

(1) Discharge from an extra-terrestrial source, by BIRKELAND (1908, 1913), ALFVÉN (1950, 1955), KARLSON (1962, 1963), BLOCK (1966).

(2) Diamagnetism, by MARIS and HULBURT (1929), HULBURT (1937).

(3) Polarization of the radiation belts, by MARTYN (1951), SHAW (1959), CHAMBERLAIN (1961), KERN (1962), KERN and VESTINE (1961), FEJER (1963, 1964).

(4) Dynamo action, by FUKUSHIMA and OGUTI (1953), OBAYASHI and JACOBS (1957), COLE (1960), SWIFT (1963).

(5) Convection, by PIDDINGTON (1962 a,b, 1963), AXFORD (1962, 1964), AXFORD and HINES (1961), HINES (1964), NISHIDA (1966), BRICE (1967).

(6) Neutral point discharge, by DUNGEY (1961, 1963, 1968).

(7) Neutral line discharge, AKASOFU and CHAPMAN (1961).

(8) Electric field acceleration, by TAYLOR and HONES (1965).

(9) Magnetospheric tail instability, by ATKINSON (1966, 1967), AXFORD (1967a,b), DUNGEY (1968), PETSCHEK (1964), PIDDINGTON (1967, 1968), SPEISER (1967a,b, 1968).

(10) Trapping boundary instability, by SWIFT, (1964, 1965, 1967a,b, 1968).

1. Discharge from an Extra-Terrestrial Source

BIRKELAND's study (1913) was based on an extensive analysis of magnetic records collected during the First International Polar Year. He correctly revealed the essential feature of the polar geomagnetic disturbances, recognizing that they consist of a successive appearance of impulsive disturbances, namely what he called the polar elementary storms. They are called here the polar magnetic substorms.

He proposed that polar magnetic substorms are caused by a beam of electrons from the sun, which is injected into the polar upper atmosphere, flows along the auroral zone, and escapes into interplanetary space. The interaction between the electrons and the upper atmosphere was thought to cause the aurora. He demonstrated his idea by using the first model experiment, terrella, which consists of a magnetized iron ball and electron gun in a vacuum tube.

Alfvén proposed that any space charge at the boundary of his forbidden region will be immediately discharged along the field lines toward the auroral zone. The discharge causes a current flow from the dayside boundary of the forbidden region to the nightside boundary along the field lines, and then along the dawn and dusk sides of the auroral zone and again along the field lines. He proposed that these currents are the auroral electrojets. It is unlikely that the electrojet results from a direct injection of solar wind particles. However, the fact that his model current system has an important similarity with the model discussed in Section 10.3, suggests that plasma *within* the magnetosphere may have a large-scale drift $(E \times B)$ motion, assuming that an appropriate electric field exists during the magnetospheric substorms. This problem was discussed by KAVANAGH et al. (1968). In Section 9.3. B and 10.2, we noted an inward motion of the plasma in the dark side.

2. Diamagnetism

MARIS and HULBURT (1929) proposed that a highly concentrated plasma sheet appears in the auroral ionosphere and causes a diamagnetic distortion of the earth's field. Such a distortion would have essentially the same characteristics as a polar magnetic substorm. Since each plasma particle has a magnetic moment $\mu = (W_{\perp}/B)$, the magnetic moment of the plasma sheet is given by $An\mu = An\kappa T/B$ (see ALFVÉN, 1950, p. 59), and thus the pole magnetic strength $An\kappa T/4\pi B$, where A denotes the bottom cross-section of the sheet.

The theory suggests that the H component disturbance field should change sign just under the East-West center line of the bottom of the plasma sheet, a negative change poleward of the line and a positive change equatorward. However, observations indicate that a negative bay is observed in the whole region of the auroral bulge.

3. Polarization of the Radiation Belts

A possible polarization of the radiation belts and its consequences have been studied by a number of workers. MARTYN (1951) proposed that a radial electric polarization field associated with the ring current (proposed by CHAPMAN and FERRARO (1933)) would repel some of the charged particles toward the auroral zone; the repelled charged particles would transmit the polarization field into the auroral ionosphere and thereby generate the auroral electrojet. He suggested that the potential difference between the inner and outer edges of the proposed ring current will be of order 10^5 volts, and thus taking into account the geometrical convergence of the two field lines from the two edges, a poleward electric field of more than 10^{-3} volts/cm can be introduced into the polar ionosphere. Martyn assumed that positive ions will drift faster than negative ions (the Hall current), so that a westward current flows in the dawn sector and an eastward current in the dusk sector. We now know, however, that in the E region of the ionosphere the Hall current is mainly carried by electrons so that his current system must be reversed.

After the discovery of the outer radiation belts, its possible axial asymmetry and the resulting polarization electric field have been extensively studied. Both CHAMBERLAIN (1961) and KERN (1962) proposed that a longitudinal gradient of the magnetic field (∇B) will be produced by an inhomogeneous plasma distribution and also by compression of the magnetosphere and that ∇B is directed towards the midnight meridian on both the evening side and morning side. The direction of the drift motion associated with ∇B is opposite for opposite charges, resulting in thin layers of space charges and thus an outward polarization field in the evening sector and inward field in the morning sector. The electric fields thus produced will be transmitted to the polar ionosphere, causing an eastward current in the evening sector and a westward current in the morning sector.

In Chapter 6, we noted that in the evening sector auroras which are excited mainly by electrons are separated from a diffuse luminous band which is excited by proton bombardment. It is this separation which has the attracted attention of many workers who are looking for an electric field which could drive the auroral electrojet. It is worthwhile to examine the existence of such an electric field by a rocket chemical release and other techniques.

The polarization of a radiation belt and its consequences was most extensively studied by FEJER (1964). His basic assumption is that the energetic proton belt, a positive space charge, is embedded in the low energy magnetospheric plasma. Ignoring the inertial term $m \, dv/dt$, charged particles in the magnetosphere drift with the velocity

$$v_G = \left(E - \frac{\mu \nabla B}{e} \right) \times B/B^2 \,.$$

For the low energy plasma, the above equation is simplified, since the ∇B drift will be negligible compared with the daily rotation with the earth, namely

$$v_G = E \times B/B^2 \,,$$

where E is given (cf. HINES, 1964) by

$$E = \Omega \times r \times B,$$

where Ω denotes the rotational velocity, r the radius vector measured from the axis of rotation.

On the other hand, on a very energetic particle, the rotation of the earth has a negligible effect which drifts with velocity

$$v_G \simeq -\frac{\mu}{e} \frac{\nabla B \times B}{B^2}.$$

Therefore, in a non-axial symmetric field (such as the magnetosphere compressed by the solar wind), the paths of the energetic protons and the low energy plasma are different, giving rise to a large space charge separation. Charge neutrality can, however, be nearly restored by currents along the geomagnetic field lines and also currents in the ionosphere.

Suppose a tube of magnetic flux carves out an area of 1 cm² from the ionosphere which is assumed to be a spherical shell. Any electrostatic field E causes a motion of tube of force, together with the low energy plasma filling it. The foot of the tube of force moves with velocity v_D with respect to the earth's rotation. Thus, if Q denotes half the space charge of the low energy plasma, the current density associated with the motion of the foot is given by Qv_D. Space charge $(-Q)$ of the energetic particles seen from an observer on the earth, will produce a current $(-Q)(-v_R) = Qv_R$, where v_R denotes the velocity of the rotation of the earth.

Recalling that the equations for the dynamo theory by CHAPMAN and BARTELS (1940) is equivalent to

$$\nabla \cdot J = \nabla \cdot \left[-\left(\int \sigma \, dz \right) \nabla \phi + \int \sigma (v \times B) \, dz \right] = 0,$$

our new situation is simply expressed by

$$\nabla \cdot (J + Qv_R + Qv_D) = 0,$$

where ϕ denotes the electrostatic potential. The above differential equation may be rewritten as a differential equation for ϕ, from which $E = -\nabla \phi$ and thus J can be obtained. Figure 163 shows an example of Fejer's calculation; the integrated Hall conductivity along the auroral zone is taken to be 45 mho $(= 4.5 \times 10^{-8}$ emu), which is about twice the midday conductivity at medium latitudes. According to his calculation, the magnetic field produced by the above mechanism is of order 80 γ. If there is a great enhancement of the proton belt and of conductivity along the auroral zone during geomagnetic storms the magnetic field would be somewhat larger.

This theory provides the fundamentals for examining consequences of an asymmetry of radiation belts and the resulting coupling between the belts and the ionosphere.

SUN

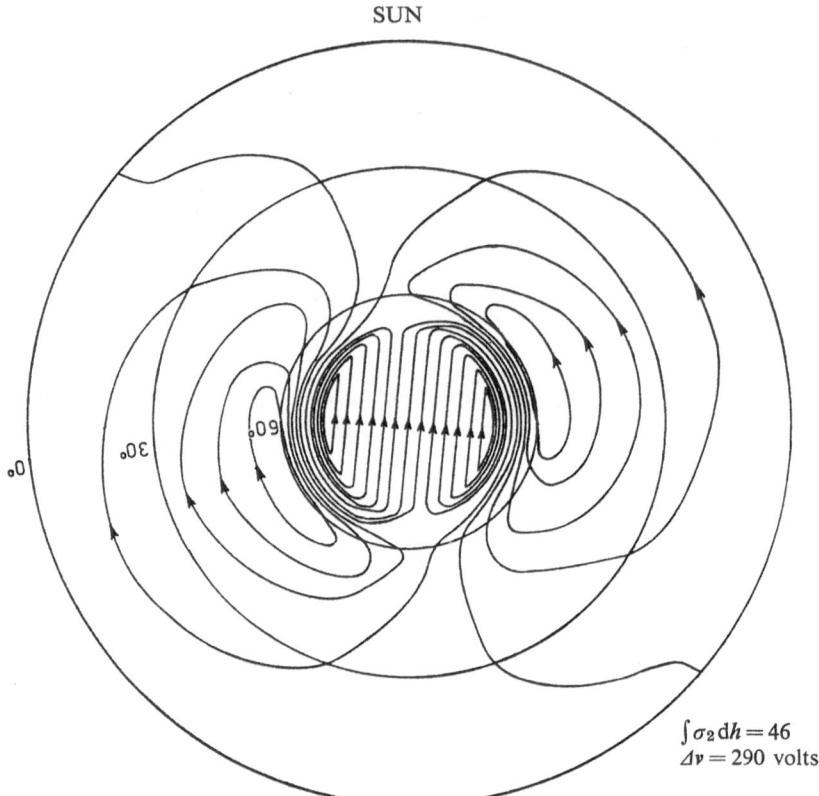

$\int \sigma_2 dh = 46$
$\Delta v = 290$ volts

Fig. 163. Ionospheric current system generated by the polarization of the radiation belt (Fejer, J. A.: *J. Geophys. Res.* **69**, 123, 1964).

4. *Dynamo Action*

The success of the dynamo theory for the solar quiet day daily variation has led some people to speculate that the dynamo action ($v \times B$) in the ionosphere is also responsible for auroral electrojets.

The dynamo theory of polar magnetic storms assumes a highly conductive belt (one in each hemisphere) along the auroral zone, so that in addition to the usual Sq current system, an extra current system is induced by the dynamo action.

FUKUSHIMA and OGUTI (1953) and OBAYASHI and JACOBS (1957) made illustrative calculations for the ionosphere with anisotropic conductivity. The differential equation for the current intensity J is the same as that for the dynamo theory of Sq,

$$\frac{\partial}{\partial \lambda}\left(\frac{\partial J}{\sin \theta \partial \lambda}\right) + \frac{\partial}{\partial \theta}\left(\sin \theta \frac{\partial J}{\partial \theta}\right) = 2G \sum_3 \left[\cot \theta \frac{\partial^2 \Psi}{\partial \lambda^2} + \frac{\partial}{\partial \theta}\left(\sin \theta \cos \theta \frac{\partial \Psi}{\partial \theta}\right)\right].$$

Here, $J = J_q + J_d$ (where J_q denotes the current intensity for the Sq variation and J_d the additional current intensity due to the increased conductivity [the Cowling conductivity, $\sum_3 = \int \sigma_3 \, dh$] along the auroral zones); the wind velocity is assumed to have a potential

$$\Psi = k_1^1 P_1^1 (\cos \theta) \sin (n\lambda + \pi/2).$$

The conductivity along the auroral zones of width 5° is assumed to be about 10 times larger than the rest of the ionosphere.

FUKUSHIMA and OGUTI (1953) found that the wind system which is responsible for the Sq variation cannot explain the orientation of the electrojet current system with respect to the sun. Further, the current intensity is much weaker than the observed one. To overcome these difficulties, it is necessary that the air motion in high latitudes differ in phase by about 100° ~150° with respect to the Sq wind system, and also that the wind speed or the conductivity along the auroral zone be greater than the values assumed. These difficulties were also pointed out by MAEDA (1957).

The most serious difficulty of the theory is, however, the fact that at geomagnetically conjugate points the Sq variations are quite different (indicating that the wind system in the Northern and Southern Hemispheres may differ greatly from each other), while the polar magnetic substorms are, in general, strikingly similar; Section 3.5.

Both COLE (1960) and SWIFT (1963) considered the dynamo action in a little different way. Assuming a highly conductive and narrow slab, they examined effects of a uniform wind blowing across it. The wind tends to carry positive ions along its direction more closely than electrons, resulting in a polarization electric field across the high conductive slab and thus an intense Hall current along it. Again, their proposed mechanism depends greatly on the direction of the ionospheric wind, so that criticism given in the above may also be applied here. Further, their model is independent of height, but the proposed mechanism should occur only in the E region of the ionosphere where the collision frequency of positive ions with neutral particles is high enough. Therefore, if the space charge is discharged vertically along the field lines, their effect will be considerably diminished.

5. Convection

PIDDINGTON (1960, 1962a), AXFORD and HINES (1961), and HINES (1964) proposed that a 'friction' (coined by the first author) or a 'viscous-like interaction' (coined by the latter authors) between the solar wind and the magnetosphere could cause a large-scale 'convective motion' in the magnetosphere. Here, the convective motion is limited to the motions (of velocity v) which satisfy

$$E + v \times B = 0, \tag{1}$$

where the electric field E is given by the gradient of a scalar function ϕ

$$E = - \nabla \phi \tag{2}$$

so that

$$v \times B = \nabla \phi. \tag{3}$$

This type of motion has been called the interchange of magnetic tubes of force (GOLD, 1959); plasma which occupies a tube of force at a particular time migrates with the tube at all times. Since both v and B should be perpendicular to $\nabla\phi$, they should lie on the surfaces of ϕ=const, the equipotential surfaces. Consider the equatorial cross-section of the equipotential surfaces in a dipolar field. The intersection lines should then be the equipotential lines and these equipotential lines should coincide with the flow lines of the plasma (or the flow lines of the equatorial crossing point of a thin tube of force).

Instead of obtaining the generated convective flow pattern from the proposed interaction between the plasma flow and the magnetosphere, PIDDINGTON (1960, 1962a) and AXFORD and HINES (1961) set up an *ad hoc* flow pattern which consists of two closed systems (Figure 164). The projections of the two closed systems onto the ionosphere coincides with the trace of the foot of convecting geomagnetic field lines (Figure 165). In the E region of the ionosphere, however, positive ions can no longer be attached to the field lines by frequent collisions with neutral particles during their gyration. The result is a flow of electrons along the convective flow path and thus a

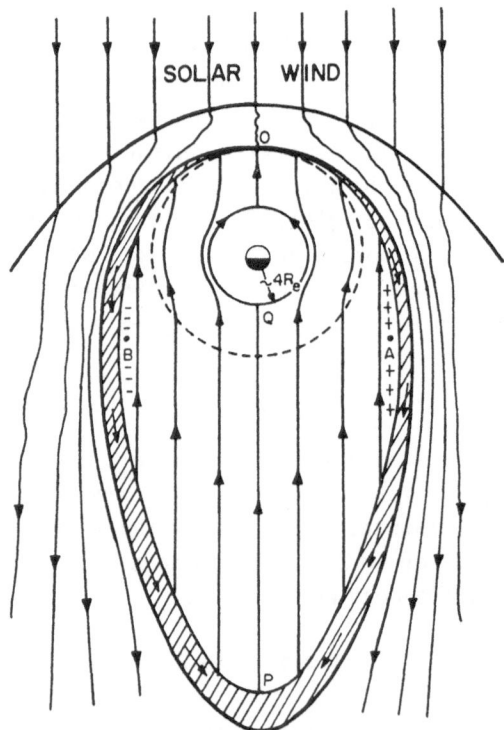

Fig. 164.   Sketch of the equatorial section of the earth's magnetosphere looking from above the North Pole. Streamlines of the solar wind are shown on the exterior, and the internal streamlines refer to the convective motion. The internal streamlines are also equipotentials of an associated electric field which may be regarded as being due to accumulations of positive and negative charges as indicated at A and B (Axford, W. I.: *Planetary Space Sci.* **12**, 45, 1964).

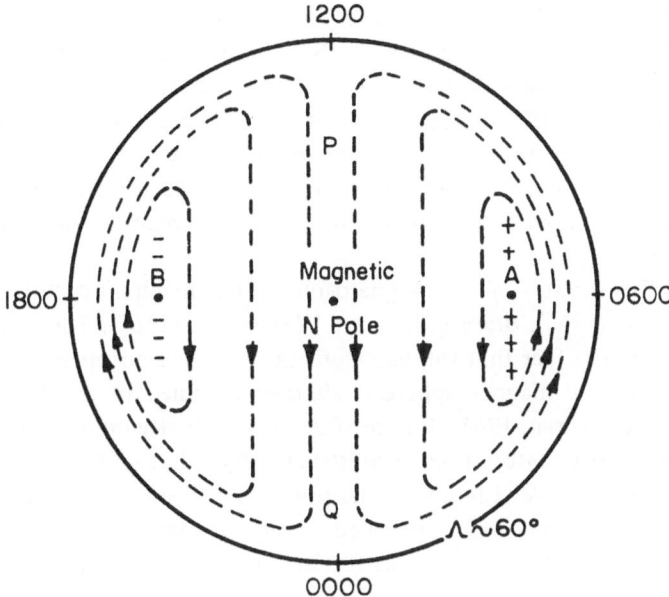

Fig. 165. Sketch of the circulation corresponding to Figure 164 at the ionospheric levels in the north polar cap (Axford, W. I.: *Planetary Space Sci.* **12**, 45, 1964).

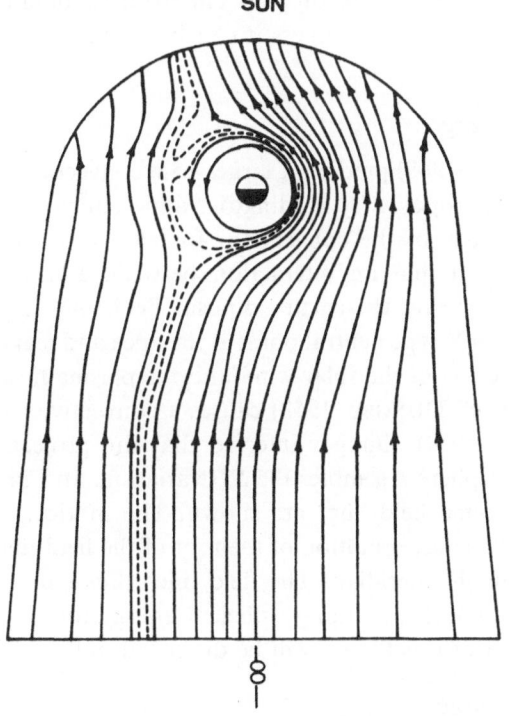

Fig. 166. Convective motion of the magnetospheric plasma suggested by Brice (Brice, N. M.: *J. Geophys. Res.* **72**, 5093, 1967).

Hall current which has an opposite direction to the electron flow. On this basis, they proposed that the *SD* current system (Section 2.1) results from this convection.

If the *SD* current system is generated by the proposed convective motion, the current lines should coincide with the equipotential lines; there should be positive space changes at the center of the current vortex in the dawn sector, and negative space charges at the centers of the current vortex in the dusk sector. Both Piddington, and Axford and Hines estimated the potential difference between the two centers to be of order 20 kilovolts.

The convective motion of the magnetospheric plasma has recently been studied further by NISHIDA (1966), BRICE (1967) and WALBRIDGE (1967). The first two authors took into account the fact that the geomagnetic field has an open tail and that the thermal plasma in the magnetosphere is sharply confined in the region called the plasmosphere (CARPENTER, 1966). Figure 166 shows the streamlines of the magneto-spheric plasma in the equatorial plane constructed by BRICE (1967); here both convection and daily rotation are combined. The potential difference between solid lines is 3 kilovolts. The dashed lines give intermediate equipotentials at 1 kilovolt intervals. The outermost closed streamline gives the location of the plasmapause (or the so-called 'knee').

In order for this theory to be a successful theory of the magnetospheric substorm, it must explain some of the most vital features of the substorm, such as its explosive phase and the formation of the proton belt. The theory should not rely on an equivalent (ionospheric) current system, unless it can be proved to be the true ionospheric current system.

6. Neutral Point Discharge

DUNGEY's theory (1963, 1968) predicts a large-scale convection of the magnetospheric plasma similar to that proposed by Piddington, and Axford and Hines. The difference is the driving mechanism of the convection. Instead of a friction or a viscous-like interaction, he proposed that an interaction between a southward oriented interplanetary magnetic field and the earth's dipolar field plays an important role. The interaction results in an X type neutral point at the apex and rear of the magnetosphere. Then, the combined effect of the solar wind and the plasma flow in the vicinity of the X type neutral point (cf. DUNGEY, 1958) causes a convective motion of the field lines (see also LEVY et al., 1964). Dungey inferred that the projection of the convection pattern onto the ionosphere resembles the *SD* variation. In Figure 167 the successive motions of the convected field lines are shown. The motions corresponding to the sequence 1 to 8 should cause a motion of the foot of the field line across the polar cap along the noon-midnight meridian. The field lines thus connected to the midnight meridian fuse or reconnect, providing sufficient energy for the magnetospheric substorm. The latter part of this theory will be discussed in 9.

7. Neutral Line Discharge

AKASOFU and CHAPMAN (1961) regarded the ribbon-like structure of some auroras to be one of the most vital points in any theory of the aurora. They proposed that there must be a specific structure (whatever it may be) in the magnetosphere to generate thin sheet beams of energetic auroral electrons. They concluded that the structure must be closely related to a deformation of the earth's magnetic field since an undeformed dipolar field possesses no distinctive regions which might modulate electron fluxes in such a way as to produce thin sheet beams.

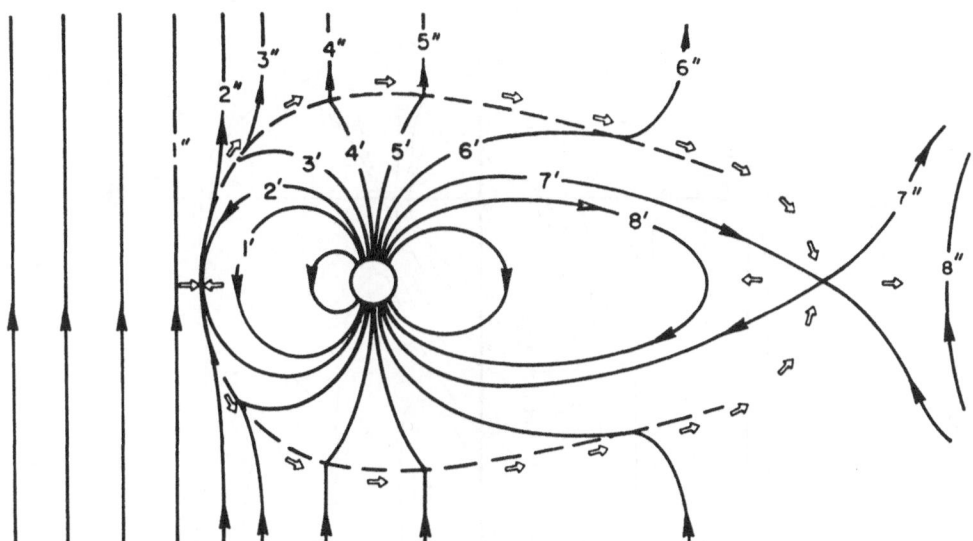

Fig. 167. Interaction between interplanetary and geomagnetic fields and the resulting motions of the plasma (open arrows) proposed by Dungey. Numbers indicate the motion of individual field lines with the motion progressing toward higher numbers (Levy, R. H., H. E. Petschek, and G. L. Siscoe: *AIAA Journal* **2**, 2065, 1964).

Further, since the ribbon-like structure has a planetary scale, the scale of the deformation must also be of large scale. Based on such a consideration, they concluded that the ring current might be one of the qualified causes of the deformation which satisfies the above requirements. The diamagnetic effect of the ring current tends to cause a 'dip' in the dipole field so that an enhanced ring current may reduce the dipole field and produce a *line* on which the field intensity may become null ($B=0$) or very small, although this requires a very large ratio of the kinetic energy density, to the magnetic field energy density ($=n\kappa T/(B^2/8\pi)$) value. They also suggested that a further enhancement of the belt might result in a reversal of the original magnetic field, forming the X and O type neutral lines. This point has, however, been criticized by several workers. PARKER (1962) states that there is no reason to believe that the over-inflation can generate new lines of force with a reversed direction.

8. *Electric Field Acceleration*

TAYLOR and HONES (1965) proposed that the interaction between the solar wind and the magnetosphere permanently maintains an electric (static) field which could drive the ionospheric current when the conductivity is sufficiently increased by particle bombardments, and that it is this electric field which accelerates auroral particles. In order to demonstrate their proposed idea, they calculated the electric field distribu-

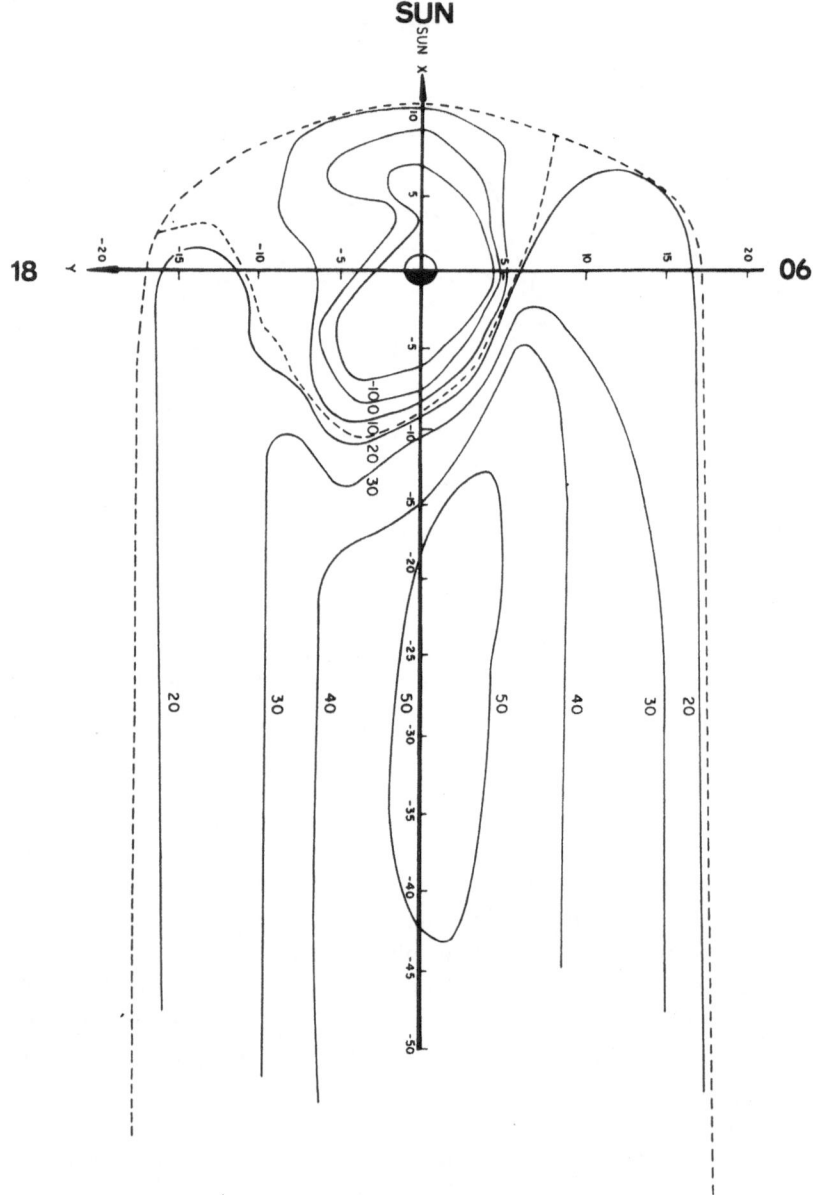

Fig. 168.   Equipotential contours (labeled in kilovolts) in the equatorial plane of the magnetosphere which is deduced on the basis of the equivalent current system (Figure 31a) (Taylor, H. E. and E. W. Hones Jr.: *J. Geophys. Res.* **70**, 3605, 1965).

tion and the equipotential contours from the ionospheric current pattern given in Figure 31a (by assuming a certain model for the distribution of conductivity). Then, by using a particular model magnetosphere, they projected the electric field distribution and the equipotential contours onto the equatorial plane. Figure 168 shows their equipotential contours (labeled in kilovolts) in the equatorial plane.

Motions of charged particles in this complex geomagnetic-geoelectric system may be studied by assuming that the motions are adiabatic. Such motions are associated with three constants of the motions:

(1) Total energy

$$K = W + eV$$

(2) Magnetic moment

$$\mu = W_\perp / B$$

(3) Integral invariant

$$J = \sqrt{2m\mu} \oint \sqrt{B_m - B}\, ds \quad \text{or} \quad J' = J/\sqrt{2m\mu} = \oint \sqrt{B_m - B}\, ds = f(B),$$

where

W = total kinetic energy,
$W_\perp$ = kinetic energy associated with the gyration,
$B_m = W/\mu$,
ds = element of the field line.

Therefore, if $\mathbf{B}(r, \theta, \phi)$ is known, a specification of K, μ and J' for a particle determines the surface on which the particle must be found.

They computed paths of solar wind electrons and protons with energies less than $W = 1$ keV for a given set of $K = \pm 18$ keV and μ by finding surfaces of constant J'. Figure 169 shows an example of their computed results for $K = \pm 18$ keV and $\mu = 5$ keV/gauss. They noted that their model can provide the following description of auroral processes:

(a) Auroral particles originate in the solar wind with energies less than ~ 1 keV.

(b) Auroral particles are trapped near the surface of the magnetosphere with $0.5 \leq J' \leq 5.0\ r_e$ gauss$^{1/2}$, and as they drift along surfaces J' their energies are increased by the electrostatic field constant from 1 to 35 keV. They will not remain in trapped orbits, but will either precipitate into the upper atmosphere and be absorbed, or drift back out of the magnetosphere.

(c) There will be a definite boundary between the region where solar wind electrons can be found and the region where solar wind protons can be found, the protons being confined equatorward of the electrons.

(d) Electron spectra should be harder equatorward of the boundary than poleward of the boundary.

(e) Electron precipitation will occur primarily in a narrow range of latitudes on the night side of the earth; protons will be precipitated over a large area in the afternoon and in a narrow region equatorward of the electrons throughout the night.

An important but unanswered question in this theory is how the electric field is set up in the magnetosphere. The theory implies (at least as a first approximation) that the mechanism which provides the electric field is independent of the magnetospheric substorm. There is now some evidence that the intensity of the electric field is greatly increased and is variable during polar magnetic substorms (HAERENDEL and LÜST (1968)).

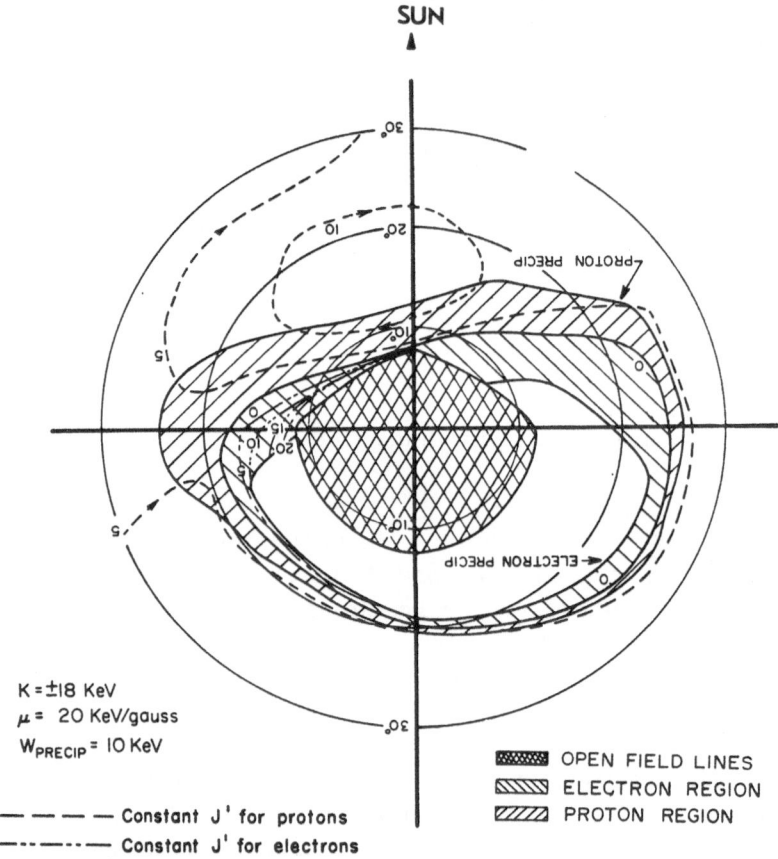

Fig. 169.   Intersection of surfaces of constant J' with the surface of the earth for solar wind particles with $K = \pm 18$ keV and $\mu = 20$ keV/gauss. The electron precipitation zone and the proton precipitation zone are shaded differently (Taylor, H. E. and E. W. Hones Jr.: *J. Geophys. Res.* **70**, 3605, 1965).

9. *Magnetospheric Tail Instability*

The magnetospheric tail has attracted the attention of a number of workers during the last several years, because the magnetic energy in the tail could be a possible energy source which may be converted into the substorm energy. Several theories have been proposed along this line, but they differ considerably in details, in particular on the mechanism of magnetospheric tail energy (cf. AXFORD, 1967a, b; DUNGEY, 1968; PIDDINGTON, 1968).

Applying the energy flow diagram discussed at the beginning of this subsection to the theories in this category we have

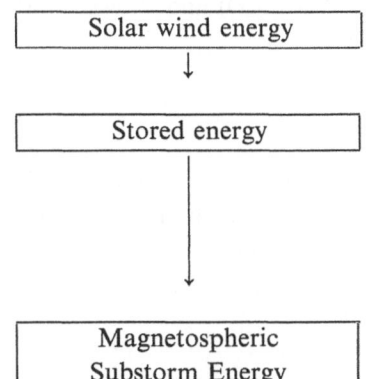

(1) = Kinetic energy of solar wind particles
(2) = X-line discharge (Dungey)
    = 'Friction' (Axford, Piddington)
(3) = Magnetic energy in the magnetospheric tail
    = Tearing mode instability (Coppi, Laval, and Pellat)
(4) = Turbulence (Axford)
    = Electron drift motion enhanced by HM waves (Piddington)

Because a similar problem has been extensively studied for a similar magnetic configuration in connection with searches for the energy source of solar flares (SWEET, 1956; PARKER, 1957; DUNGEY, 1958; PETSCHEK, 1964), the same or similar arguments can be applied to the problem of the conversion of the magnetic energy in the tail into the substorm energy (cf. AXFORD *et al.*, 1965; see also AXFORD, 1967a).

In particular, PETSCHEK's theory (1964) predicts that magnetic field lines in the tail region drift toward the current sheet in the equatorial plane with a speed of approximately one-tenth of the Alfvén wave speed, namely $0.1\,V_A$ $(=B/\sqrt{4\pi\rho})$, which must be the same as the local $(E \times B)$ drift speed, namely,

$$E/B = 0.1B/\sqrt{4\pi\rho}.$$

AXFORD *et al.* (1965) showed that $\phi = 30$ kV for $L = 2 \times 10^5$ km, $\rho = 2 \times 10^{-24}$ g/cm³ and $B \simeq 10\ \gamma$.

As pointed out by STURROCK and COPPI (1966), Petschek's mechanism (as well as Sweet's and Dungey's) is not an instability mechanism and is concerned with the steady-state dynamical behavior of the plasma and magnetic field configuration in the vicinity of a current sheet. Therefore, the above discussion does not seem to apply for the auroral substorm.

The current sheet may have three types of instability modes: the rippling mode, tearing mode, and gravitational mode. They have been extensively studied by MURTY (1961), FURTH (1963, 1964), FURTH *et al.* (1963), JOHNSON *et al.* (1963) and others under the name of finite-resistivity instability. Our particular interest is concerned with the tearing mode. This is essentially the disintegration of a sheet current into a number of filaments by the pinch effect. Further, since the medium has a finite resistance, a decoupling between the field lines and plasma occurs, and the field energy can be converted into thermal energy by Ohmic heating. The existence of this type of instability has been proved experimentally by BODIN (1963) and EBERHAGEN and GLASER (1964). FURTH (1964) also studied the growth of the tearing instability in a collisionless plasma, since effects of collisions (leading to the resistivity) can be replaced by electron inertia. He examined the stability of the neutral sheet configuration obtained by HARRIS (1962) for a perturbation (exp $(\omega t + ikz)$) and showed that the stability condition can be written in a form

$$\tfrac{1}{2}An_0r_c\left(\frac{v_s}{v}\right)^2 > 1,$$

where $A = 8\pi/k(h+k)$; $r_c = e^2/mc^2$; $h^2 = 2\pi n_0 r_c (v_s/v)^2 \simeq$ (the sheet-pinch thickness)$^{-2}$; v_s = the directed particle velocity; and v = the root mean square thermal velocity. The above condition is analogous to the Bennett pinch condition

$$I^2 > 2A_s nmv^2$$
$$\left(\text{or} \quad \tfrac{1}{2}A_s nr_c\left(\frac{v_s}{v}\right)^2 > 1\right),$$

where I denotes the current of a cylindrical particle stream and A_s its area. Thus Furth suggests that the area A_s of each first-order pinch be identified with A, which is roughly the product of the sheet-pinch thickness $2/h$ and the instability wavelength $2/k$. Therefore, $k = h$ represents marginal stability and $k < h$ represents instability. For the growth rate, Furth noted that in a special case the hydromagnetic and Vlasov-equation approaches to this problem give essentially the same result.

A more detailed examination of the growth rate using the Vlasov-equation approach has been made by LAVAL and PELLAT (1964), based on Harris' solution of the plasma-magnetic field configuration in the vicinity of the neutral sheet. The application of their results to auroral phenomena was discussed by COPPI et al. (1965). Their study shows that the growth time τ of the instability is given by

$$\tau = \frac{1}{1/2}\left(\frac{2\lambda}{R_{Le}}\right)^{3/2}\frac{\lambda}{v_{th}}\frac{\theta_e}{\theta_e + \theta_i},$$

where λ denotes the sheet thickness, R_{Le} the gyro-radius of electrons; v_{th} the electron thermal velocity, and θ_e and θ_i are κT_e and κT_i, respectively. Taking $2\lambda \simeq 600$ km, $B \simeq 1.6 \times 10^{-4}$ gauss and $\theta_i \simeq 1$ keV, one gets

$$\tau \simeq 15 \text{ sec} \quad \text{if} \quad \theta_e \simeq 10 \text{ keV}$$

and

$$\tau \simeq 5 \text{ sec} \quad \text{if} \quad \theta_e \simeq 1 \text{ keV}.$$

As mentioned earlier, the tearing mode is essentially the pinch of a sheet current, resulting in the disintegration of the sheet current into a number of filamentary currents. Each filament produces its own magnetic field around it. Figure 170 shows schematically the situations before and during the instability.

AXFORD (1967b) pointed out that the mechanism proposed by COPPI et al. (1965) cannot explain the multiplicity of auroral arcs and suggested that the entire band of the auroral oval precipitation corresponds to the region between the trapping boundary

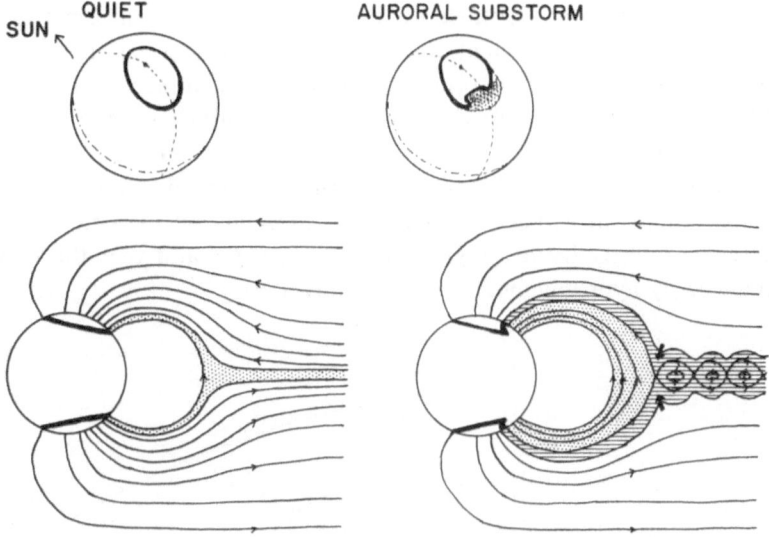

Fig. 170.   Tearing mode instability and the resulting changes of the magnetic field configuration.

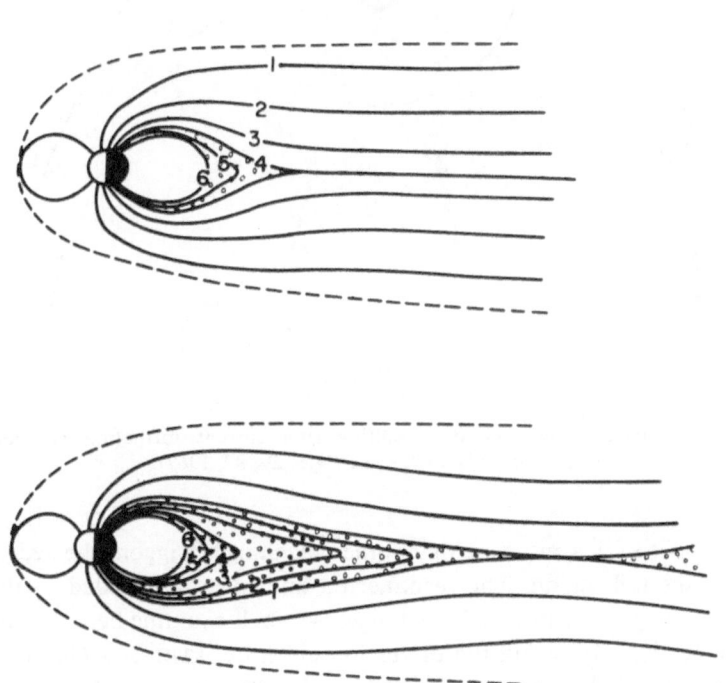

Fig. 171.   Instability in the magnetospheric tail and the resulting changes of the magnetic field configuration (Axford, W. I.: *Aurora and Airglow, Proc. NATO Advanced Study Institute*, 1966).

and the boundary of closed field lines. If this is the case, auroral electrons must be produced by some mechanism which operates in the trapping region, rather than in the tail region. AXFORD (1967b) proposed also that the reconnection of the tail field lines would result in changes of the magnetic field configuration in the way illustrated schematically in Figure 171.

PIDDINGTON (1968) has recently pointed out that the reconnection cannot be due to electron-ion collisions and suggested that it is accomplished by a flow of electrons toward the current sheet from above or below the sheet and that the flow can be greatly enhanced by hydromagnetic waves (Figure 172).

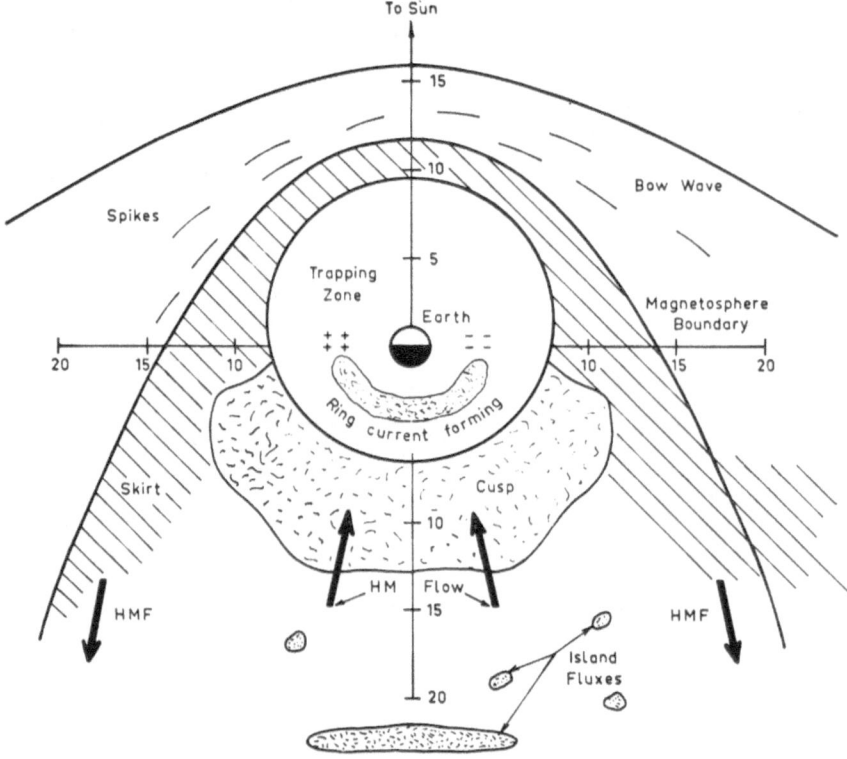

Fig. 172.   Hydromagnetic flow and the resulting ring current formation (Piddington, J. H.: *J. Atmospheric Terrest. Phys.* **29**, 87, 1967).

Let us suppose for a moment that some mechanism triggers the reconnection of field lines in the tail region. The reconnected flux tubes are added to the nightside of the trapping region, forming a large bulge. We shall examine here the consequences of the reconnection, following the discussion given by ATKINSON (1966). We assume that the intensity of the tail magnetic field is $B = 30$ γ and that the particle number density is $2/\text{cm}^3$. Since the magnetic flux in a flux tube is constant, areas in the tail region are mapped onto the polar region (where $B = 50000$ γ) in a ratio 1600:1.

Lengths and velocities are mapped in a ratio 40:1. If the field lines reconnect with a speed of 0.1 $V_A \simeq 10^7$ cm/sec in the tail region, this maps as 2.5×10^5 cm/sec = 2.5 km/ sec in the auroral zone. ATKINSON (1966) estimated the total energy deposited in the auroral bulge to be of order 3×10^{20} ergs. The volume of the tail field which must be annihilated to produce this energy has a linear dimension of $l = 4 \times 10^9$ cm $\simeq 6a$; $((B^2/8\pi) \times l^3 = 3 \times 10^{20}$ ergs). This length maps as 1000 km ($\simeq 10°$) to the ionosphere and corresponds to the North-South extent of the auroral bulge.

As mentioned in Section 10.4. A, it is unlikely that a large portion of the magnetospheric tail is involved during an early phase of the magnetospheric substorm. Since the first indication of the substorm occurs near the trapping boundary, the above estimate suggests the annihilation of the magnetic field line in a rather large volume just outside the trapping boundary.

As an inevitable consequence of the reconnection, the reconnected field lines contract violently toward the trapping boundary. The contraction will be associated with an inward flow of the plasma (PIDDINGTON, 1967); the Lorentz force ($\mathbf{J} \times \mathbf{B}$) drives such a flow, since at the point of reconnection $\nabla \times \mathbf{B} = 4\pi \mathbf{J}$ is very large. All the theories in this category suggest that the contraction will accelerate the plasma attached to the contracting field lines by the betatron acceleration process; this is because as the contraction proceeds, the magnetic field intensity B at the equatorial crossing point increases and (transverse kinetic energy)/B = constant. Therefore, if the contraction is associated with an increase of B of factor 10 at the equatorial crossing point of each field line; a 1 keV proton could be accelerated to an energy of 10 keV at the end of the contraction. This mechanism will provide ring current protons in the midnight sector (or in the late evening sector by allowing for their westward drift motion).

However, as we noted in Section 10.2. C, the ring current protons appear extensively in the sunlit sector. Furthermore, during the early phase of an intense substorm, the population of the ring current particles is greatly reduced in an extensive longitude range around the midnight meridian (20 LT – 06 LT). It seems rather difficult to explain the behavior of the ring current protons by the contraction of the field lines.

The basic problem associated with the conversion of the magnetic energy ($B^2/8\pi$) into the substorm energy may be summarized as follows: a large-scale magnetic field configuration in the magnetospheric tail results from a large-scale local ($\nabla \times \mathbf{B} \neq 0$) electric current system, and thus from a large-scale ordered motion (v) of electrons and ions $\mathbf{J} = \Sigma env$. The annihilation of the magnetic energy is then equivalent to destroying this ordered motion of electrons and protons. The energy associated with the ordered motion, $(\Sigma(1/2) mnv^2)$, may be converted into the energy of the thermal random motion $(\Sigma(1/2) mnv_t^2)$, so that $v = 0$, $J = 0$, $B = 0$ and thus $(B^2/8\pi) = 0$. (If particle-particle collisions are the basic process in this conversion, the process is nothing but the Ohmic heating.) The theories grouped in this category have not demonstrated convincingly that this will occur in an explosive way in a volume of linear dimension of at least $6a$ within a rather limited period, namely during the explosive phase $\simeq 30$ minutes.

10. *Interchange Instability*

SWIFT (1965) suggested that the auroral substorm results from a fluting instability on the outer boundary of the ring current belt. Thus, the necessary energy is first accumulated in the ring current belt as internal energy and is then converted into dynamical energy by fluting (in a manner analogous to the conversion of thermal to kinetic energy of air motion when an automobile tire has a blowout).

Applying the energy flow diagram, we may illustrate the sequence of the processes as follows:

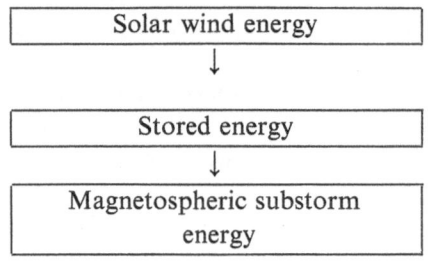

(1) = Kinetic energy of the neutral component of the solar wind?
(2) = Direct (and asymmetric) injection
(3) = Ring current particle energy
(4) = Interchange-type surface instability

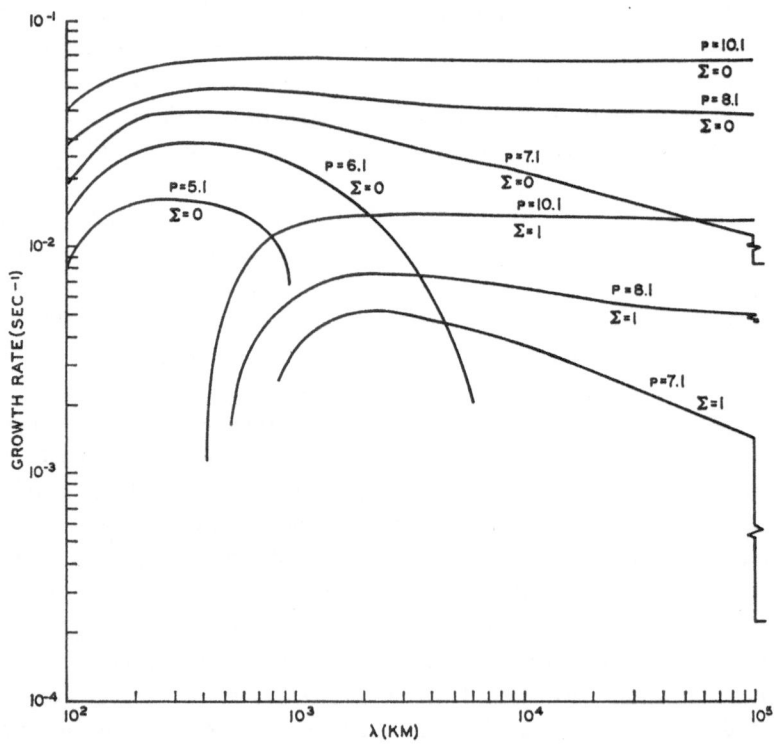

Fig. 173. Growth rate of the interchange instability as a function of wavelength (λ) near the outer boundary of the ring current belt (Swift, D. W.: *Planetary Space Sci.* **15**, 1225, 1967).

Swift considers that the electrostatic field associated with the asymmetric growth of the ring current tends to steepen the gradient of the ring current particle distribution in the midnight sector (number density $\propto r_e^{-p}$). The auroral substorm is thus considered to be a relaxation process for such a stress. This may be the explanation for the fact that substorms tend to occur when the asymmetric ring current is rapidly growing. A quiet and homogeneous arc may be explained as an effect of discharge from a thin surface boundary layer of the ring current belt. However, if the gradient of the particle distribution exceeds a certain value, the flute instability (or the interchange instability) tends to grow, but it is prevented from further growth by a short-circuiting effect of the ionosphere underneath (cf. COLE, 1963). Swift pointed out, that if this short-circuiting current flowing along the field lines becomes intense enough, ion acoustic waves can develop, and interaction between the waves and the current particles causes a sudden increase in the longitudinal resistance.

The increased resistance will free the magnetospheric plasma from a close tie with the ionosphere and allow the flute instability to grow rapidly. At the same time, the ion acoustic wave energizes electrons. He estimated that the growth rate of the instability will then be of order of a few seconds. A sudden enhancement of VLF emissions at the time of the onset of the substorm is considered to be an indication of the growth of ion acoustic waves.

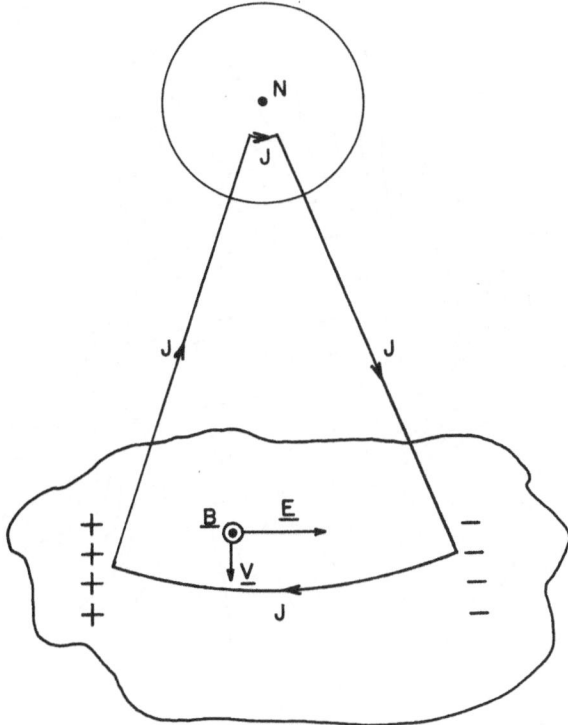

Fig. 174.   Schematic diagram showing the coupling between the magnetospheric plasma and the ionosphere.

Figure 173 shows the growth rates of the proposed interchange instability as a function of wavelength (measured in the equatorial plane), assuming that the equatorial number density of the protons (average energy 3.2 keV) falls off as r_e^{-p} The curves labeled $\Sigma = 0$ are calculated assuming that the ionosphere has zero conductivity and the curves with $\Sigma = 1$ assuming that the ionosphere has a conductivity of 1 mho.

SWIFT (1967b) suggested that a very early phase of the auroral substorm can be represented by the growth rate with $p = 7.1$ and $\Sigma = 1$ in Figure 173 for his particular model. This is the period when the growth of the instability is constrained by the depolarizing effects of the ionosphere. The rapid expansive phase would then correspond to the period of the development of the instability. This becomes possible when the resistivity of the field lines is greatly increased by the growth of the ion acoustic waves.

The fact that the magnetospheric particles are not necessarily independent from the ionospheric plasma, can be seen from the following example. Suppose a group of protons and electrons in the equatorial plane of the magnetosphere. This 'cloud' of particles will polarize by itself, because protons tend to drift westward and electrons eastward, resulting in an eastward electric field. The $(E \times B)$ drift motion of the cloud will then be outward. If the field line and the ionosphere are conductive, such an out-

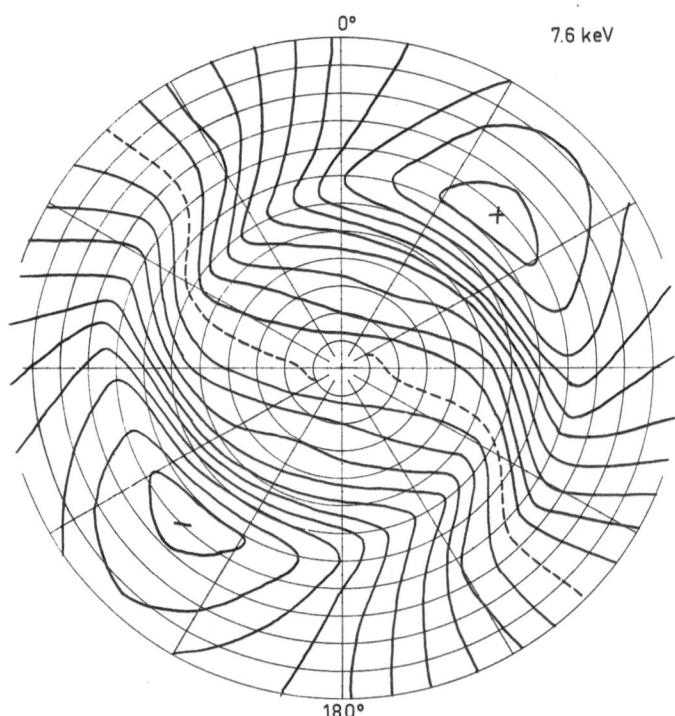

Fig. 175.   Electric potential field in the geomagnetic equatorial plane associated with ring current particles whose average energy is 7.8 keV; a highly conductive strip is assumed to exist along the auroral zone (Swift, D. W.: *Planetary Space Sci.* **15**, 835, 1967).

ward motion will be checked, because the polarization field is short-circuited by the field line-ionosphere circuit (Figure 174). The short-circuiting current flows westward, so that the resulting Lorentz force ($J \times B$) is directed inward.

SWIFT (1968) also examined consequences of an asymmetric injection of the ring current particles in a self-consistent way by following the line of thought first suggested by FEJER (1961). Fejer's study was described in detail in Section 10.3. Swift showed that the combined effect of the electric field generated by the asymmetric injection and the differential motions of charged particles causes a new space charge system near the boundary layer of the ring current and that the electrostatic field thus developed drives an electric current system similar to the model examined in Section 10.3. Figure 175 illustrates the potential field developed in the equatorial plane of the magnetosphere for an asymmetric injection of protons of energy 7.6 keV; the intensity of the injection is a function of longitude (λ) with respect to the noon meridian, $(1 + \cos \lambda)$ for $-\pi/2 < \lambda < \pi/2$. The ionospheric current generated by the potential field in Figure 174 is shown in Figure 175. Comparing both Figures 175 and 176, it can be inferred that there are field aligned currents from the positive center of the

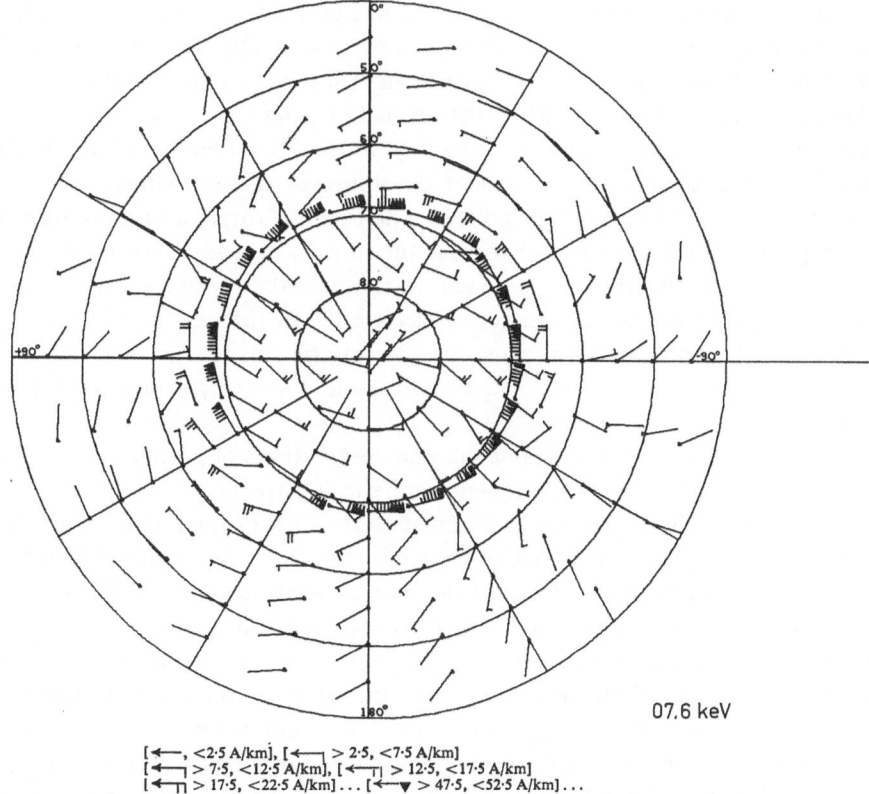

07.6 keV

[⟵—, <2·5 A/km], [⟵⊓ > 2·5, <7·5 A/km]
[⟵⊓ > 7·5, <12·5 A/km], [⟵⊤⌐ > 12·5, <17·5 A/km]
[⟵⊤⊓ > 17·5, <22·5 A/km]...[⟵⟶▼ > 47·5, <52·5 A/km]...

Fig. 176. Ionospheric component of the current associated with the potential field in Figure 174. The arrows indicate the direction of the current. The magnitude of the current (in amperes per kilometer) is indicated by the tails on the arrows (Swift, D. W.: *Planetary Space Sci.* **16**, 329, 1968).

equipotential field into the 09 LT sector of the auroral zone. The current is then divided into two, one flows eastward along the auroral zone in the day sector and the other flows westward along the auroral zone in the night sector. The current flows out along the field lines from the ionosphere to the negative center of the equipotential field in the equatorial plane. In addition to this current system, the Hall current flows around the feet of the field lines along which the currents flow. We have shown in Chapter 3 and in Section 10.3. that such a current system can explain reasonably well the observed distribution of the magnetic disturbance fields below dp lat 55°.

10.5. Concluding Remarks

It may be worthwhile to review briefly what we have discussed in this monograph. For each polar upper atmospheric phenomenon, we first examined the daily variation at different latitudes on the basis of typical daily records. This study enabled us to understand the meaning of the statistical daily variation pattern which expresses, on the dipole latitude-time coordinates, how characteristics of active feature of each phenomenon vary. However, we noted that the daily variation does not arise from the fact that the earth rotates under such a fixed (with respect to the sun-earth line) pattern once a day. A more appropriate way of expressing the actual active condition would be to say that such a pattern appears intermittently with a lifetime of order $1 \sim 3$ hours, several times a day, when the sun is fairly active.

However, the pattern varies considerably during the lifetime of a substorm. Therefore, we constructed the pattern of development of each substorm by using that of the auroral substorm (or the polar magnetic substorm) as a frame of reference; the development of the auroral substorm is the most accurately known to date.

If the cause of each substorm is known in terms of the precipitation of energetic particles or in terms of the appearance of electric fields, the pattern of the development of each substorm will tell us how the precipitation of the particles varies over the polar cap during the substorm. The construction of such a pattern was attempted in Section 10.2.

In Chapter 9, we reviewed available satellite observations of particles and magnetic fields in the magnetosphere during magnetospheric substorms.

Then, on the basis of the precipitation pattern on the polar region and *in situ* satellite observations, we attempted to construct the three dimensional distribution of electron fluxes during three epochs of the magnetospheric substorm (Section 10.3). The steps we have taken may be expressed by the chart on p. 247.

We have concluded in Section 10.2 that the magnetospheric substorm is an internal process of the magnetosphere. This is because onset time of the magnetospheric substorm at a geocentric distance of $r_e \simeq 10a$ is $T = 2$ or 3 min, where T is reckoned from the first indication of the polar substorm and because T becomes larger as r_e increases. Therefore, it is rather unlikely that the major part of the magnetospheric tail is of prime importance in the magnetospheric substorm; it is known that the length of the tail is at least of order $1000a$.

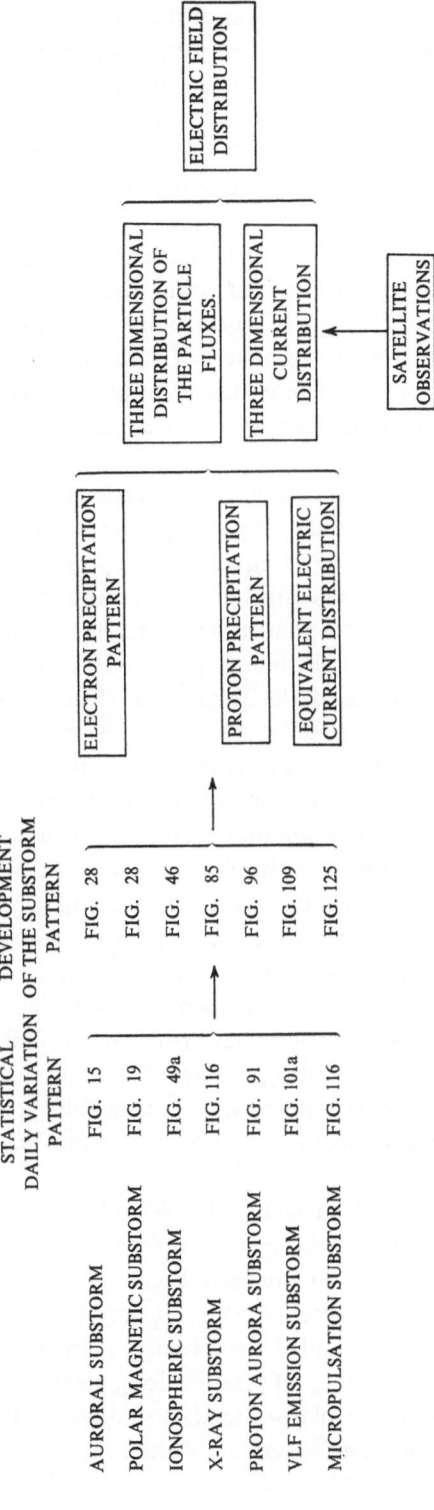

We have noted that the first indication of the auroral substorm at $T=0$ is a sudden increase of the brightness of the arc which is the equatorward boundary of the auroral oval. It is thus of vital importance to determine the exact relationship between the equatorward boundary of the auroral oval and the corresponding structure in the magnetosphere. Since the equatorward boundary of the auroral oval and the proton aurora lie close together, it is worthwhile to examine their relationship a little before and during the auroral substorm.

In this connection, it is worthwhile to look for any indication of the polar substorm before the time defined as the onset time $T=0$ in this monograph. Recently, ROSTOKER (1968) presented evidence to suggest that a polar magnetic substorm occurs in two stages. In many cases, the major excursion of a bay (negative bay) is preceded by a small excursion (trigger bay) that often has a duration of less than 20 minutes. Both the trigger bay and the main bay are associated with the Pi-2 micropulsations (Chapter 8).

We suggested in Section 10.2 that the asymmetric growth of the ring current plays a key role in the development of the magnetospheric substorm. Therefore, the formation mechanism of the asymmetric ring current may provide a clue in understanding the magnetospheric substorm. One of the most important features of the asymmetric ring current in this connection is that it is formed in the sunlit sector.

At this point, it may be worthwhile to point out a close phenomenological similarity between the substorm and the solar flare. The latter is associated with the growth of a strong electric current (MORETON and SEVERNY, 1968) and the energization of energetic particles in the lower corona or in the upper chromosphere. Some energetic electrons thus produced are injected downward, resulting in the bremsstrahlung X-rays, and the ionization and excitation of the chromospheric gas. Since hydrogen is the main constituent in the chromosphere, the resulting visible emissions are the Balmer lines. A strong Hα emission occurs along narrow strips in the chromosphere. Most flares show an explosive increase of the intensity of the Hα emission ($\sim$ of time constant of order 10^3 sec) and a slow decay (~ 1 hr).

In much the same way, the magnetospheric substorm is associated with the sudden growth of an intense electric current (the ring current and the auroral electrojet) and the generation of energetic particles. Some of the energetic particles thus produced are injected into the polar upper atmosphere, resulting in the bremsstrahlung X-rays and the atmospheric emissions (from N_2^+, O, etc.) along narrow strips (auroras).

Recently, JACOBSEN and CARLQVIST (1964) and ALFVÉN and CARLQVIST (1967) called attention to the fact that there is a critical intensity of electric current in low pressure discharge. If the current intensity becomes greater than the critical value, the magnetic energy ε_M of the discharge circuit may be dissipated explosively. Here, $\varepsilon_M = (1/2) L J^2$; $L=$inductance and $J=$total current intensity. In particular, if there is a gradient of the number density of plasma along a magnetic flux tube (along which the electric current flows), the plasma tends to adjust itself in such a way that the current intensity across any cross-sections of the tube is constant.

However, if the current density exceeds the critical value given by $en\sqrt{\kappa T_i/m_e}$, the plasma tends to move toward the region of higher density, creating a gap in the plasma distribution in a lower density region and thus resulting in a disruption of the current circuit. The voltage drop V over the gap may reach the value $V = L\,dJ/dt$. In this situation, the magnetic energy ε_M stored in the circuit is consumed in the gap region, in accelerating both electrons and ions to the energy V electronvolts.

The electric current system associated with the magnetospheric substorm consists of the asymmetric ring current belt in the day-evening sector, the current along the field lines from the morning end of the ring current to the Northern and Southern ionospheres, the ionospheric current along the auroral oval in the dark sector (that is, the auroral electrojet, one in each hemisphere) and the current along the field lines from the Western ends of the auroral electrojet to the ring current; Figure 160.

Indeed, it was shown by AKASOFU (1968) that the magnetic energy of the (inductive) ring current circuit $\varepsilon_M = (1/2)LJ^2 = (0.5) \times (1.01 \times 10^2 \, H) \times (2 \times 10^6 \, \text{amp})^2 = 2.02 \times 10^{14}$ joules $= 2.02 \times 10^2$ ergs can be dissipated in a time scale of order $\tau = L/R = 1.01 \times 10^2 H/0.05\Omega = 2.02 \times 10^3$ sec $= 34$ min. Here, the inductance L of the circuit (assuming a toroidal ring) is taken to be $1.01 \times 10^2 \, H$. The main contribution of the resistance of the whole circuit comes from the ionosphere; the resistance of the Northern ionosphere R_N may be estimated to be $R_N = V/J_N = 10^5$ volts/10^6 amp $= 0.1 \, \Omega$. The unstable discharge occurs only when the North-South thickness of the sheet current becomes of order 1.2 km or less; that is the thickness of the aurora. In the sheet, protons and electrons can be accelerated to have an energy of order $V = L\,d \, (J/2)/dt \simeq L(J/2)/\tau = (1.01 \times 10^2 \, H) \times (10^6 \, \text{amp})/2.02 \times 10^3$ sec $= 50$ kV and a flux of order 3.4×10^{10} particles/cm^2 sec. In the evening sector, the electrons are accelerated downward and protons upward, while in the morning sector the electrons are accelerated upward and protons downward. This pattern can provide the basic East-West asymmetry of the magnetospheric substorm, although subsequent processes would undoubtedly change the basic pattern: For example, the accelerated electrons in the equatorial region in the morning sector might be dumped by various types of instability; indeed, the precipitation of electrons in the morning sector shows considerable fluctuations in time and space. Further, in the equatorial plane, the accelerated particles might have an energy density which is much greater than the local magnetic energy density ($B^2/8\pi = 9 \times 10^{-8}$ ergs/cm^3). Therefore, a large-scale outward motion of the magnetospheric plasma can be expected; the poleward motion of active auroras and the outward flow of the plasma in the tail region may be manifestations of such a dynamical effect.

As the chart indicates, our goal is to find the magnetospheric electric fields, which are responsible for the magnetospheric substorm, and their causes. On the basis of the review of magnetospheric substorm theories in Section 10.4, we may say that there are two lines of thought on the cause of the electric field associated with the magnetospheric substorm. The first is that the electric field is associated with the convective $(E \times B)$ motion of the plasma in the magnetosphere (Section 10.4 (5, 6, 9)), although most of the theories in this category are not necessarily explicit on the cause of the electric field; see ALFVÉN (1968), KAVANAGH et al. (1968).

The other line of thought is that the asymmetric growth of the ring current and the electric field associated with it play a key role in the development of the magnetospheric substorm (Sections 10.4.10). The ring current is discharged in the form of a thin sheet current (along the surface generated by rotating a field line around the dipole axis) from the morning end of the ring current to the ionosphere and also from the ionosphere to the evening end of the ring current; the auroral electrojet (one in each hemisphere) is the ionospheric portion of the discharge circuit.

In the above we have briefly reviewed the two lines of thought on the nature of the electric field which causes the magnetospheric substorm. They differ greatly in their interpretation of a number of phenomena associated with the magnetospheric substorm. For example, in the convection model the ring current is thought to be merely a by-product of the convection, but in the asymmetric ring current model the ring current plays an essential role in magnetospheric substorm processes. Here, it is interesting to note that both models face a major stumbling block. That is, both models do not explain the way in which the solar wind energy is introduced into the magnetosphere.

We all agree, however, that the magnetospheric substorm offers difficult but challenging problems. Further, it produces a variety of phenomena in the polar upper atmosphere. In particular, the auroral substorm is the most mysterious and beautiful sight in the polar sky.

References

GENERAL

HESS, W. N.: 1968, *The radiation belt and magnetosphere*, Blaisdell Pub. Co., Waltham, Mass.

REFERRED TO IN TEXT

AKASOFU, S.-I.: 1968, *Magnetospheric substorm as a discharge process* (in preparation).
AKASOFU, S.-I. and CHAPMAN, S.: 1961, 'A neutral line discharge theory of the auroral polaris', *Phil. Trans. Roy. Soc.* **253**, 359–406.
ALFVÉN, H.: 1950, *Cosmical electrodynamics*, Clarendon Press, Oxford, England.
ALFVÉN, H.: 1955, 'On the electric field theory of magnetic storms and aurorae', *Tellus* **7**, 50–64.
ALFVÉN, H.: 1968, 'Some properties of magnetospheric neutral surfaces', *J. Geophys. Res.* **73**, 4379–4381.
ALFVÉN, H. and CARLQVIST, P.: 1967, 'Currents in the solar atmosphere and a theory of solar flares, *Solar Phys.* **1**, 220–228.
ATKINSON, G.: 1966, 'A theory of polar substorms'. *J. Geophys. Res.* **71**, 5157–5164.
ATKINSON, G.: 1967, 'Polar magnetic substorms', *J. Geophys. Res.* **72**, 1491–1494.
AXFORD, W. I.: 1962, 'The interaction between the solar wind and the earth's magnetosphere', *J. Geophys. Res.* **67**, 3791–3796.
AXFORD, W. I.: 1964, 'Viscous interaction between the solar wind and the earth's magnetosphere', *Planetary Space Sci.* **12**, 45–54.
AXFORD, W. I.: 1967a, 'Magnetic storm effects associated with the tail of the magnetosphere', *Space Sci. Rev.* 149–157.
AXFORD, W. I.: 1967b, 'The interaction between the solar wind and the magnetosphere', in *Aurora and Airglow* (ed. by B. M. McCormac), Reinhold Pub. Co., New York, pp. 499–509.
AXFORD, W. I. and HINES, C. O.: 1961, 'A unifying theory of high-latitude geophysical phenomena and geomagnetic storms', *Canadian J. Phys.* **39**, 1433–1464.

AXFORD, W. I., PETSCHEK, H. E., and SISCOE, G. L.: 1965, 'Tail of the magnetosphere', *J. Geophys. Res.* **70**, 1231–1236.

BIRKELAND, K.: 1913, *The Norwegian aurora polaris expedition 1902–1903*, Vol. 1, Section 2, H. Aschehoug Co., Christiania.

BLOCK, L. P.: 1966, 'On the distribution of electric fields in the magnetosphere', *J. Geophys. Res.* **71**, 858–864.

BODIN, H. A. B.: 1963, 'Observations of resistive instabilities in a Theta pinch', *Nuclear Fusion* **3**, 215–217.

BOSTRÖM, R.: 1967a, 'Desirable magnetic-field measurements in the high-latitude magnetosphere', *Space Sci. Rev.* **7**, 191–197.

BOSTRÖM, R.: 1967b, 'Auroral electric fields', in *Aurora and airglow* (ed. by B. M. McCormac), Reinhold Pub. Co., New York, pp. 293-303.

BRICE, N. M.: 1967, 'Bulk motion of the magnetosphere', *J. Geophys. Res.* **72**, 5193–5211.

CAHILL, L. J. Jr.: 1966, 'Inflation of the inner magnetosphere during a magnetic storm', *J. Geophys. Res.* **71**, 4505–4519.

CARPENTER, D. L.: 1966, 'Whistler studies of the plasmapause in the magnetosphere. 1. Temporal variations in the position of the knee and some evidence on plasma motions near the knee', *J. Geophys. Res.* **71**, 693–709.

CHAMBERLAIN, J. W.: 1961, 'Theory of auroral bombardment', *Astrophys. J.* **134**, 401–424.

CHAPMAN, S. and BARTELS, J.: 1940, *Geomagnetism*, The Clarendon Press, London.

CHAPMAN, S. and FERRARO, V. C. A.: 1933, 'A new theory of magnetic storms. Part II. The main phase', *Terr. Magn. and Atmos. Elect.* **38**, 79–96.

COLE, K. D.: 1960, 'A dynamo theory of the aurora and magnetic disturbance', *Australian J. Phys.* **13**, 484–497.

COLE, K. D.: 1963, 'Damping of magnetospheric motions by the ionosphere', *J. Geophys. Res.* **68** 3231–3235.

COPPI, B., LAVAL, G., and PELLAT, R.: 1965, 'A model for the influence of the earth magnetic tail on geomagnetic phenomena', International Center for Theoretical Physics Pub.

CUMMINGS, W. D. and COLEMAN, P. J., Jr.: 1968, 'Simultaneous magnetic field variations at the earth's surface and at synchronous, equatorial distance', *Radio Sci.* **3**, 758–761.

DESSLER, A. J. and PARKER, E. N.: 1959, 'Hydromagnetic theory of geomagnetic storms', *J. Geophys. Res.* **64**, 2239–2252.

DUNGEY, J. W.: 1958, *Cosmic electrodynamics*, Cambridge Univ. Press, England.

DUNGEY, J. W.: 1961, 'Interplanetary magnetic field and the auroral zones', *Phys. Rev. Letters* **6**, 47–48.

DUNGEY, J. W.: 1963, 'The structure of the exosphere or adventures in velocity space', in *Geophysics, the earth's environment* (ed. by C. DeWitt, J. Hieblot, and A. Lebeau), Gordon and Breach, New York, pp. 503-550.

DUNGEY, J. W.: 1968, 'The reconnection model of the magnetosphere', in *Earth's particles and fields* (ed. by B. M. McCormac), Reinhold, N.Y., pp. 385–392.

EBERHAGEN, A. and GLASER, H.: 1964, 'Studies on macroinstabilities in a Theta pinch with antiparallel magnetic field', *Nuclear Fusion* **4**, 296–299.

FEJER, J. A.: 1961, 'The effects of energetic trapped particles on magnetospheric motions and ionospheric currents', *Canadian J. Phys.* **39**, 1409–1417.

FEJER, J. A.: 1963, 'Theory of auroral electrojets', *J. Geophys. Res.* **68**, 2147–2157.

FEJER, J. A.: 1964, 'Theory of the geomagnetic daily disturbance variations', *J. Geophys. Res.* **69**, 123–137.

FRANK, L. A.: 1967, 'On the extraterrestrial ring current during geomagnetic storms', *J. Geophys. Res.* **72**, 3753–3767.

FUKUSHIMA, N. and OGUTI, T.: 1953, 'Polar magnetic storms and geomagnetic bays. Appendix 1. A theory of S_D-field', *Rep. Ionos. Space Res. Japan.* **7**, 137–146.

FURTH, H. P.: 1963, 'Prevalent instability of nonthermal plasmas', *Phys. Fluids* **6**, 48–57.

FURTH, H. P.: 1964, 'Instabilities due to finite resistivity of finite current-carrier mass', Advanced plasma theory, Proc. Intern. School of Phys. Course XXV, Academic Press.

FURTH, H. P., KILLEEN, J., and ROSENBLUTH, M. N.: 1963, 'Finite-resistivity instabilities of a sheet pinch', *Phys. Fluids* **6**, 459–484.

GOLD, T.: 1959, 'Plasma and magnetic fields in the solar system', *J. Geophys. Res.* **64**, 1665–1674.

HAERENDEL, G. and LÜST, R.: 1968, 'Electric fields in the upper atmosphere', in *Earth's particles and fields* (ed. by B. M. McCormac), Reinhold Pub. Co., New York, pp. 271–285.

HARGREAVES, J. K., HONES, E. W. Jr., and SINGER, S.: 1968, 'Relations between bursts of energetic electrons at 17 earth-radii in the magnetotail and radio absorption events in the ionospheric D-region', *Planetary Space Sci.* **16**, 567–580.

HARRIS, E. G.: 1962, 'On a plasma sheath separating regions of oppositely directed magnetic field' *Nuovo Cimento* **23**, 115–121.

HARTZ, T. R. and BRICE, N. M.: 1967, 'The general pattern of auroral particle precipitation', *Planetary Space Sci.* **15**, 301–329.

HINES, C. O.: 1964, 'Hydromagnetic motions in the magnetosphere', *Space Sci. Rev.* **3**, 342–379.

HULBURT, E. O.: 1937, 'Terrestrial magnetic variations and aurorae', *Rev. Mod. Phys.* **9**, 44–68.

JACOBSEN, C. and CARLQVIST, P.: 1964, 'Solar flares caused by circuit interruptions', *Icarus* **3**, 270–272.

JOHNSON, J. L., GREENE, J. M., and COPPI, B.: 1963, 'Effect of resistivity on hydromagnetic instabilities in multipolar systems', *Phys. Fluids* **6**, 1169–1183.

KARLSON, E. T.: 1962, 'Motion of charged particles in an inhomogeneous magnetic field', *Phys. Fluids* **5**, 476–486.

KARLSON, E. T.: 1963, 'Streaming of a plasma through a magnetic dipole field', *Phys. Fluids* **6**, 708–722.

KAVANAGH, L. D., Jr., FREEMAN, J. W., Jr., and CHEN, A. J.: 1968, 'Plasma flow in the magnetosphere', *J. Geophys. Res.* **73**, 5511–5519.

KERN, J. W.: 1962, 'A charge separation mechanism for the production of polar auroras and electrojets', *J. Geophys. Res.* **67**, 2649–2665.

KERN, J. W. and VESTINE, E. H.: 1961, 'Theory of auroral morphology', *J. Geophys. Res.* **66**, 713–723.

LAVAL, G. and PELLAT, R.: 1964, 'Méthode d'étude de la stabilité de certaines solutions de l'équation de Vlasov', *Compt. Rend.* **259**, 1706–1709.

LEVY, R. H., PETSCHEK, H. E., and SISCOE, G. L.: 1964, 'Aerodynamic aspects of the magnetospheric flow', *AIAA Journal* **2**, 2065–2076.

MAEDA, H.: 1957, 'Wind systems for the geomagnetic S_D field', *J. Geomag. Geoelec.* **9**, 119–121.

MARIS, H. B. and HULBURT, E. O.: 1929, 'A theory of auroras and magnetic storms', *Phys. Rev.* **33**, 412–431.

MARTYN, D. F.: 1951, 'The theory of magnetic storms and auroras', *Nature* **167**, 92–94.

MORETON, G. E. and SEVERNY, A. B.: 1968, 'Magnetic fields and flares in the CMP, 20 September 1963', *Solar Phys.* **3**, 715–719.

MURTY, G. S.: 1961, 'Instability of a conducting fluid slab carrying uniform current in the presence of a homogeneous magnetic field', *Arkiv Fysik* **19**, 499–510.

NAGATA, T. and KOKUBUN, S.: 1960, 'Polar magnetic storms, with special reference to relation between geomagnetic disturbances in the northern and southern auroral zones', *Rep. Ionos. Space Res. Japan* **14**, 273–290.

NISHIDA, A.: 1966, 'Formation of plasmapause, or magnetospheric plasma knee, by the combined action of magnetospheric convection and plasma escape from the tail', *J. Geophys. Res.* **71**, 5669–5679.

OBAYASHI, T. and JACOBS, J. A.: 1957, 'Sudden commencements of magnetic storms and atmospheric dynamo action', *J. Geophys. Res.* **62**, 589–616.

PARKER, E. N.: 1957, 'Sweet's mechanism for merging magnetic fields in conducting fluids', *J. Geophys. Res.* **62**, 509–520.

PARKER, E. N.: 1962, 'Dynamics of the geomagnetic storm', *Space Sci. Rev.* **1**, 62–99.

PETSCHEK, H. E.: 1964, 'Magnetic field annihilation', AAS-NASA Symposium on the Physics of Solar Flares (ed. by W. N. Hess), NASA SP-50, pp. 425–439.

PIDDINGTON, J. H.: 1960, 'Geomagnetic storm theory', *J. Geophys. Res.* **65**, 93–106.

PIDDINGTON, J. H.: 1962a, 'A hydromagnetic theory of geomagnetic storms', *Geophys. J.* **7**, 183–193.

PIDDINGTON, J. H.: 1962b, 'A hydromagnetic theory of geomagnetic storms and auroras', *Planetary Space Sci.* **9**, 947–957.

PIDDINGTON, J. H.: 1963, 'Connexions between geomagnetic and auroral activity and trapped ions', *Planetary Space Sci.* **11**, 451–462.

PIDDINGTON, J. H.: 1967, 'A theory of auroras and the ring current', *J. Atmospheric Terrest. Phys.* **29**, 87–105.

PIDDINGTON, J. H.: 1968, 'The growth and decay of the geomagnetic tail', *Earth's particles and fields* (ed. by B. M. McCormac), Reinhold, N.Y., pp. 417–427.

ROSTOKER, G.: 1968, 'Macrostructure of geomagnetic bays', *J. Geophys. Res.* **73**, 4217–4229.

SHAW, J. E.: 1959, 'Outline of a theory of magnetic separation of auroral particles and the origin of the S_D field', *Planetary Space Sci.* **2**, 49–55.

SPEISER, T. W.: 1965, 'Particle trajectories in model current sheets. 1. Analytical solutions', *J. Geophys. Res.* **70**, 4219–4226.

SPEISER, T. W.: 1967a, 'Plasma density and acceleration in the tail from the reconnection model', *Earth's particles and fields* (ed. by B. M. McCormac), Reinhold, N.Y., pp. 393–402.

SPEISER, T. W.: 1967b, 'Particle trajectories in model current sheets, 2. Applications to auroras using a geomagnetic tail model', *J. Geophys. Res.* **72**, 3919–3932.

SPEISER, T. W.: 1968, 'On the uncoupling of parallel and perpendicular particle motion in a neutral sheet', *J. Geophys. Res.* **73**, 1112–1113.

STURROCK, P. A. and COPPI, B.: 1966, 'A new model of solar flares', *Astrophys. J.* **143**, 3–22.

SWEET, P. A.: 1956, 'The neutral point theory of solar flares', *I.A.U. Symposium No. 6* (Stockholm, 1956) (ed. by B. Lennert), Cambridge Univ. Press. pp. 123–134.

SWEET, P. A.: 1963, 'Instability problems in the origin of solar flares', AAS-NASA Symposium on the physics of solar flares (ed. by W. N. Hess), NASA-SP-50, pp. 409–413.

SWIFT, D. W.: 1963, 'The generation and effect of electrostatic fields during an auroral disturbance', *J. Geophys. Res.* **68**, 2131–2140.

SWIFT, D. W.: 1964, 'The connection between the ring current belt and the auroral substorm', *Planetary Space Sci.* **12**, 945–960.

SWIFT, D. W.: 1965, 'A mechanism for energizing electrons in the magnetosphere', *J. Geophys. Res.* **70**, 3061–3073.

SWIFT, D. W.: 1967a, 'Possible consequences of the asymmetric development of the ring current belt', *Planetary Space Sci.* **15**, 835–862.

SWIFT, D. W.: 1967b, 'The possible relationship between the auroral breakup and the interchange instability of the ring current', *Planetary Space Sci.* **15**, 1225–1237.

SWIFT, D. W.: 1968, 'Further possible consequences of the asymmetric development of the ring current belt – effect of variations in ionospheric conductivity', *Planetary Space Sci.* **16**, 329–342.

TAYLOR, H. E. and HONES Jr., E. W.: 1965, 'Adiabatic motion of auroral particles in a model of the electric and magnetic fields surrounding the earth', *J. Geophys. Res.* **70**, 3605–3628.

VASYLIUNAS, V. M.: 1968, 'A survey of low-energy electrons in the evening sector of the magnetosphere with OGO 1 and OGO 3', *J. Geophys. Res.* **73**, 2839–2884.

WALBRIDGE, E.: 1967, 'The limiting of magnetospheric convection by dissipation in the ionosphere', *J. Geophys. Res.* **72**, 5213–5230.

APPENDIX

APPENDIX

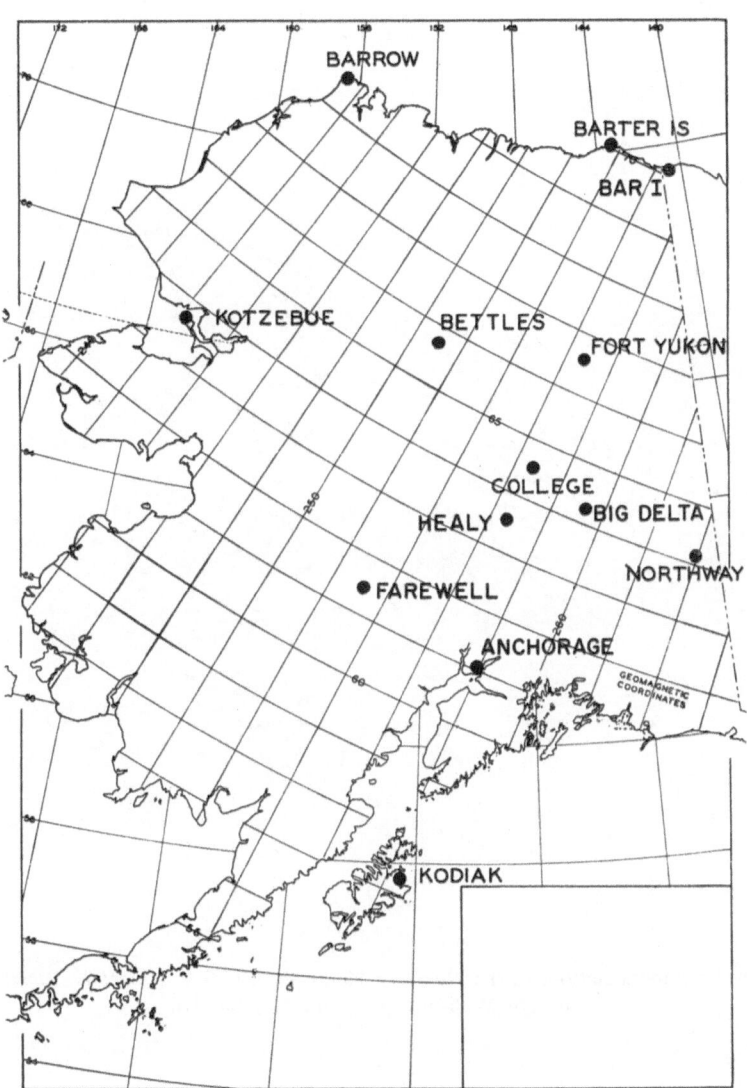

Fig. A.   Map of Alaska and the location of the Alaskan geophysical observatories.

Fig. B.   All-sky camera stations in the Northern Hemisphere (*I.G.Y. Ascaplots, Annals of the IGY* (ed. by W. Stoffregen), vol. XX, part 1.).

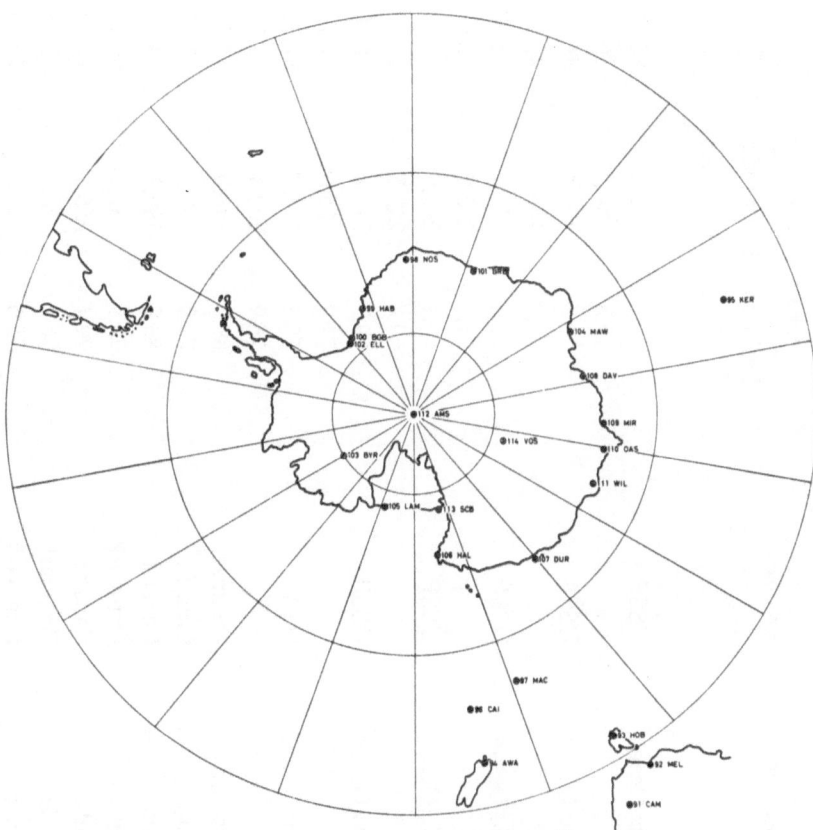

Fig. C.   All-sky camera stations in the Southern Hemisphere (*I.G.Y. Ascaplots, Annals of the IGY* (ed. by W. Stoffregen), vol. XX, part 1.).

LIST OF GEOMAGNETIC OBSERVATORIES COMPILED BY THE WORLD DATA CENTER A

CSAGI No.	Observatory	Location	Sponsor	Geomagnetic Latitude	Geomagnetic Longitude	Geographic Latitude °'	Geographic Longitude °'
A001	North Pole 6	Arctic Ocean	USSR	70.6	197.4	80 56 N	150 02 E
				76.8	164.2	86 15 N	39 23 E
A003	North Pole 7	Arctic Ocean	USSR	78.5	198.4	86 21 N	149 38 W
				81.9	160.5	85 17 N	34 05 W
	North Pole 8	Arctic Ocean	USSR	71.0	230.0	75 32 N	162 47 W
				71.5	213.7	79 06 N	179 23 E
	North Pole 10	Arctic Ocean	USSR	66.1	208.9	75 11 N	160 52 E
				69.8	211.7	78 02 N	171 54 E
	North Pole 12	Arctic Ocean	USSR	69.7	217.8	76 50 N	179 59 E
				74.9	215.0	81 19 N	194 26 E
A019	Arctic Ice Floe A	Arctic Ocean	USA	75.2	214.5	81 38 N	164 34 W
				76.9	200.4	85 23 N	169 20 W
A021	Thule	Greenland	Denmark	89.0	358.0	77 29 N	69 10 W
A005	Alert	Canada	Canada	85.9	168.2	82 30 N	62 30 W
A030	Resolute Bay	Canada	Canada	83.0	289.3	74 42 N	94 54 W
A049	Godhavn	Greenland	Denmark	79.9	32.5	69 14 N	53 31 W
A028	Mould Bay	Canada	Canada	79.1	256.4	76 12 N	119 24 W
A070	Sukkertoppen	Greenland	Denmark	76.1	28.7	65 25 N	52 54 W
A044	Kap Tobin	Greenland	Denmark	75.6	81.6	70 25 N	21 58 W
A010	Murchison Bay	Norway	Sweden	75.2	137.5	80 03 N	18 15 E
A018	Barentsburg	Norway	USSR	74.6	132.5	78 38 N	16 23 E

List of geomagnetic observatories (continued)

CSAGI No.	Observatory	Location	Sponsor	Geomagnetic Latitude °	Geomagnetic Longitude °	Geographic Latitude ° ′	Geographic Longitude ° ′
A099	Baker Lake	Canada	Canada	73.8	315.2	64 20 N	96 02 W
A009	Tikhaya Bay	USSR	USSR	71.5	153.2	80 18 N	52 48 E
A009	Heiss Island	USSR	USSR	71.3	156.1	80 37 N	58 03 E
A031	Bear Island	Norway	Norway	71.1	124.5	74 30 N	19 12 E
A132	Julianehaab	Greenland	Denmark	70.8	35.5	60 43 N	46 02 W
A104	Leirvogur	Iceland	Iceland	70.2	71.0	64 11 N	21 42 W
A045	Barter Island	Alaska	USA	70.0	253.1	70 08 N	143 40 W
A122	Yellowknife	Canada	Canada	69.0	293.3	62 26 N	114 24 W
A145	Fort Churchill	Canada	Canada	68.7	322.8	58 48 N	94 06 W
A039	Barrow	Alaska	USA	68.5	241.1	71 18 N	156 45 W
A047	Tromso	Norway	Norway	67.1	116.7	69 40 N	18 57 E
A148	Great Whale River	Canada	Canada	66.6	347.4	55 16 N	77 47 W
A073	Fort Yukon	Alaska	USA	66.6	256.8	66 34 N	145 18 W
A020	Cape Chelyuskin	USSR	USSR	66.3	176.5	77 43 N	104 17 E
A054	Abisko	Sweden	Sweden	66.0	115.0	68 21 N	18 49 E
A059	Utsjoki	Finland	Finland	65.8	122.9	69 45 N	27 02 E
A060	Kiruna	Sweden	Sweden	65.3	115.6	67 50 N	20 25 E
A180	Ivalo	Finland	Finland	64.7	121.9	68 36 N	27 29 E
A092	College	Alaska	USA	64.6	256.5	64 52 N	147 50 W
A102	Big Delta	Alaska	USA	64.3	259.3	64 00 N	145 44 W
A118	Northway	Alaska	USA	64.1	263.8	62 58 N	141 57 W

List of geomagnetic observatories (continued)

CSAGI No.	Observatory	Location	Sponsor	Geomagnetic Latitude	Geomagnetic Longitude	Geographic Latitude ° '	Geographic Longitude ° '
A065	Sodankyla	Finland	Finland	63.8	120.0	67 22 N	26 38 E
A069	Kotzebue	Alaska	USA	63.7	242.1	66 53 N	162 36 W
A107	Healy	Alaska	USA	63.5	256.6	63 51 N	148 58 W
A074	Murmansk	USSR	USSR	63.5	125.8	68 15 N	33 05 E
A033	Dixon Island	USSR	USSR	63.0	161.6	73 33 N	80 34 E
A057	Lovozero	USSR	USSR	62.9	127.0	67 58 N	35 01 E
A140	Lerwick	Scotland	United Kingdom	62.5	88.6	60 08 N	1 11 W
A123	Dombas	Norway	Norway	62.3	100.1	62 04 N	9 07 E
A154	Meanook	Canada	Canada	61.8	301.0	54 37 N	113 20 W
A077	Cape Wellen	USSR	USSR	61.8	237.1	66 10 N	169 50 W
A130	Anchorage	Alaska	USA	61.0	258.1	61 14 N	149 52 W
A037	Tixie Bay	USSR	USSR	60.4	191.4	71 35 N	129 00 E
A149	Sitka	Alaska	USA	60.0	275.3	57 04 N	135 20 W
B038	Eskdalemuir	Scotland	United Kingdom	58.5	82.9	55 19 N	3 12 W
B009	Lovo	Sweden	Sweden	58.1	105.8	59 21 N	17 50 E
A134	Nurmijarvi	Finland	Finland	57.9	112.6	60 31 N	24 39 E
B349	Stonyhurst	England	United Kingdom	56.9	82.7	53 51 N	2 28 W
B098	Valentia	Ireland	Ireland	56.6	73.5	51 56 N	10 15 W
B259	Leningrad	USSR	USSR	56.2	117.3	59 57 N	30 42 E
B114	Rude Skov	Denmark	Denmark	55.9	98.5	55 51 N	12 27 E
B249	Agincourt	Canada	Canada	55.1	347.0	43 47 N	79 16 W

List of geomagnetic observatories (continued)

CSAGI No.	Observatory	Location	Sponsor	Geomagnetic Latitude	Geomagnetic Longitude	Geographic Latitude ° ′	Geographic Longitude ° ′
	Newport	USA	USA	55.1	300.0	48 16 N	117 07 W
B119	Hartland	England	United Kingdom	54.6	79.0	51 00 N	4 29 W
B058	Wingst	German Fed Rep	German Fed Rep	54.6	94.1	53 45 N	9 04 E
B326	Warkenhagen	German Dem Rep	German Dem Rep	54.4	96.1	54 01 N	11 04 E
B159	Victoria	Canada	Canada	54.2	293.0	48 31 N	123 25 W
B071	Witteveen	Netherlands	Netherlands	54.1	91.2	52 49 N	6 40 E
B269	Weston	USA	USA	53.9	357.1	42 23 N	71 19 W
B044	Hel	Poland	Poland	53.4	103.7	54 36 N	18 49 E
A121	Srednikan	USSR	USSR	53.1	210.6	62 26 N	152 19 E
B014	Borok	USSR	USSR	53.0	123.2	58 02 N	38 58 E
B106	Gottingen	German Fed Rep	German Fed Rep	52.3	93.7	51 32 N	9 58 E
B095	Niemegk	German Dem Rep	German Dem Rep	52.2	96.6	52 04 N	12 41 E
B132	Manhay	Belgium	Belgium	52.0	88.9	50 18 N	5 41 E
B136	Dourbes	Belgium	Belgium	52.0	87.7	50 06 N	4 36 E
B261	Casper	USA	USA	51.5	314.5	42 51 N	106 18 W
B325	Collm	German Dem Rep	German Dem Rep	51.5	96.5	51 19 N	13 00 E
	Minsk	USSR	USSR	51.5	110.4	54 06 N	26 31 E
A124	Yakutsk	USSR	USSR	51.0	193.8	62 01 N	129 43 E
B035	Moscow	USSR	USSR	50.9	120.5	55 29 N	37 19 E
B089	Swider	Poland	Poland	50.6	104.6	52 07 N	21 15 E
B161	Chambon-la-Forêt	France	France	50.5	84.4	48 01 N	2 16 E

List of geomagnetic observatories (continued)

CSAGI No.	Observatory	Location	Sponsor	Geomagnetic Latitude	Geomagnetic Longitude	Geographic Latitude ° '	Geographic Longitude ° '
	Nantes	France	France	50.5	80.1	47 15 N	1 33 W
B101	Belsk	Poland	Poland	50.4	104.0	51 50 N	20 48 E
B143	Pruhonice	Czechoslovakia	Czechoslovakia	49.9	97.3	49 59 N	14 33 E
B269	Carrollton	USA	USA	49.6	330.4	39 22 N	93 28 W
B318	Fredericksburg	USA	USA	49.6	349.8	38 12 N	77 22 W
B182	Garchy	France	France	49.6	84.9	47 18 N	3 06 E
B028	Kazan	USSR	USSR	49.3	130.4	55 50 N	48 51 E
B137	Raciborz	Poland	Poland	49.3	100.8	50 05 N	18 11 E
B012	Beloit	USA	USA	49.2	324.9	39 29 N	98 08 W
B328	Budkov	Czechoslovakia	Czechoslovakia	49.1	96.4	49 04 N	14 01 E
B295	Boulder	USA	USA	49.0	316.5	40 08 N	105 14 W
B163	Furstenfeldbruck	German Fed Rep	German Fed Rep	48.8	93.3	48 10 N	11 17 E
	Regensburg	Switzerland	Switzerland	48.7	90.3	47 29 N	8 27 E
B054	Shatsk	USSR	USSR	48.7	123.7	53 59 N	41 51 E
B019	Sverdlovsk	USSR	USSR	48.5	140.7	56 44 N	61 04 E
C401	Burlington	USA	USA	48.5	320.1	39 17 N	102 16 W
B145	Lvov	USSR	USSR	48.0	105.9	49 54 N	23 45 E
B305	Leadville	USA	USA	48.0	315.5	39 17 N	106 17 W
B162	Wien-Kobenzl	Austria	Austria	47.9	98.2	48 16 N	16 19 E
B440	Price	USA	USA	47.7	310.3	39 37 N	110 47 W
B150	Kiev	USSR	USSR	47.6	112.2	50 43 N	30 18 E

List of geomagnetic observatories (continued)

CSAGI No.	Observatory	Location	Sponsor	Geomagnetic Latitude	Geomagnetic Longitude	Geographic Latitude ° '	Geographic Longitude ° '
B100	Adak	Aleutian Islands	USA	47.2	240.0	51 52 N	176 39 W
B168	Nagycenk	Hungary	Hungary	47.2	98.3	47 38 N	16 43 E
B172	Hurbanovo	Czechoslovakia	Czechoslovakia	47.2	99.8	47 52 N	18 11 E
B232	Castel Tesino	Italy	Italy	46.7	92.8	46 03 N	11 39 E
B191	Tihany	Hungary	Hungary	46.3	99.1	46 54 N	17 54 E
B267	Logrono	Spain	Spain	46.1	77.2	42 27 N	2 30 W
B022	Tomsk	USSR	USSR	45.9	159.6	56 28 N	84 56 E
B236	Monte Capellino	Italy	Italy	45.8	89.5	44 33 N	8 57 E
B280	Roburent	Italy	Italy	45.8	88.4	44 18 N	7 53 E
B239	Castellaccio	Italy	Italy	45.7	89.4	44 26 N	8 56 E
B321	San Miguel	Azores Islands	Portugal	45.6	50.9	37 46 N	25 39 W
C329	Baja	Hungary	Hungary	45.4	99.9	46 11 N	19 00 E
C093	Coimbra	Portugal	Portugal	45.0	70.3	40 13 N	8 25 W
	Jassy	Rumania	Rumania	44.7	108.2	47 11 N	27 32 E
C001	Petropavlovsk	USSR	USSR	44.7	218.6	53 06 N	158 38 E
C406	Espanola	USA	USA	44.6	316.6	35 49 N	106 04 W
C086	Del Ebro	Spain	Spain	43.9	79.7	40 49 N	0 30 E
C098	Toledo	Spain	Spain	43.9	74.7	39 53 N	4 03 W
C018	Odessa	USSR	USSR	43.7	111.1	46 47 N	30 53 E
C028	Grocka	Yugoslavia	Yugoslavia	43.6	100.9	44 38 N	20 46 E
	Castle Rock	USA	USA	43.5	298.6	37 14 N	122 08 W

List of geomagnetic observatories (continued)

CSAGI No.	Observatory	Location	Sponsor	Geomagnetic		Geographic	
				Latitude	Longitude	Latitude °　′	Longitude °　′
C405	Dallas	USA	USA	43.0	327.7	32 59 N	96 45 W
C063	L'Aquila	Italy	Italy	42.9	92.9	42 23 N	13 19 E
C085	La Maddalena	Italy	Italy	42.5	88.7	41 14 N	9 24 E
C215	Surlari	Rumania	Rumania	42.5	106.1	44 41 N	26 15 E
C082	Ponza	Italy	Italy	41.5	92.1	40 55 N	12 57 E
C026	Simferopol	USSR	USSR	41.2	113.3	44 50 N	34 04 E
C360	Aloushta	USSR	USSR	41.0	113.6	44 41 N	34 25 E
C143	San Fernando	Spain	Spain	41.0	71.3	36 28 N	6 12 W
C078	Capri	Italy	Italy	40.9	93.2	40 33 N	14 13 E
C059	Panagyurishte	Bulgaria	Bulgaria	40.8	103.4	42 31 N	24 11 E
B102	Irkutsk	USSR	USSR	40.7	174.8	52 10 N	104 27 E
C137	Almeria	Spain	Spain	40.6	75.3	36 51 N	2 28 W
C236	Tucson	USA	USA	40.4	312.2	32 15 N	110 50 W
C124	Gibilmanna	Italy	Italy	38.5	92.2	37 59 N	14 01 E
	Kandilli	Turkey	Turkey	38.5	107.4	41 04 N	29 04 E
C016	Yuzhno-Sakhalinsk	USSR	USSR	36.9	206.7	46 57 N	142 43 E
C364	Tbilisi	USSR	USSR	36.7	122.1	42 05 N	44 42 E
C133	Pendeli	Greece	Greece	36.6	101.5	38 03 N	23 52 E
C439	Ulan Bator	Mongolia	Mongolia	36.4	176.5	47 51 N	106 45 E
C264	Tenerife	Canary Islands	Spain	35.0	58.6	28 29 N	16 17 W
C425	Centro Geofisico	Cuba	Cuba	34.1	345.3	22 58 N	82 09 W

List of geomagnetic observatories (continued)

CSAGI No.	Observatory	Location	Sponsor	Geomagnetic Latitude	Geomagnetic Longitude	Geographic Latitude	Geographic Longitude
C034	Memambetsu	Japan	Japan	34.0	208.4	43 54 N	144 12 E
C050	Alma Ata	USSR	USSR	33.4	150.7	43 15 N	76 55 E
C051	Vladivostock	USSR	USSR	32.8	198.1	43 41 N	132 10 E
C361	Tashkent	USSR	USSR	32.3	144.0	41 20 N	69 37 E
C126	Ashkhabad	USSR	USSR	30.5	133.1	37 57 N	58 06 E
C416	Ksara	Lebanon	Lebanon	30.2	111.7	33 49 N	35 53 E
C300	San Juan	Puerto Rico	USA	29.6	3.1	18 07 N	66 09 W
C287	Teoloyucan	Mexico	Mexico	29.6	327.0	19 45 N	99 11 W
C163	Tehran	Iran	Iran	29.4	126.5	35 44 N	51 23 E
C339	Nitsanim	Israel	Israel	28.4	110.0	31 44 N	34 36 E
C117	Onagawa	Japan	Japan	28.3	206.8	38 26 N	141 28 E
C256	Helwan	UAR	UAR	27.2	106.4	29 52 N	31 20 E
C239	Misallat	UAR	UAR	26.9	105.9	29 31 N	30 54 E
C069	Seoul	Rep of Korea	Rep of Korea	26.5	194.2	37 35 N	127 03 E
C170	Gilgit	Pakistan	Pakistan	26.4	147.3	35 55 N	74 18 E
C147	Kakioka	Japan	Japan	26.0	206.0	36 14 N	140 11 E
C273	Tamanrasset	Algeria	Algeria	25.4	79.6	22 48 N	5 32 E
	Kanozan	Japan	Japan	25.0	205.9	35 15 N	139 58 E
E743	Midway	Midway Islands	USA	24.1	246.3	28 13 N	177 22 W
C214	Simosato	Japan	Japan	23.0	202.4	33 34 N	135 56 E
C220	Hachijojima	Japan	Japan	22.9	206.0	33 08 N	139 48 E

List of geomagnetic observatories (continued)

CSAGI No.	Observatory	Location	Sponsor	Geomagnetic Latitude	Geomagnetic Longitude	Geographic Latitude	Geographic Longitude
C223	Aso	Japan	Japan	22.0	198.1	32 53 N	131 01 E
	Araira	Venezuela	Venezuela	21.9	2.7	10 27 N	66 29 W
C251	Quetta	Pakistan	Pakistan	21.6	139.7	30 11 N	66 57 E
C311	M'Bour	Senegal	Senegal	21.3	55.0	14 24 N	16 58 W
C277	Honolulu	Hawaiian Islands	USA	21.1	266.5	21 19 N	158 00 W
C250	Sabhawala	India	India	20.5	149.7	30 20 N	77 48 E
C245	Kanoya	Japan	Japan	20.5	198.1	31 25 N	130 53 E
E575	Paramaribo	Surinam	Netherlands	17.0	14.3	5 49 N	55 13 W
E578	Fuquene	Colombia	Colombia	16.9	355.1	5 28 N	73 44 W
E532	Sokoto	Nigeria	Nigeria	15.9	77.1	13 03 N	5 15 E
E558	Freetown	Sierra Leone	Sierra Leone	14.8	57.8	8 28 N	13 13 W
E488	Lunping	Rep of China	Rep of China	13.7	189.5	25 00 N	121 10 E
E485	Zaria	Nigeria	Nigeria	13.6	79.1	11 09 N	7 39 E
E597	Kontagora	Nigeria	Nigeria	13.3	76.8	10 24 N	5 27 E
E525	Chittagong	Pakistan	Pakistan	11.4	161.9	22 21 N	91 49 E
E528	Cha-Pa	Vietnam	Vietnam	10.9	173.3	22 21 N	103 50 E
E571	Ibadan	Nigeria	Nigeria	10.7	74.7	7 26 N	3 54 E
E590	Tatuoca	Brazil	Brazil	9.6	20.8	1 12 S	48 31 W
E472	Legon	Ghana	Ghana	9.6	70.2	5 38 N	0 11 W
E538	Alibag	India	India	9.5	143.6	18 38 N	72 52 E
E542	Hyderabad	India	India	7.6	148.9	17 25 N	78 33 E

List of geomagnetic observatories (continued)

CSAGI No.	Observatory	Location	Sponsor	Geomagnetic		Geographic	
				Latitude	Longitude	Latitude °'	Longitude °'
E593	Talara	Peru	Peru-USA	6.6	347.7	4 38 S	81 18 W
E586	Moca	Fernando Poo Is	Spain	5.7	78.6	3 21 N	8 40 E
E568	Addis Abeba	Ethiopia	Ethiopia	5.4	109.2	9 02 N	38 46 E
E574	Palmyra Island	Palmyra Island	USA	5.2	265.8	5 53 N	162 05 W
E547	Baguio	Philippine Rep	Philippine Rep	5.1	189.2	16 25 N	120 36 E
E583	Bangui	Central Afr Rep	Central Afr Rep	4.8	88.5	4 26 N	18 34 E
E595	Chiclayo	Peru	Peru-USA	4.5	349.2	6 48 S	79 48 W
E556	Guam	Mariana Islands	USA	4.0	212.9	13 35 N	144 52 E
E585	Fanning Island	Gilbert-Ellice Is	UK-USA	3.7	268.8	3 54 N	159 23 W
E553	Muntinlupa	Philippine Rep	Philippine Rep	3.0	189.7	14 22 N	121 01 E
E596	Chimbote	Peru	Peru-USA	2.2	350.5	9 06 S	78 36 W
E562	Annamalainagar	India	India	1.5	149.4	11 24 N	79 41 E
E561	Majuro	Marshall Islands	USA	1.3	239.6	7 05 N	171 23 E
E566	Kodaikanal	India	India	0.6	147.1	10 14 N	77 28 E
E611	Bunia	Congo	Congo	-0.3	99.3	1 32 N	30 11 E
E618	Jarvis Island	Jarvis Island	USA	-0.6	269.1	0 23 S	160 02 W
E646	Huancayo	Peru	Peru	-0.6	353.8	12 03 S	75 20 W
	Cebu	Philippine Rep	Philippine Rep	-0.9	192.7	10 18 N	123 54 E
E603	Trivandrum	India	India	-1.1	146.4	8 29 N	76 57 E
E651	Cuzco	Peru	Peru	-2.1	357.1	13 32 S	71 58 W
E631	Binza	Congo	Congo	-3.1	83.6	4 16 S	15 22 E

List of geomagnetic observatories (continued)

CSAGI No.	Observatory	Location	Sponsor	Geomagnetic		Geographic	
				Latitude	Longitude	Latitude ° '	Longitude ° '
E606	Koror	Palau Islands	USA	−3.2	203.3	7 20 N	134 30 E
E624	Lwiro	Congo	Congo	−3.8	97.2	2 15 S	28 48 E
	Davao	Philippine Rep	Philippine Rep	−4.1	194.5	7 05 N	125 35 E
E740	Yauca	Peru	Peru-USA	−4.1	354.5	15 32 S	74 40 W
E621	Nairobi	Kenya	Kenya	−4.4	105.3	1 19 S	36 49 E
E666	Arequipa	Peru	Peru	−5.0	357.6	16 28 S	71 29 W
E676	Santa Cruz	Bolivia	Bolivia	−6.4	5.5	17 48 S	63 10 W
E640	Luanda	Angola	Portugal	−7.2	80.6	8 55 S	13 10 E
E702	La Quiaca	Argentina	Argentina	−10.6	3.2	22 06 S	65 36 W
E629	Vassouras	Brazil	Brazil	−11.9	23.9	22 24 S	43 39 W
E625	Hollandia	Dutch New Guinea	Indonesia	−12.5	210.3	2 34 S	140 31 E
E644	Elizabethville	Congo	Congo	−12.7	94.0	11 38 S	27 25 E
E672	Tahiti	Society Islands	France	−15.3	282.7	17 33 S	149 37 W
E653	Apia	Samoa Islands	Western Samoa	−16.1	260.2	13 48 S	171 47 W
E634	Kuyper	Indonesia	Indonesia	−17.5	175.6	6 02 S	106 44 E
	Tangerang	Indonesia	Indonesia	−17.6	175.4	6 10 S	106 38 E
E712	Easter Island	Easter Island	Chile	−18.2	322.6	27 10 S	109 25 W
E686	Tsumeb	Southwest Africa	German Fed Rep	−18.2	82.8	19 13 S	17 42 E
E642	Port Moresby	New Guinea	Australia	−18.6	217.9	9 24 S	147 09 E
C932	Pilar	Argentina	Argentina	−20.2	4.6	31 40 S	63 53 W
C901	Tananarive	Malagasy Rep	Malagasy Rep	−23.7	112.5	18 55 S	47 33 E

List of geomagnetic observatories (continued)

CSAGI No.	Observatory	Location	Sponsor	Geomagnetic Latitude	Geomagnetic Longitude	Geographic Latitude ° ′	Geographic Longitude ° ′
C982	Las Acacias	Argentina	Argentina	−23.7	10.1	35 00 S	57 41 W
	Mauritius	Indian Ocean	United Kingdom	−26.6	124.4	20 06 S	57 33 E
C899	Plaisance	Indian Ocean	United Kingdom	−27.0	122.5	20 26 S	57 40 E
C914	Lourenco Marques	Mozambique	Portugal	−27.7	95.8	25 55 S	32 35 E
C988	Trelew	Argentina	Argentina	−31.8	3.2	43 15 S	65 19 W
C957	Hermanus	Rep of S Africa	Rep of S Africa	−33.3	80.5	34 25 S	19 14 E
C838	Gough Island	Gough Island	Rep of S Africa	−33.8	51.9	40 21 S	9 53 W
C918	Brisbane	Australia	Australia	−35.8	226.9	27 32 S	152 55 E
C925	Watheroo	Australia	Australia	−41.8	185.7	30 19 S	115 53 E
C993	Gnangara	Australia	Australia	−43.2	185.8	31 47 S	115 57 E
B966	Toolangi	Australia	Australia	−46.7	220.8	37 32 S	145 28 E
B979	Amberley	New Zealand	New Zealand	−47.7	252.5	43 09 S	172 43 E
B996	Marion Island	Marion Island	Rep of S Africa	−49.0	94.3	46 51 S	37 52 E
A962	Orcadas Del Sur	Orcadas	Argentina	−50.0	18.2	60 44 S	44 47 W
	G Gonzales Videla	Antarctica	Chile	−53.4	4.4	64 49 S	62 51 W
A973	Argentine Islands	Antarctica	United Kingdom	−53.8	3.4	65 15 S	64 16 W
B998	Port-aux-Français	Indian Ocean	France	−57.3	128.0	49 21 S	70 12 E
A961	Macquarie Island	Antarctica	Australia	−61.1	243.1	54 30 S	158 57 E
A987	Sanae Station	Antarctica	Rep of S Africa	−63.6	44.2	70 18 S	2 21 W
A987	Norway Station	Antarctica	Norway	−63.8	43.9	70 30 S	2 32 W
A952	Eights	Antarctica	USA	−63.8	355.3	75 14 S	77 10 W

List of geomagnetic observatories (continued)

CSAGI No.	Observatory	Location	Sponsor	Geomagnetic Latitude	Geomagnetic Longitude	Geographic Latitude ° '	Geographic Longitude ° '
A989	Halley Bay	Antarctica	United Kingdom	−65.8	24.3	75 31 S	26 37 W
A951	Novolazarevskaya	Antarctica	USSR	−66.2	53.6	70 46 S	11 49 E
A994	General Belgrano	Antarctica	Argentina	−67.3	15.8	77 58 S	38 48 W
A986	Roi Baudouin	Antarctica	Belgium	−68.0	63.2	70 26 S	24 18 E
A984	Syowa Base	Antarctica	Japan	−69.7	77.7	69 00 S	39 35 E
A997	Byrd	Antarctica	USA	−70.6	336.3	80 00 S	119 30 W
A980	Mawson	Antarctica	Australia	−73.1	102.9	67 36 S	62 53 E
A995	Little America	Antarctica	USA	−74.0	312.0	78 11 S	162 12 W
A988	Cape Hallett	Antarctica	New Zealand-USA	−74.7	278.2	72 19 S	170 13 E
A979	Dumont D'Urville	Antarctica	France	−75.7	230.8	66 40 S	140 00 E
A978	Mirny	Antarctica	USSR	−77.0	146.8	66 33 S	93 01 E
	Plateau	Antarctica	USA	−77.2	52.5	79 15 S	40 30 E
A976	Oasis	Antarctica	USSR	−77.5	160.7	66 18 S	100 43 E
A977	Wilkes	Antarctica	Australia-USA	−77.7	179.2	66 15 S	110 35 E
A999	South Pole	Antarctica	USA	−78.5	0.3	89 57 S	13 19 W
A985	Charcot	Antarctica	France	−78.3	234.5	69 23 S	139 01 E
A991	Scott Base	Antarctica	New Zealand	−79.0	294.4	77 51 S	166 47 E
A959	Pionerskaya	Antarctica	USSR	−80.3	146.5	69 44 S	95 30 E
A996	Vostok	Antarctica	USSR	−89.2	91.4	78 27 S	106 52 E

Time Comparison Table for College, Alaska

College Local Time (150°WMT)⇌UT

College LT	0	1	2	3	4	5	6	7	8	9	10	11
UT	10	11	12	13	14	15	16	17	18	19	20	21
College LT	12	13	14	15	16	17	18	19	20	21	22	23
UT	22	23	0	1	2	3	4	5	6	7	8	9

College Dipole Midnight
$$\begin{cases} = 0142 \text{ LT} = 1142 \text{ UT in May} \\ = 0115 \text{ LT} = 1115 \text{ UT in December} \end{cases}$$

INDEX OF NAMES

INDEX OF SUBJECTS